Springer Textbooks in Earth Sciences, Geography and Environment

The Springer Textbooks series publishes a broad portfolio of textbooks on Earth Sciences, Geography and Environmental Science. Springer textbooks provide comprehensive introductions as well as in-depth knowledge for advanced studies. A clear, reader-friendly layout and features such as end-of-chapter summaries, work examples, exercises, and glossaries help the reader to access the subject. Springer textbooks are essential for students, researchers and applied scientists.

Martin H. Trauth

Python Recipes for Earth Sciences

 Springer

Martin H. Trauth
Institute of Geosciences
University of Potsdam
Potsdam, Brandenburg, Germany

ISSN 2510-1307   ISSN 2510-1315   (electronic)
Springer Textbooks in Earth Sciences, Geography and Environment
ISBN 978-3-031-07721-0   ISBN 978-3-031-07719-7   (eBook)
https://doi.org/10.1007/978-3-031-07719-7

Responsible Editor: Dr. Annett Büttner
This Springer imprint is published by the registered company Springer Nature Switzerland AG
The registered company address is: Gewerbestrasse 11, 6330 Cham, Switzerland

Preface

The book *Python Recipes for Earth Sciences* is designed to help undergraduate and postgraduate students, doctoral students, post-doctoral researchers, and professionals alike find quick solutions to common data analysis problems in the earth sciences. It provides a minimal amount of theoretical background and demonstrates the application of all described methods via examples. The book is based on the popular *MATLAB Recipes for Earth Sciences* book that I wrote many years ago, which is now available in its 5th edition (Springer 2021). The idea for this new book was to translate the text of the MATLAB book to another, increasingly popular programming language in the spirit of the classic *Numerical Recipes* book by W.H. Press, S.A. Teukolsky, W.T. Vetterling, and B.P. Flannery (Cambridge University Press 2007).

The present book contains Python scripts that can be used to solve typical problems in the earth sciences via simple statistics, time series analysis, geostatistics, and image processing. It also demonstrates the application of selected advanced techniques of data analysis, such as nonlinear time series analysis, adaptive filtering, bootstrapping, and terrain analysis. The book's supplementary electronic material (available online through Springer Link) includes recipes that contain all the example data and all the Python commands featured in the book. The Python codes can be easily modified for use with the reader's own data and projects. The Python software, which has existed for more than 30 years, is used since it not only provides numerous ready-to-use algorithms for most methods of data analysis, but also allows the existing routines to be modified and expanded and additionally enables new software to be developed. However, most methods are introduced with very simple codes before any Python-specific functions are used. As a result, users of alternatives to Python will also find value in this book.

In order to derive the maximum benefit from this book, the reader will need to have access to the Python software and be able to execute the recipes while reading the book. The Python recipes yield various graphs on the screen that are not shown in the printed book. The tutorial-style book does, however, contain numerous figures, thereby making it possible to go through the text without actually running Python on a computer. I developed the recipes using Python 3.8.8, though most recipes will also work with earlier software releases, but not with Python 2. While undergraduates participating in a course on data analysis might wish to go through the entire book, more experienced readers may choose to refer to only one particular method in order to solve a specific problem. The concept of the book and the contents of its chapters are therefore outlined below in order to make it easier to use for readers with a variety of different requirements.

- Chapter 1—This chapter introduces some fundamental concepts of samples and populations. It also links the various types of data and questions that can be answered using the data to the methods described in the subsequent chapters.
- Chapter 2—This chapter offers a tutorial-style introduction to Python designed for earth scientists. Readers who are already familiar with the software are advised to proceed directly to the subsequent chapters.
- Chapters 3 and 4—These two chapters cover fundamentals in univariate and bivariate statistics. They contain basic concepts in statistics and also introduce advanced topics, such as resampling schemes and cross-validation. Readers who are already familiar with basic statistics might want to skip these chapters.
- Chapter 5 and 6—Readers who wish to work with time series are advised to read both of these chapters. Time series analysis and signal processing are closely linked. A good knowledge of statistics is required to work successfully with these methods.
- Chapters 7 and 8—I recommend reading through both of these chapters since the processing methods used for spatial data and for images have much in common. Moreover, spatial data and images are often combined in the earth sciences, for instance, when projecting satellite images onto digital elevation models.
- Chapter 9—Data sets in the earth sciences often have many variables and many data points. Multivariate methods are applied to a great variety of large data sets, including satellite imagery data. Any reader who is particularly interested in multivariate methods is advised to read Chaps. 3 and 4 before proceeding to this chapter.
- Chapter 10—Methods for analyzing circular and spherical data are widely used in the earth sciences. Structural geologists, for example, measure and analyze the orientation of slickensides (or striae) on a fault plane, and the statistical analysis of circular data is used in paleomagnetic applications. Microstructural investigations include the analysis of grain shapes and quartz c-axis orientations in thin sections.

When using this book, it is important to keep in mind that it is based on a book that uses MATLAB for data analysis in the earth sciences. An attempt has been made to remain as close as possible to the original didactic concept and text of the MATLAB book, even if it may seem unusual or awkward for some Python users. On the other hand, this didactic concept will be familiar to the many users who, for whatever reason, have made switch from MATLAB to Python. These users, like me, are familiar with MATLAB and first look for similar solutions in Python instead of programming Python code from scratch. This happens very often and is reflected in many Python packages, such as *NumPy* and *Matplotlib*, which are highly similar to MATLAB toolboxes, right down to their identical function names and input arguments.

Readers who place both books side by side will notice that I have directly translated the code in order to make the numerical and graphical output as similar as possible. However, since different random generators are used in MATLAB and Python, the results do not always match with one another. In these cases, after generating synthetic data, the sample files generated with MATLAB are used in the Python examples in order to reproduce the results. There may also be some discrepancies in cases for which MATLAB produces a result of zero but for which Python produces a result slightly different from zero, such as $-3.0552e{-}17$. Such discrepancies are likely due to the less stringent floating-point error mitigation in many Python packages.

The attentive reader will also notice that in some cases, there are minor discrepancies between the graphs generated with Python and the illustrations in the first half of the book. This is also due to the fact that the random generators used here provide different results and that not all of the used random generators can be reset. Therefore, some results (e.g., the bootstrapping results) are different between individual runs. However, the figures at least qualitatively reflect the results of statistics, time series analysis, and signal processing. For the second half of the book, the graphics were newly created to reflect the peculiarities in the output of graphics packages such as *Matplotlib* and *PyGMT,* which certainly led to some very nice results.

I wrote the Python scripts, which are based on the MATLAB scripts, with the best of my knowledge and with the support of the colleagues mentioned below. However, I cannot offer any warranty, especially when using software such as Python and the Python packages, which themselves offer no warranty. For example, I have found a number of discrepancies between results from Python and MATLAB, which I attribute to minor floating-point issues with some Python packages, such as the use of very small values instead of (correctly) zero in the calculation. I have flagged these discrepancies in the text. I recommend that the reader always test the code (be it with Python or MATLAB) with synthetic examples before applying it to real examples because it is also possible that users have made their own mistakes when using floating-point numbers.

This book has benefited from the comments of many people, in particular from my contributing authors, Robin Gebbers and Norbert Marwan, as well as from many colleagues and friends worldwide who have read the work, pointed out mistakes, and made suggestions for improvement. I wish thank Ed Manning and Ryan DeLaney for their professional proofreading, which has greatly improved the readability of my book. I also very much appreciate the expertise and patience of Elisabeth Sillmann at *blaetterwaldDesign*, who has supported and advised me with the graphic design. I am very grateful to Annett Büttner, Marion Schneider,

Helen Rachner, Nirmal Iyer, Stefan Kreickenbaum, and their team at *Springer* for their continuing interest and support in publishing my books. I am very grateful to Nadine Berner, Bodo Bookhagen, Matt Hall, Hauke Krämer, Pawel Lachovicz, Norbert Marwan, and Maurizio Petrelli for answering my Python questions. I would also like to thank Brunhilde Schulz, Andreas Bohlen, Ingo Orgzall, and their team at *UP Transfer GmbH* for organizing short courses on the book throughout the past two decades at the universities of Kiel, Bremen, Bratislava, Ghent, Munich, Nairobi, Cologne, Stockholm, Amsterdam, Aberystwyth, and Potsdam as well as at UA Barcelona, UC London, Brown University Providence, the BGR Hannover, the BGI Bayreuth, the NHM Vienna, and GNS Science Wellington.

I additionally wish to thank the *NASA/GSFC/METI/ERSDAC/JAROS* and the *U.S./Japan ASTER Science Team* as well as director Mike Abrams for allowing me to include the ASTER images in this book. I am grateful to Stuart W. Frye and his team at the NASA Goddard Space Flight Center (GSFC) for allowing me to include EO-1 data in the book as well as for our fruitful discussions while working on the section about Hyperion images.

Potsdam   Martin H. Trauth
April 2022

Contents

1   Data Analysis in the Earth Sciences 1
    1.1   Introduction 1
    1.2   Data Collection 1
    1.3   Types of Data.. 3
    1.4   Methods of Data Analysis 6
    Recommended Reading 8

2   Introduction to Python 9
    2.1   Introduction 9
    2.2   Getting Started...................................... 11
    2.3   Python Syntax 12
    2.4   Array Manipulation 19
    2.5   Data Types in Python................................. 25
    2.6   Data Storage and Handling 37
    2.7   Control Flow 43
    2.8   Scripts and Functions................................ 47
    2.9   Basic Visualization Tools............................. 50
    2.10   Generating Code to Recreate Graphics 54
    2.11   Publishing and Sharing MATLAB Code 54
    2.12   Creating Graphical User Interfaces 54
    Recommended Reading 55

3   Univariate Statistics 57
    3.1   Introduction 57
    3.2   Empirical Distributions 58
    3.3   Examples of Empirical Distributions.................. 63
    3.4   Theoretical Distributions 71
    3.5   Examples of Theoretical Distributions................ 79
    3.6   Hypothesis Testing................................... 81
    3.7   The t-Test... 84
    3.8   The F-Test 88
    3.9   The χ^2-Test 92

3.10   The Kolmogorov–Smirnov Test.   95
3.11   Mann–Whitney Test.   98
3.12   The Ansari–Bradley Test   101
3.13   Distribution Fitting.   104
3.14   Error Analysis   108
Recommended Reading   115

4   Bivariate Statistics.   117
4.1   Introduction   117
4.2   Correlation Coefficients.   118
4.3   Classical Linear Regression Analysis   129
4.4   Analyzing the Residuals   132
4.5   Bootstrap Estimates of the Regression Coefficients   134
4.6   Jackknife Estimates of the Regression Coefficients   136
4.7   Cross-Validation.   138
4.8   Reduced Major Axis Regression   139
4.9   Curvilinear Regression.   141
4.10   Nonlinear and Weighted Regression   143
4.11   Classical Linear Regression of Log-Transformed Data.   147
Recommended Reading   150

5   Time Series Analysis   151
5.1   Introduction   151
5.2   Generating Signals.   152
5.3   Auto-Spectral and Cross-Spectral Analysis.   157
5.4   Examples of Auto-Spectral and Cross-Spectral Analysis   161
5.5   Interpolating and Analyzing Unevenly Spaced Data.   171
5.6   Evolutionary Power Spectrum   177
5.7   Lomb–Scargle Power Spectrum.   180
5.8   Wavelet Power Spectrum   186
5.9   Detecting Abrupt Transitions in Time Series.   191
5.10   Aligning Stratigraphic Sequences   191
5.11   Nonlinear Time Series Analysis (by N. Marwan)   192
Recommended Reading   210

6   Signal Processing.   213
6.1   Introduction   213
6.2   Generating Signals.   214
6.3   Linear Time-Invariant Systems   216
6.4   Convolution, Deconvolution, and Filtering   219
6.5   Comparing Functions for Filtering Data Series.   223
6.6   Recursive and Nonrecursive Filters   225
6.7   Impulse Response   227
6.8   Frequency Response   230
6.9   Filter Design.   238
6.10   Adaptive Filtering   241
Recommended Reading   249

7   Spatial Data 251
     7.1   Introduction 251
     7.2   The Global Geography Database GSHHG 252
     7.3   The 1 Arc-Minute Gridded Global Relief Data ETOPO1 254
     7.4   The 30 Arc-Second Elevation Model GTOPO30 259
     7.5   The Shuttle Radar Topography Mission SRTM 262
     7.6   Exporting 3D Graphics to Create Interactive Documents 268
     7.7   Gridding and Contouring 268
     7.8   Comparison of Methods and Potential Artifacts 275
     7.9   Statistics of Point Distributions 281
     7.10   Analysis of Digital Elevation Models (by R. Gebbers) 291
     7.11   Geostatistics and Kriging (by R. Gebbers) 301
     Recommended Reading 319

8   Image Processing 321
     8.1   Introduction 321
     8.2   Data Storage 322
     8.3   Importing, Processing, and Exporting Images 327
     8.4   Importing, Processing, and Exporting Landsat Images 332
     8.5   Importing and Georeferencing Terra ASTER Images 338
     8.6   Processing and Exporting EO-1 Hyperion Images 344
     8.7   Digitizing from the Screen 350
     8.8   Image Enhancement, Correction, and Rectification 350
     8.9   Color Intensity Transects Across Varved Sediments 355
     8.10   Grain Size Analysis from Microscopic images 356
     8.11   Quantifying Charcoal in Microscopic images 361
     8.12   Shape-Based Object Detection in Images 364
     8.13   The Normalized Difference Vegetation Index NDVI 371
     Recommended Reading 379

9   Multivariate Statistics 381
     9.1   Introduction 381
     9.2   Principal Component Analysis 383
     9.3   Independent Component Analysis (by N. Marwan) 401
     9.4   Discriminant Analysis 406
     9.5   Cluster Analysis 411
     9.6   Multiple Linear Regression 416
     9.7   Aitchison's Log-Ratio Transformation 423
     Recommended Reading 428

10   Directional Data 429
     10.1   Introduction 429
     10.2   Graphical Representation of Circular Data 431
     10.3   Empirical Distributions of Circular Data 433
     10.4   Theoretical Distributions of Circular Data 436
     10.5   Test for Randomness of Circular Data 440

10.6   Test for the Significance of a Mean Direction   441
10.7   Test for the Difference Between Two Sets of Directions   444
10.8   Graphical Representation of Spherical Data   447
10.9   Statistics of Spherical Data   449
Recommended Reading   452

Data Analysis in the Earth Sciences

1

1.1 Introduction

Earth scientists make observations and gather data about the natural processes that operate on planet Earth. They formulate and test hypotheses on the forces that have acted on a particular region to create its structure, and they also make predictions about future changes to the planet. All of these steps in exploring the Earth involve acquiring and analyzing numerical data. An earth scientist therefore needs to have a firm understanding of statistical and numerical methods as well as the ability to utilize relevant computer software packages in order to be able to analyze the acquired data.

This book introduces some of the most important methods of data analysis employed in the earth sciences and illustrates their use through examples with the Python software package. These examples can then be used as recipes for analyzing the reader's own data after the reader has understood the application of the examples with synthetic data. This introductory chapter deals with data acquisition in Sect. 1.2, the various types of data in Sect. 1.3, and the appropriate methods for analyzing earth science data in Sect. 1.4. We therefore first explore the characteristics of typical data sets and subsequently investigate the various ways of analyzing data using Python.

1.2 Data Collection

Most data sets in the earth sciences have a very limited sample size and also contain a large number of uncertainties. Such data sets are typically used to describe rather large natural phenomena, such as a granite body, a large landslide, or a widespread sedimentary unit. The methods described in this book aim to find a

© The Author(s), under exclusive license to Springer Nature Switzerland AG 2022
M. H. Trauth, *Python Recipes for Earth Sciences*, Springer
Textbooks in Earth Sciences, Geography and Environment,
https://doi.org/10.1007/978-3-031-07719-7_1

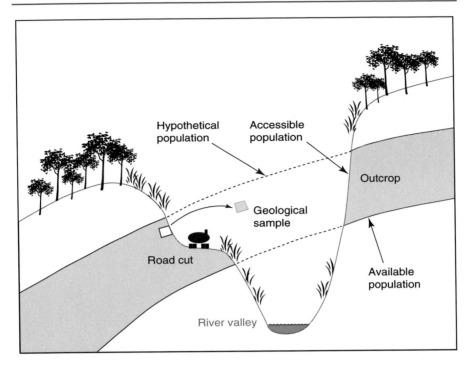

Fig. 1.1 Samples and populations. Deep valley incision has eroded parts of a sandstone unit (*hypothetical population*). The remaining sandstone (*available population*) can only be sampled from outcrops, such as road cuts and quarries (*accessible population*). Note the difference between a statistical sample as a representative of a population and a geological sample as a piece of rock. Modified from Swan and Sandilands (1995).

way of predicting the characteristics of a larger *population* from a much smaller *sample* (Fig. 1.1).

It is important to note that depending on the population of interest in the research context, a sample obtained in the field (e.g., a piece of granite) is not necessarily identical to what is later described as a statistical sample. If the object of study is the chemical composition of a granite sample, then the sample is the collection of replicate measurements taken on subsamples of the granite. If, however, the researcher is more interested in the chemical composition of the entire granite body, then the sample consists of the chemical composition of several (preferably representative) rock samples from this granite body.

In any case, an appropriate sampling strategy is the first step toward obtaining a good data set. Developing a successful strategy for field sampling requires making decisions on the *sample size* and the *spatial sampling scheme*. The sample size (again for the rock sample) is determined by the sample's volume and/or weight or (for the statistical sample) by the number of rock samples collected from a granite body. If the population is heterogeneous, the sample needs to be large enough to represent the population's variability; however, samples should also be as small as possible in order to minimize the time and costs involved in their analysis. Collecting smaller pilot samples is recommended prior to defining a suitable sample size.

The design of the spatial sampling scheme is dependent on the availability of outcrops or other materials that are suitable for sampling (Fig. 1.2). Sampling in quarries typically leads to clustered data, whereas sampling along road cuts, shoreline cliffs, or steep gorges results in one-dimensional traverse sampling schemes. A more uniform sampling pattern can be designed in which there is 100% exposure or in which there are no financial limitations. A regular sampling scheme results in a gridded distribution of sample locations, whereas a uniform sampling strategy includes the random location of a sampling point within a grid square. Although these sampling schemes might be expected to provide superior methods for sampling collection, evenly spaced sampling locations tend to miss small-scale variations in the area, such as thin mafic dykes within a granite body or the spatially restricted presence of a particular type of fossil.

The correct sampling strategy depends on the objectives of the investigation, the type of analysis required, and the desired level of confidence in the results. Once a suitable sampling strategy has been chosen, the quality of the sample can be influenced by several factors, thereby resulting in samples that are not truly representative of the larger population. Chemical or physical alteration, contamination by other material, or displacement by natural or anthropogenic processes may all result in erroneous results and interpretations. It is therefore recommended that the quality of the samples, the employed method of data analysis, and the validity of the conclusions drawn from the analysis be checked at each stage of the investigation.

1.3 Types of Data

Most earth science data sets consist of numerical measurements, although some information, such as fossils and minerals, can also be represented by a list of names (Fig. 1.3). The available methods for data analysis may require certain types of data in the earth sciences. These types of data include

- *Nominal data*—Information in the earth sciences is sometimes presented as a list of names, such as the various fossil species collected from a limestone bed or the minerals identified in a thin section of rock. In some studies, these data are converted to a binary representation, that is, *one* for present and *zero* for absent. Special statistical methods are available for analyzing such data sets.
- *Ordinal data*—These are numerical data that represent observations that can be ranked but for which the intervals along the scale are irregularly spaced. The hardness scale created by German mineralogist Friedrich Mohs (1773 to 1839) is one example of an ordinal scale. The hardness value indicates the material's resistance to scratching. Diamond has a hardness of 10, whereas the value for talc is 1, but in terms of absolute hardness, diamond (hardness 10) is four times

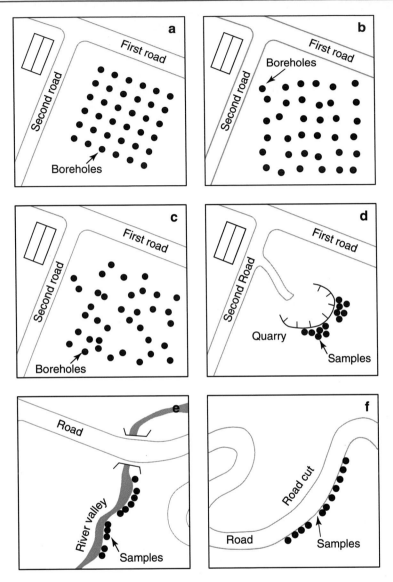

Fig. 1.2 Sampling schemes. **a** *Regular sampling* on an evenly spaced rectangular grid, **b** *uniform sampling* by obtaining samples that are randomly located within regular grid squares, **c** *random sampling* using uniformly distributed *xy*-coordinates, **d** *clustered sampling* constrained by limited access in a quarry, and *traverse sampling* along **e** river valleys and **f** road cuts. Modified from Swan and Sandilands (1995).

harder than corundum (hardness 9) and six times harder than topaz (hardness 8). The Modified Mercalli Scale, which attempts to categorize the effects of earthquakes, is another example of an ordinal scale and ranks earthquakes from intensity I (barely felt) to intensity XII (total destruction) (Richter 1958).

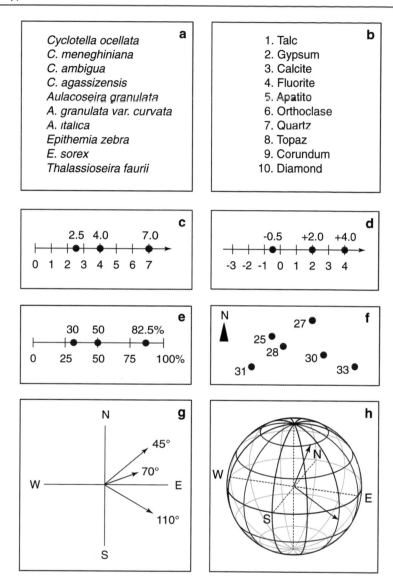

Fig. 1.3 Types of earth science data. **a** *Nominal data*, **b** *ordinal data*, **c** *ratio data*, **d** *interval data*, **e** *closed data*, **f** *spatial data*, and **g–h** *directional data*. All of these data types are described in this book. Modified from Swan and Sandilands (1995)

- *Ratio data*—These data are characterized by a constant length of successive intervals and therefore offer a great advantage over ordinal data. The zero point is the natural termination of the data scale. This type of data allows for either discrete or continuous data sampling. Examples of such data sets include length and weight data.

- *Interval data*—These are ordered data that have a constant length of successive intervals but for which the data scale is not terminated by zero. Temperatures in °C and °F represent an example of this data type even though arbitrary zero points exist for both scales. This type of data may be sampled continuously or in discrete intervals.

In addition to these standard data types, earth scientists frequently encounter special kinds of data, such as

- *Closed data*—These data are expressed as proportions and that add to a fixed total, such as 100%. Compositional data (e.g., the element compositions of rock samples) represent the majority of closed data.
- *Spatial data*—These data are collected in a 2D or 3D study area. The spatial distribution of a certain fossil species, the spatial variation in the thickness of a sandstone bed, and the distribution of tracer concentrations in groundwater are examples of this type of data, which is likely the most important data type in the earth sciences.
- *Directional data*—These data are expressed in angles. Examples include the strike and dip of bedding, the orientation of elongated fossils, and the flow direction of lava. This is another very common type of data in the earth sciences.

Most of these different types of data require specialized methods of analysis, some of which are outlined in the next section below.

1.4   Methods of Data Analysis

Data analysis uses precise characteristics of small samples to hypothesize about the general phenomenon of interest. Determining which particular method to use to analyze the data depends on the data type and the project requirements. The various methods available include

- *Univariate methods*—Each variable is assumed to be independent of the others and is explored individually. The data are presented as a list of numbers that represent a series of points on a scaled line. Univariate statistical methods include collecting information about the variable, such as the minimum and maximum values, the average, and the dispersion about the average. This information is then used to attempt to infer the underlying processes that are responsible for variations in the data. Examples include the effects of chemical weathering on the sodium content of volcanic glass shards and the influence of specific environmental factors on the size of snail shells within a sediment layer.
- *Bivariate methods*—Two variables are investigated together to detect relationships between them. For example, the correlation coefficient may be calculated

in order to investigate whether there is a linear relationship between two variables. Alternatively, bivariate regression analysis may be used to find an equation that describes the relationship between the two variables. An example of a bivariate plot is the *Harker Diagram*, which is one of the oldest methods of visualizing geochemical data from igneous rocks and simply plots the concentrations of oxides of chemical elements against SiO_2 (Harker 1909).

- *Time series analysis*—These methods investigate data sequences as a function of time. The time series is decomposed into a long-term trend, a systematic (periodic, cyclic, rhythmic) component, and an irregular (random, stochastic) component. A widely used technique for describing the cyclic components of a time series is spectral analysis. Examples of the application of these techniques include investigating cyclic climatic variations in sedimentary rocks and analyzing seismic data.

- *Signal processing*—This includes all techniques used for manipulating a signal in order to minimize the effects of noise with the aim of correcting all kinds of unwanted distortions or of separating various components of interest. Signal processing includes designing and realizing filters as well as their application to the data. These methods are widely used in combination with time series analysis, for example, to increase the signal-to-noise ratio in climate time series, in digital images, or in geophysical data.

- *Spatial analysis*—This is the analysis of variables in 2D or 3D space, and two or three of the required variables are hence coordinate numbers. Spatial analysis includes descriptive tools for investigating the spatial patterns of geographically distributed data. Other techniques involve using spatial regression analysis to detect spatial trends. Spatial analysis also includes 2D and 3D interpolation techniques, which help to estimate surfaces that represent the predicted continuous distribution of the variable throughout an area. Examples include analyzing drainage systems, identifying old landscape forms, and analyzing lineament in tectonically active regions.

- *Image processing*—Processing and analyzing images has become increasingly important in the earth sciences. These methods involve importing & exporting, compressing & decompressing, and displaying images. Image processing also aims to enhance images for improved intelligibility and to manipulate images in order to increase the signal-to-noise ratio. Advanced techniques are used to extract specific features or to analyze shapes and textures, such as for counting mineral grains or fossils in microscopic images. Another important application of image processing lies in the use of satellite remote sensing to map certain types of rocks, soils, and vegetation, as well as to measure other variables, such as soil moisture, rock weathering, and erosion.

- *Multivariate analysis*—These methods involve observing and analyzing more than one statistical variable at a time. Since graphically representing multidimensional data sets is difficult, most of these methods include dimension reduction. Multivariate methods are widely used with geochemical data, such as in tephrochronology, in which volcanic ash layers are correlated via the geochemical fingerprinting of glass shards. Another important use lies in comparing species assemblages in ocean and lake sediments for reconstructing paleoenvironments.

- *Analysis of directional data*—Methods of analyzing circular and spherical data are widely used in the earth sciences. Structural geologists measure and analyze the orientation of slickensides (or striae) on a fault plane. Moreover, circular statistical methods are common in paleomagnetic studies, and microstructural investigations include the analysis of grain shapes and quartz *c*-axis orientations in thin sections of rocks.

Some of these methods of data analysis require applying numerical methods, such as interpolation techniques. While the following text deals mainly with statistical techniques, it also introduces several numerical methods that are commonly used in the earth sciences.

Recommended Reading

Harker A (1909) The natural history of igneous rocks. Macmillan, New York
Richter CF (1958) Elementary seismology. W.H. Freeman, San Francisco
Swan ARH, Sandilands M (1995) Introduction to geological data analysis. Blackwell Sciences, Oxford

Further Reading

Davis JC (2002) Statistics and data analysis in geology, 3rd edn. John Wiley and Sons, New York
Petrelli M (2021) Introduction to python in earth science data analysis—from descriptive statistics to machine learning. Springer, Berlin Heidelberg New York
Press WH, Teukolsky SA, Vetterling WT, Flannery BP (2007) Numerical recipes: the art of scientific computing, 3rd edn. Cambridge University Press, Cambridge
Trauth MH (2021a) MATLAB Recipes for Earth Sciences, 5th edn. Springer, Berlin Heidelberg New York
Trauth MH (2021b) Signal and noise in geosciences, MATLAB Recipes for Data Acquisition inEarth Sciences. Springer, Berlin Heidelberg New York
Trauth MH, Sillmann E (2018) Collecting, processing and presenting geoscientific information, MATLAB and design recipes for earth sciences, 2nd edn. Springer, Berlin Heidelberg New York

Introduction to Python

2

2.1 Introduction

Python is an interpreted programming language that was developed by Guido van Rossum in the late 1980s and first publicly released in 1991 (http://www.python. org). Python is a free and open-source language that is promoted, projected, and advanced by the *Python Software Foundation* (PSF), which was founded in 2001 (http://www.python.org/psf). According to its website, the PSF is a non-profit corporation that holds the intellectual property rights to the Python programming language. Members of the PSF are individuals who have made significant contributions to language's development, including the original developer and PSF President G. von Rossum. In addition, numerous companies are both members and sponsors of the PSF, including Google, Microsoft, Amazon, Meta/Facebook, Netflix, and Huawei (https://www.python.org/psf/sponsors).

Python is used in many application domains, including in web development, in performing mathematical calculations, in analyzing and visualizing data, and in facilitating the writing of new software programs. The advantage of this software is that it combines comprehensive math and graphics functions with a powerful high-level language. Python users have developed numerous extensions (*packages*) during the past several decades (https://pypi.org). While other packages suffer from buggy codes and dubious authorship, these Python packages have

Supplementary Information The online version contains supplementary material available at https://doi.org/10.1007/978-3-031-07719-7_2.

Sections 2.10–2.12 can be found in the MATLAB version of this book but could not be translated to Python. However, they are listed here with corresponding section numbers in order to create identical chapter numbering (and thereby also identical figure numbering and computer scripts) between the two versions of the book. The reader can thus use both books side by side to trace the translation of the scripts from MATLAB to Python (and back).

generally been of relatively high quality depending on the development time, the size of the developer community, and the package distribution. It is important to note that neither the Python Software Foundation nor the developers of the packages offer a warranty and that there is no professional support and no claim to complete or up-to-date documentation. The PSF makes Python available to licensees on an *as-is* basis (https://docs.python.org/3/license.html); however, numerous companies offer commercial support and consulting for Python applications, most notably Anaconda Inc., which is also known for providing a free Python installation package called *Anaconda Distribution* (https://www.anaconda.com).

Over the last few years, Python has become an increasingly popular tool in the earth sciences. It has been used for finite element modeling, for processing seismic data, for analyzing satellite imagery, and for generating digital elevation models from satellite data. The continuing popularity of the software is also apparent in published scientific literature, with many conference presentations also having referenced Python. Universities and research institutions alike have recognized the need for Python training for their staff and students, and many earth science departments across the world thus now offer Python courses for undergraduates.

The following sections contain a tutorial-style introduction to Python that covers its setup on the computer in Sect. 2.2, Python syntax in Sects. 2.3 and 2.4, data input and output in Sects. 2.5 and 2.6, programming in Sects. 2.7 and 2.8, and visualization in Sect. 2.9. We recommend that the reader go through the entire chapter in order to acquire solid foundational knowledge of the software before proceeding to the following chapters of the book. A more detailed introduction can be found in the Python Beginner's Guide on the Python website, where we can also find the documentation, a list of numerous Python-based books, and access to the Python community. Another important resource for solving Python problems is Stackoverflow (https://stackoverflow.com), where users of Python and other programming languages discuss problems and possible solutions.

In this book, we use *Python 3.8.12* from the free *Anaconda Individual Edition*, which includes an installer for Python and a manager for the most popular packages. We also use the integrated development environment (IDE) *Spyder 5.1.5*, the web-based *JupyterLab 3.2.1*, and the web-based interactive computing notebook environment *Jupyter Notebook 6.4.6*. In the following chapters, we use Spyder as a comfortable development environment because it contains a graphical user interface with an *Editor*, a *Console*, a graphics window, and several tools, for example, for displaying variables in the workspace. From the large collection of available packages, we limit ourselves to a few very popular and sufficiently tested packages that are necessary to perform the data analysis applications in this book.

Our collection of packages includes *IPython* (https://ipython.org). According to its website, *IPython* provides, inter alia, an interactive shell, the kernel for Jupyter, and support both for interactive data visualization and for the use of graphical user interface (GUI) toolkits. The second important package is *NumPy* for scientific computing with multidimensional arrays, which includes an assortment of routines for fast operations on arrays, basic linear algebra, basic statistical operations, random simulation, and much more (https://numpy.org). Another popular package for data analysis and manipulation is *pandas* (https://pandas.pydata.org), which we do not use because many users have advised against it. The third important package is *Matplotlib*, which is a comprehensive library for creating static, animated, and interactive visualizations in Python (https://matplotlib.org). We also use *SciPy*, which is a package for mathematics, science, and engineering (https://scipy.org), and we additionally use *scikit-learn* for machine learning in Python (https://scikit-learn.org). The packages *scikit-image* (https://scikit-image.org), *OpenCV* (https://pypi.org/project/opencv-python), and *pyhdf* (https://pypi.org/project/pyhdf) are used for image processing in Python.

In addition to these multidisciplinary packages, a variety of Python packages are available for geoscience applications, such as the ones listed on the well-curated websites Awesome Open Geoscience and Python Resources for Earth Sciences. From this collection, we use *GDAL* (https://gdal.org/api/python.html), *Rasterio* (https://rasterio.readthedocs.io), and *PyGMT* (https://www.pygmt.org), the latter of which requires *xarray* (https://docs.xarray.dev), and for manipulating geospatial data, we use *Cartopy* (https://scitools.org.uk/cartopy).

2.2   Getting Started

The Python website includes, *inter alia*, extensive documentation, audio/visual talks, a beginner's guide, answers to frequently asked questions, and a list of books about Python. Since this extensive material provides all the information required to use the software, the present introduction concentrates on the most relevant software components and tools used in the subsequent chapters of this book.

After installing the Anaconda Individual Edition, the Anaconda Navigator is launched by clicking on the application's shortcut icon. On the navigator's graphical user interface, we start the Scientific PYthon Development EnviRonment (Spyder) by double-clicking on the Spyder icon. We can also launch the software without using the Spyder desktop on computers with UNIX-based operating systems by typing

```
python
```

in the operating system prompt. If we launch Spyder with its desktop, it comes up with several window panels and tabs (Fig. 2.1). The default desktop layout includes the *Editor* for coding Python scripts and viewing data in text format and the *Files* tab, which lists the files in the directory that is currently being used. The *Console* presents the interface between the software and the user, that is, it accepts Python commands that are typed after the prompt In [1]: appears. The *Variable Explorer* lists the variables in the Python workspace, which is empty when starting a new software session. The *Help* tab is used to get help with any object either in the Editor or in the Console. Help with a particular function can also be found by writing the function name in parentheses after the Help command in the Console. Graphs are displayed in the *Plots* tab. All of these panels and tabs are part of the Spyder desktop by default but can also be undocked from it.

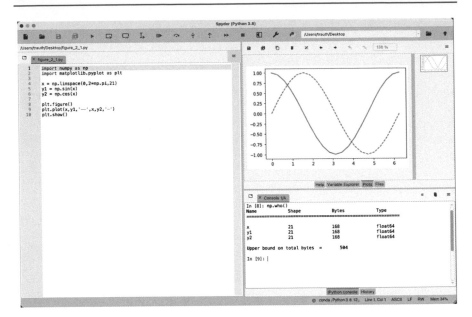

Fig. 2.1 Screenshot of the Spyder default desktop layout, including the *Editor* (left in the figure), the *Console* (lower right), and the *Plots* (right) panels. On the right side, we can also see the *Help*, *Variable Explorer*, and *Files* tabs behind the Plots panel as well as the *History* tab behind the Console tab.

By default, the software stores all of our Python-related files in our home directory. Alternatively, we can create a personal working directory in which to store our Python-related files. We should then make this new directory the working directory in the Spyder preferences. The software uses a *search path* to find Python-related files, which are organized into directories on the hard disk. The default search path includes only the directory that has been created by the installer in the applications folder and the default working directory. To see which directories are in the search path or to add new directories, we use the *PYTHONPATH Manager* from the *Tools* menu in the Spyder desktop. The modified search path is then saved in the software preferences on our hard disk, and the software then directs Python to use our custom search path.

2.3   Python Syntax

Although Python was not developed primarily for use in scientific data analysis, it has achieved great popularity in this area thanks to the development and distribution of packages such as *NumPy*. In this respect, it is not surprising that from the large number of available data types in Python, we mainly use the *n*-dimensional array *ndarray*, which is also called the *NumPy* array in reference to its rectangular two-dimensional *array* of numbers. A simple 1-by-1 array is a *scalar*. Arrays with one column or row are *vectors*, time series, or other one-dimensional data fields. An *m*-by-*n* array can be used for a digital elevation model or a grayscale image. Red, green, and blue (RGB) color images are usually stored as three-dimensional arrays, that is, the three colors are represented by an *m*-by-*n*-by-3 array.

Before we start with our Python project, we clear the workspace by typing

```
%reset -f
```

after the prompt in the Console and then press the *return* or *enter* key, which resets the *IPython* kernel without restarting it. Clearing the workspace is always recommended before working on a new Python project in order to avoid name conflicts with previous projects. We can also go a step further and close all plots by pressing the *Remove all plots* button in the *Plots* tab to avoid confusing the graphical representation of the results with that of the previous results. We can then use

```
clear
```

to clear the content from the Console. In Python, the packages (e.g., *NumPy*) and thus also the functions contained within them must be imported before using them, for example, by typing

```
import numpy as np
```

When importing the *NumPy* package, we define an abbreviation for the functions it contains (e.g., np), and this abbreviation is prefixed to all functions. For example, if we want to create a *NumPy* array in the workspace, we use array() from *NumPy*, which is then called by np.array(), as shown below:

```
A = np.array([[2,4,3,7],[9,3,-1,2],[1,9,3,7],[6,6,3,-2]])
```

While we must use the np prefix in the Python examples, we omit it in the explanatory text, that is, we use array() instead of np.array(). The function array() defines a variable A and then lists the elements of the array (separated by commas) in square brackets, with the rows of the array again being enclosed by square brackets separated by commas. After pressing *return* or *enter*, the variable A is actually created in the workspace but is not displayed in the Console. To display the variable in the Console, we must use the print() command,

```
print(A)
```

which yields

```
[[ 2  4  3  7]
 [ 9  3 -1  2]
 [ 1  9  3  7]
 [ 6  6  3 -2]]
```

Displaying the elements of A using print() could be problematic for very large arrays, such as digital elevation models consisting of thousands or millions of elements. The array A is now stored in the workspace, and we can carry out some basic operations with it, such as computing the sum of elements using A as the *input variable* or *input argument* of the *function* sum(), as in

```
print(sum(A))
```

which results in the display

```
[18 22  8 14]
```

The example above illustrates an important point about Python: The software prefers to work with the columns of arrays. The four results of sum(A) are obviously the sums of the elements in each of the four columns of A. In order to sum all elements of A and store the result in a scalar b, we simply need to type

```
print(sum(sum(A)))
```

which sums first the columns of the array and then the elements of the resulting vector. Since we did not specify an *output variable* or *output argument*, such as A for the array entered above, Python does not store the results of the calculation. In order to sum all the elements of A and store the result in a scalar b, we simply need to type

```
b = sum(sum(A))
print(b)
```

The attentive reader may have noticed that sum() is not prefixed by np in the Python code above. We are obviously calling not a function from the *NumPy* package, but a function from Python's standard collection, that is, a so-called *built-in function* that is contained in Python's module *builtins* as the output of

```
help(sum)
```

which suggests

```
Help on built-in function sum in module builtins:

sum(iterable, /, start=0)
    Return the sum of a 'start' value (default: 0) plus
    an iterable of numbers

    When the iterable is empty, return the start value.
    This function is intended specifically for use with
    numeric values and may reject non-numeric types.
```

However, the *NumPy* package also contains `sum()`, but it returns a different result, namely the sum of all elements in the first application, as shown below:

```
b = np.sum(A)
print(b)
```

We now have two variables A and b stored in the workspace. We can easily check this by typing

```
%whos
```

which lists the names of all the variables in the workspace together with information about their type and size (or dimension):

```
Variable   Type          Data/Info
-------------------------------
A          ndarray       4x4: 16 elems, type `int64`, 128 bytes
b          int64         62
np         module        <module 'numpy' from '/Us<...>kages/
                         numpy/__init__.py'>
```

The corresponding *NumPy* command is

```
np.who()
```

which yields

```
Name            Shape            Bytes            Type
===============================================================

A               4 x 4            128              int64

Upper bound on total bytes  =       128
```

which only lists the *NumPy* arrays together with their shape (or dimension), their number of bytes, and their type (see Sect. 2.5 for more details). Note that Python is case sensitive by default such that A and a define two different variables. In this context, it has long been recommended that uppercase letters be used for arrays (or matrices) that have two dimensions or more and that lowercase letters be used for one-dimensional arrays (or vectors) and for scalars. However, it is also common to use variables with mixed uppercase and lowercase letters. This is particularly important when using descriptive variable names, that is, variables whose names contain information concerning their meaning or purpose (e.g., the variable `CatchmentSize`) rather than a single-character variable (e.g., a). We can now delete the contents of variable b by typing

```
del b
```

Next, we learn how specific array elements can be accessed or exchanged. In so doing, it is important to bear in mind that the first element in array A has index (0,0) and not (1,1) as, for example, in MATLAB (https://de.mathworks.com) and Julia (https://julialang.org). Furthermore, the indices in Python are written in square brackets, that is, the element A[0,0] corresponds to A(1,1) in MATLAB and Julia. In our example, in order to access A(3,2) in MATLAB, we need to type

```
print(A[2,1])
```

in Python together with print() in order to actually display the array element located in the third row and second column, which is 9. The array indexing therefore follows the rule *(row-1, column-1)*. Python's indexing, which starts with 0 instead of 1, has been controversial for many years, especially among users of other languages like MATLAB and Julia, whose indexing starts with 1. Both indexing conventions have advantages and disadvantages, but it will become apparent throughout the present book that the indexing used by Python poses greater risks, for example, when selecting spectral bands from the hyperspectral Hyperion sensor (Sect. 8.6). If, for example, the user wants to access Hyperion Bands 29, 23, and 16 in order to create a pseudocolor image, this is not done with indices 29, 23, and 16 as in MATLAB and Julia, but with indices 28, 22, and 15 in Python. This fact may be familiar to Python users but may lead to errors among users who have switched to Python from other languages. As an example, we type

```
A[2,1] = 30
print(A)
```

in order to replace element A[2,1] with 30 and to display the entire array, as

```
[[ 2   4   3   7]
 [ 9   3  -1   2]
 [ 1  30   3   7]
 [ 6   6   3  -2]]
```

shown below:

If we wish to replace several elements at one time, we can use the *colon operator*. Typing

```
A[2,0:4] = [1 3 3 5]
print(A)
```

or

```
A[2,0:] = np.array([1,3,3,5])
print(A)
```

replaces all elements in the third row of array A. Here, another peculiarity of Python becomes obvious because the interval [0:4] is open to the right, that is, [0, 4) using mathematical notation, which is why we actually access the 4th column with index 4 and not with index 3 as would be expected when starting indexing with 0. The interval [0:] without an index to the right of the : sign means that all columns are accessed, that is, the missing index corresponds to end in other programming languages like MATLAB and Julia. Therefore, A(3,1:end-1) in MATLAB and Julia corresponds to A[2,0:-1] in Python. A third variant is the command, which (similar to in MATLAB and Julia) completely omits the specification of indices for the columns if all columns are to be accessed, as shown below:

```
A[2,:] = np.array([1,3,3,5])
print(A)
```

NumPy provides several ways to create evenly spaced values within a given interval, for instance, as a shortcut for entering array elements. The first option is to use arange(start,stop,step) by typing

```
c = np.arange(0,10)
print(c)
```

which creates a vector or a one-dimensional array with a single row that contains all integers in the given interval but that uses the half-open interval (*start, stop*) (using mathematical notation), that is, the interval includes *start* but excludes *stop*. Therefore, we actually get the integers from 0 to 9 and not from 0 to 10, as shown below:

```
[0 1 2 3 4 5 6 7 8 9]
```

If wish to include 10, we must choose stop=11. An alternative function for generating evenly spaced values within a given interval is linspace(start,stop,num), but for this function, we need to know the number num of values: for example, 11 values between 0 and 10:

```
c = np.linspace(0,10,11)
print(c)
```

We can use both functions to create evenly spaced values with increments of 0.5 instead of 1 by typing

```
c = np.arange(1,10+0.5,0.5)
print(c)
```

or

```
c = np.linspace(1,10,19)
print(c)
```

which both yield

```
[ 1.    1.5  2.   2.5  3.   3.5  4.    4.5  5.   5.5  6.
  6.5  7.   7.5  8.   8.5  9.   9.5  10. ]
```

which autowraps the lines that are longer than the width of the Console, that is, it introduces a line break after 6. in the first line. The display of the values of a variable can be interrupted by pressing *Ctrl+C* (*Control+C*) on the keyboard. This interruption affects only the output in the Console, whereas the actual command is processed before displaying the result.

Python provides standard arithmetic operators for addition (+) and subtraction (–). The at sign @ denotes matrix multiplication that involves inner products between rows and columns. For instance, to multiply matrix A by a new matrix B, as follows

```
B = np.array([[4,2,6,5],[7,8,5,6],[2,1,-8,-9],[3,1,2,3]])
```

we type

```
C = A @ np.transpose(B)
print(C)
```

or

```
C = np.matmul(A,np.transpose(B))
print(C)
```

using the *NumPy* function matmul(), in which transpose() yields the complex conjugate transpose, which turns rows into columns and columns into rows. This generates the output

```
[[ 69 103 -79  37]
 [ 46  94  11  34]
 [117 304 -55  60]
 [ 44  93  12  24]]
```

In linear algebra, matrices are used to keep track of the coefficients of linear transformations. Multiplying two matrices represents a combination of two linear transformations in a single transformation. Please note that neither matrix multiplication nor matrix division is commutative.

In the earth sciences, however, matrices are often simply used as two-dimensional arrays of numerical data rather than as matrices *sensu stricto* when representing a linear transformation. Arithmetic operations on such arrays are carried out element by element. While this does not make any difference in addition or subtraction, it does affect multiplicative operations. As an example, multiplying A and B element by element is performed by typing

```
C = A * B
print(C)
```

or

```
C = np.multiply(A,B)
print(C)
```

which generates the output

```
[[  8   8  18  35]
 [ 63  24  -5  12]
 [  2  30 -24 -63]
 [ 18   6   6  -6]]
```

2.4 Array Manipulation

Python provides a wide range of functions with which to manipulate arrays. This section introduces the most important functions for array manipulation, and these functions are used later in the book. We begin by clearing the workspace, importing the *NumPy* package as *np* and the *NumPy* module for masked arrays as *ma* in order to be able work with arrays with missing or invalid entries, and creating two arrays A and B by typing

```
%reset -f

import numpy as np
import numpy.ma as ma

A = np.array([[2,4,3],[9,3,-1]])
print(A)

B = np.array([[1,9,3],[6,6,3]])
print(B)
```

which yields

```
[[ 2   4   3]
 [ 9   3  -1]]

[[1 9 3]
 [6 6 3]]
```

When we work with arrays, we sometimes need to concatenate two or more arrays into one single array. We can use `concatenate(arrays,axis=0)` to concatenate arrays A and B along the first dimension (i.e., along the rows) by using `axis=0`. Alternatively, we can use `vstack()` to concatenate arrays A and B vertically. By typing either

```
C = ma.concatenate([A,B],axis=0)
print(C)
```

or

```
C = ma.vstack([A,B])
print(C)
```

we obtain (in both cases)

```
[[ 2   4   3]
 [ 9   3  -1]
 [ 1   9   3]
 [ 6   6   3]]
```

Similarly, we can concatenate arrays horizontally, that is, along the second dimension (i.e., along the columns) using `axis=1`. Typing

```
D = ma.concatenate([A,B],axis=1)
print(D)
```

or instead using `hstack()`

```
D = ma.hstack([A,B])
print(D)
```

yields (in both cases)

```
[[ 2   4   3   1   9   3]
 [ 9   3  -1   6   6   3]]
```

When working with satellite images, we often concatenate three spectral bands into three-dimensional arrays using the colors red, green, and blue (RGB) (Sects. 2.5 and 8.4). Since `concatenate()` allows the use of `axis=2`, we use `dstack()` to concatenate arrays A and B along the third dimension by typing

```
E = ma.dstack([A,B])
print(E)
```

which yields

```
[[[2 1]
  [4 9]
  [3 3]]

 [[9 6]
  [3 6]
  [-1 3]]]
```

Typing

```
np.whos()
```

yields

Name	Shape	Bytes	Type
A	2 x 3	48	int64
B	2 x 3	48	int64
D	2 x 6	96	int64
C	4 x 3	96	int64
E	2 x 3 x 2	96	int64

which indicates that we have now created a three-dimensional array, as revealed by the resulting size of 2-by-3-by-2. Alternatively, we can use

```
print(E.shape)
```

which yields

```
(2, 3, 2)
```

which tells us that the array has 2 rows, 3 columns, and 2 layers in the third dimension. Using `len` instead of `shape()`.

```
print(len(E))
```

yields

2

which tells us the first dimension of the array only. Hence, `len()` is normally used to determine the length of a one-dimensional array (or vector), such as the evenly spaced time axis c, which was created in Sect. 2.3.

Python uses an array indexing that includes the `[0,0]` element in the upper-left corner. Other types of data that are imported into Python may follow a different indexing convention. As an example, digital terrain models (introduced in Sects. 7.3 to 7.5) are often indexed differently and therefore need to be flipped in an up–down direction, or in other words, about a horizontal axis. Alternatively, we can flip arrays in a left–right direction (i.e., about a vertical axis). We can use `flipud()` for flipping in an up–down direction and `fliplr()` for flipping in a left–right direction, as in

```
F = np.flipud(A)
print(F)

F = np.fliplr(A)
print(F)
```

which yields

```
[[ 9   3  -1]
 [ 2   4   3]]

[[ 3   4   2]
 [-1   3   9]]
```

In more complex examples, we can use `roll(a,shift,axis=1)` to circularly shift (i.e., rotate) arrays by `shift` positions along the dimension `axis=1`. For example, we can shift array A by one position along the 2nd dimension (i.e., along the rows) by typing

```
G = np.roll(A,1,axis=1)
print(G)
```

which yields

```
[[ 3   2   4]
 [-1   9   3]]
```

We can also use `reshape()` to completely reshape the array. The result is an m-by-n array H whose elements are taken column-wise from A. As an example, we can create a 3-by-2 array from A by typing

```
H = np.reshape(A,np.array([3,2]),order='F')
print(H)
```

which yields

```
[[ 2  3]
 [ 9  3]
 [ 4 -1]]
```

where order='F' is an instruction for reading/writing the elements using Fortran-like index order, with the first index changing fastest and the last index changing slowest.

Another important way of manipulating arrays is by sorting their elements. As an example, we can use sort(C,axis=0) with axis=0 to sort the elements of C in ascending order along the first array dimension (i.e., the rows). Typing

```
print(C)
I = np.sort(C,axis=0)
print(I)
```

yields

```
[[2 4 3]
 [9 3 -1]
 [1 9 3]
 [6 6 3]]

[[1 3 -1]
 [2 4 3]
 [6 6 3]
 [9 9 3]]
```

The function argsort() can be used to sort the rows of C according to the second column C[:,1]. Typing

```
print(C)
J = C[C[:,1].argsort()]
print(J)
```

yields

```
[[2 4 3]
 [9 3 -1]
 [1 9 3]
 [6 6 3]]

[[9 3 -1]
 [2 4 3]
 [6 6 3]
 [1 9 3]]
```

Array manipulation also includes making comparisons of arrays, for example, by checking whether elements in A are also found in B by using isin(). Typing

```
print(A)
print(B)

K = np.isin(A,B)
print(K)
```

yields

```
[[False False  True]
 [ True  True False]]
```

with False indicating that the element A[i,j] is not in B and True indicating that the element A[i,j] is in B. We can also locate elements within A for which a statement is true. For example, we can locate elements with values of less than zero and replace them with NaNs by typing

```
L = A
L = L.astype(float)
L[np.where(L<0)] = np.NaN
print(L)
```

which yields (in both cases)

```
[[ 2.  4.  3.]
 [ 9.  3. nan]]
```

This is very useful when working with digital elevation models for which values below sea level are not relevant. Alternatively, we can replace data voids other than NaNs, such as -32,768, which is often used as a void with digital terrain models (Sects. 7.3 to 7.5). We can then determine which elements of an array are NaNs by typing

```
M = np.isnan(L)
print(M)
```

which yields

```
[[False False False]
 [False False  True]]
```

with NaNs indicated by True and non-NaN values indicated by False. Which of the elements in array A are unique can be determined by typing

```
N = np.unique(A)
print(N)
```

which returns

```
[-1  2  3  4  9]
```

which are the same values as in A but with no repetitions. The value of 3 occurs twice in A, and the number of elements in N is therefore one less than in A.

2.5   Data Types in Python

On a computer, we describe numbers with a finite (e.g., 32 or 64) total of zeros and ones, that is, with the so-called binary system. The total of representable numbers is therefore finite, in contrast to the total of real numbers. In a 64 bit system, there are 2^{64} (or ~$1.8447 \cdot 10^{19}$) different numbers. These *floating-point numbers* are arranged such that the density is greatest around zero and decreases in the direction of minus infinity and plus infinity in order to support a trade-off between precision and range.

To demonstrate the difference between real numbers and floating-point numbers, consider the value of $x = 4/3$, which we can write as 1.3333333... (i.e., with an infinitely repeating decimal of 3). We can write 4/3 in Python by typing

```
%reset -f

import numpy as np
import matplotlib.pyplot as plt
import matplotlib.image as mpimg
from datetime import date

print(4/3)
```

which yields

```
1.3333333333333333
```

As we can see, Python rounds the value of 4/3 to the next floating-point number. Real numbers are continuous, whereas floating-point numbers are not. The result of our calculation can therefore fall within the gap between two floating-point numbers; it is then rounded to the next floating-point number and is therefore not exact.

Converting decimals to binaries (and back) is not difficult. As an example, let us convert 25.1 to a binary representation. This number has an integer part of 25 and a fractional part of 0.1, which are converted separately. First, let us convert the integer part by typing

```
2^7 2^6 2^5 2^4 2^3 2^2 2^1 2^0
128 64  32  16  8   4   2   1
0   0   0   1   1   0   0   1      binary number
                                  (power of 2)
            16  8           1 = 25 decimal number
                                  (power of 10)
```

As we can see, we divide the number 25 into a sum of powers of two instead of powers of ten as in the decimal system. Alternatively, we can calculate the binary representation by using

```
25/2 = 12 remainder 1
12/2 =  6 remainder 0
 6/2 =  3 remainder 0
 3/2 =  1 remainder 1
 1/2 =  0 remainder 1 (results in a zero therefore stop)
```

The remainders are the binaries, which are arranged in reverse order such that the first remainder becomes the least significant digit (*LSD*) and the last remainder becomes the most significant digit (*MSD*). In our example, the integer part of the number 25.1 corresponds to the binary representation 11001, or to a normalized floating-point number of $1.1001 \cdot 2^4$.

Let us now convert 0.1 to a binary representation. Unfortunately, 0.1 is one of the many examples that have an infinitely repeating binary representation. This is because 0.1 lies between two floating-point numbers. In base 10, it is $1.0 \cdot 10^{-1}$, but in base 2, it is

```
.0001100110011001100110011001100110011...
```

which is calculated by multiplying 0.1 by two, taking the decimal as the digit, taking the fraction as the starting point for the next step, repeating the process until we get either to 0 or a periodic number, and then reading the binary number starting from the top, as shown below:

```
0.1 * 2 = 0.2 -> 0
0.2 * 2 = 0.4 -> 0
0.4 * 2 = 0.8 -> 0
0.8 * 2 = 1.6 -> 1
0.6 * 2 = 1.2 -> 1
0.2 * 2 = 0.4 -> 0
0.4 * 2 = 0.8 -> 0
0.8 * 2 = 1.6 -> 1
0.6 * 2 = 1.2 -> 1
```

The result is $000011(0011)$ periodic, or $1.10011001\ldots \cdot 2^{-4}$ as a normalized floating-point number. Merging the integer with the fractional part yields

```
11001.0001100110011001100110011001100110011...
```

as the binary representation of 25.1. On a computer that describes numbers with a finite (e.g., 32 or 64) total of zeros and ones, we therefore use an approximate representation of 25.1.

In Python, we use `finfo(np.float64).eps` or `np.spacing(1)` to calculate the *floating-point relative accuracy* in order to estimate the distance from x to the next-largest double-precision number. As an example, we can calculate the distance from 1 to the next-largest double-precision floating-point number by typing

```
print(np.finfo(np.float64).eps)
```

or

```
print(np.spacing(1))
```

both of which yield

```
2.220446049250313e-16
```

To calculate the distance from 5 to the next-largest floating-point number, we type

```
print(np.spacing(5))
```

which yields

```
8.881784197001252e-16
```

which demonstrates the decreasing density of floating-point numbers with increasing distance from zero. The range of available floating-point numbers in a computer is limited to between the smallest and largest possible floating-point numbers. Using

```
print(np.finfo(float).tiny)
print(np.finfo(float).max)
```

yields

```
2.2250738585072014e-308
1.7976931348623157e+308
```

There are problems associated with our work with floating-point numbers. The commutativity of addition $[a + b = b + a]$ and multiplication $[a \cdot b = b \cdot a]$ still holds, whereas associativity $[(a+(b+c)=(a+b)+c]$ does not, as we can see in the following example. Typing

```
print((0.5 + 0.2) - 0.7)
print(0.5 + (0.2 - 0.7))
```

yields

```
0.0
5.551115123125783e-17
```

However, the result would be the same (i.e., zero) with real numbers. Distributivity $[a \cdot (b + c) = a \cdot b + a \cdot c]$ is also violated, as the following example demonstrates. Typing

```
print(0.2 * (0.5 - 0.7) + 0.04)
print(0.2 * 0.5 - 0.2 * 0.7 + 0.04)
```

yields different values,

```
6.938893903907228e-18
2.0816681711721685e-17
```

whereas the correct value is again zero. Below is a slightly more complex example in which b is halved at every step. With real numbers, this would never stop, but with floating-point numbers, the halving stops when b becomes smaller than the floating-point relative accuracy, which is then considered to be zero. Typing

```
i = 1
a = 1
b = 1
```

```
while a + b != a:
    b = b/2
    i = i+1

print(b)
print(i)
```

yields

```
1.1102230246251565e-16
54
```

which suggests that the `while` loop indeed terminates after 54 loops.

The following example helps to explain how violating associativity – combined with overflow (larger than the largest possible floating-point number) or underflow (smaller than the smallest possible floating-point number) – can actually affect our work. The result of `((1+x)-1)/x` should be 1, but typing

```
x = 1e-15
print(((1+x)-1)/x)
```

yields

```
1.1102230246251565
```

which is ~0.11, or ~11% more than the correct value of one.

Let us now look at some examples of arrays in order to familiarize ourselves with the different data types in Python. There is no default data type in Python; instead, the data type is determined by the input of numerical values. For example, `A=1` creates a variable of the data type `int` (for *integer*), while `A=1.0` (or `A=1.`, for short) creates a variable of the data type `float` (for *floating-point*). Python's built-in `float` data type is *double precision* (or simply *double*), which stores data in a 64 bit array of floating-point numbers. A double-precision array allows the sign of a number to be stored (bit 63) together with the exponent (bits 62 to 52) and with roughly 16 significant decimal digits (bits 51 to 0). For the first example, we create a 3-by-4 *NumPy* array of random numbers with double precision by typing

```
np.random.seed(0)
A = np.random.rand(3,4)
print(A)
```

We use `random.rand()`, which generates uniformly distributed pseudorandom numbers within the closed interval [0, 1]. To obtain identical data values, we use `random.seed(0)` to reset the random number generator by using the integer 0 as the *seed* (see Chap. 3 for more details on random number generators and types of distributions). The function random.rand() and its sister function random.randn() (which are used generate normally distributed pseudorandom numbers) together with random.seed(0) (which is used to define the seed) are convenience functions for users who port code from MATLAB. The *NumPy* documentation instead recommends using random() and standard_normal() together with random.default_rng(0), and we follow this recommendation as far as possible. However, in the following example as well as in a few other examples in this book, we instead use the MATLAB-like function and generate the following output:

```
[[0.5488135  0.71518937 0.60276338 0.54488318]
 [0.4236548  0.64589411 0.43758721 0.891773  ]
 [0.96366276 0.38344152 0.79172504 0.52889492]]
```

By default, the output is in a scaled fixed-point format with 8 digits (e.g., 0.5488135 for the [0,0] element of A), but the 64 bit precision of the computation results remains unchanged after rounding to 8 digits. However, the precision is affected by converting the data type from *double* to 32 bit *single precision*. Typing

```
B = np.single(A)
print(B)
```

yields

```
[[0.5488135  0.71518934 0.60276335 0.5448832 ]
 [0.4236548  0.6458941  0.4375872  0.891773  ]
 [0.96366274 0.3834415  0.79172504 0.5288949 ]]
```

Typing

```
np.who()
```

lists variables A and B with information on their sizes (or dimensions), their number of bytes, and their classes, as shown below:

```
Name              Shape              Bytes              Type
==========================================================

A                 3 x 4              96                 float64
B                 3 x 4              48                 float32

Upper bound on total bytes  =        144
```

The data type float64 is used in all Python operations in which the physical memory of the computer is not a limiting factor, whereas float32 is used when working with large data sets. The double-precision variable A (whose size is 3-by-4 elements) requires $3 \cdot 4 \cdot 64 = 768$ bits (or 768/8 = 96 bytes) of memory, whereas B requires only 48 bytes and thus has half the memory requirement of A. Introducing at least one complex number to A doubles the memory requirement since both real and imaginary parts are double precision by default. Typing

```
A = np.complex128(A)
A[1,3] = 4j + 3
print(A)
```

yields

```
[[0.5488135 +0.j 0.71518937+0.j 0.60276338+0.j 0.54488318+0.j]
 [0.4236548 +0.j 0.64589411+0.j 0.43758721+0.j 3.        +4.j]
 [0.96366276+0.j 0.38344152+0.j 0.79172504+0.j 0.52889492+0.j]]
```

and the variable listing is now

Name	Shape	Bytes	Type
A	3 x 4	192	complex128
B	3 x 4	48	float32

which indicates the data type complex128. Please note that j (and not i) is the Python imaginary unit (or unit imaginary number) defined by the property that its square is -1.

Python also works with even smaller data types (e.g., 1 bit, 8 bit, and 16 bit data) in order to save memory. These data types are used to store digital elevation models or images (see Chaps. 7 and 8). For example, m-by-n pixel RGB true-color images are usually stored as three-dimensional arrays, that is, the three colors are represented by an m-by-n-by-3 array (see Chap. 8 for more details on RGB composites and true-color images). Such multi-dimensional arrays can be generated by concatenating three two-dimensional arrays that represent the m-by-n pixels of an image. First, we generate a 100-by-100 array of uniformly distributed random numbers in the range of [0, 1]. We then multiply the random numbers by 255 to get values between 0 and 255, as shown below:

```
np.random.seed(0)
I1 = 255 * np.random.rand(100,100)
I2 = 255 * np.random.rand(100,100)
I3 = 255 * np.random.rand(100,100)
```

The command dstack() concatenates the three two-dimensional arrays (8 bits each) into a three-dimensional array ($3 \cdot 8\text{bits} = 24\text{bits}$) by typing

```
I = np.dstack([I1,I2,I3])
```

Since RGB images are represented by integer values between 0 and 255 for each color, we convert the 64 bit double-precision values to unsigned 8 bit integers using `uint8()` (Sect. 8.2):

```
I = np.uint8(I)
```

The function `uint8()` rounds the values in `I` to the nearest integer. Any values that are outside the range of [0, 255] are rolled around from 0 to 255 (i.e., `np.uint8 (-3)` is `253`). This behavior is different from that of the corresponding MATLAB function `uint8`, which assigns values outside the range of [0, 255] to the nearest endpoint (i.e., either 0 or 255). To avoid rolling, we use `rescale_intensity()` from the *exposure* module of the *scikit-image* package instead (see Sect. 8.4). Typing

```
np.who()
```

yields

```
Name                Shape               Bytes           Type
============================================================

I1                  100 x 100           80000           float64
I2                  100 x 100           80000           float64
I3                  100 x 100           80000           float64
I                   100 x 100 x 3       30000           uint8
```

Since 8 bits can be used to store 256 different values, this data type can be used to store integer values between 0 and 255, whereas using `int8` to create signed 8 bit integers generates values between −128 and +127. The value of zero requires one bit, and there is therefore no space left in which to store +128. Finally, `imshow()` can be used to display the three-dimensional array as a true-color image:

```
plt.figure()
plt.imshow(I)
plt.show()
```

In so doing, the *Matplotlib* function `figure()` creates a new figure, `imshow()` initializes the display of the image, `show()` actually displays the current figure, and *IPython* returns the prompt indicating readiness to accept further commands.

In the following chapters, we often have to save character strings (e.g., location names or sample names) in arrays. Python provides a number of options for storing text data, including in *NumPy* arrays with character strings. For example, we can create such an array named `location_1` that contains the districts of northern Potsdam. It thus makes no difference whether we use single quotation marks

```
locations_1 = np.array([['Bornim'],
                        ['Nedlitz'],
                        ['Bornstedt'],
                        ['Sacrow'],
                        ['Eiche'],
                        ['Grube'],
                        ['Golm']])
```

or double quotation marks,

```
locations_1 = np.array([["Bornim"],
                        ["Nedlitz"],
                        ["Bornstedt"],
                        ["Sacrow"],
                        ["Eiche"],
                        ["Grube"],
                        ["Golm"]])
```

but the choice should be made consistently throughout the Python code. In this book, we prefer to use single quotation marks. In the code above, the strings do not fit on one line, and we thus insert a new line after each comma. Typing

```
np.who()
```

yields

```
Name                Shape                Bytes                Type
================================================================

locations_1        7 x 1                252                  str288

Upper bound on total bytes  =          252
```

As we can see, the shape of the character array location_1 corresponds to the number of district names times one. Storing strings require 32 bits per character, and the string with the largest number of characters (e.g., 'Bornstedt', with 9 characters) defines the number of bits as $32 \cdot 9 = 288$ for each string in locations_1, as indicated by the type str288. The array location_1 therefore requires $7 \cdot 9 \cdot 32/8 = 252$ bytes of memory.

We next look at the possibilities that Python offers for storing mixed data types in a single array using *dictionaries* (or *dicts*, for short). These data containers are particularly useful for storing any kind of information about a sample in a single variable. For example, we can generate a structure array sample_1 that includes the image array I (which we defined in the previous example) by typing

```
np.random.seed(0)
I = np.uint8(255 * np.random.rand(100,100,3))
```

Moreover, we can also generate other types of information about a sample, such as the name of the sampling location, the date of sampling, or geochemical measurements (which are stored in a 10-by-10 array), by typing

```
sample_1 = {
    'location':'Plougasnou',
    'date':date.today(),
    'image':I,
    'geochemistry':np.random.rand(10,10)
}
```

The first layer (`location`) of the dict `sample_1` contains a character string with the location name `Plougasnou`. The second layer (`date`) contains a character string generated by `date.today()`, which yields a string containing the current date in dd-mm-yyyy format. We access this particular layer in `sample_1` by typing

```
print(sample_1['date'])
```

which yields

```
2022-02-02
```

as an example. The third layer of `sample_1` contains the image, while the fourth layer contains a 10-by-10 array of uniformly distributed pseudorandom numbers. All layers of `sample_1` can be listed by typing

```
print(sample_1)
```

or

```
sample_1
```

The dict `sample_1` contains *NumPy* arrays, which are listed using `who()`, but not the other things contained in the dict, such as location names and dates. We can use the command `list()` to list the variables in the dict. Furthermore, `items()` lists all variables in the dict and also displays part of the data:

```
np.who(sample_1)
list(sample_1)
sample_1.items()
```

We can also create an array with different types of data. However, this is not advised due to the output of the following line:

```
C = np.array(['Plougasnou',
    date.today(),
    I,
    np.random.rand(10,10)])
```

Instead, we should use a list, which is another data type in Python, by typing

```
C = ['Plougasnou',
    date.today(),
    I,
    np.random.rand(10,10)]
```

and we should access the content using

```
print(C)
```

To access a single element of the list, we use

```
print(C[1])
```

which yields

```
2022-02-02
```

It is important to bear in mind that indexing in Python starts with zero.

Another very convenient way to store mixed data types in a single array involves concatenating *NumPy* arrays. As an example, we create a simple table with two columns. The first column again contains the names of the districts of northern Potsdam:

```
Districts = np.array([['Bornim'],
                ['Nedlitz'],
                ['Bornstedt'],
                ['Sacrow'],
                ['Eiche'],
                ['Grube'],
                ['Golm']])
```

The second column contains the number of inhabitants in December 2018 (source https://www.potsdam.de):

```
Inhabitants = np.array([[3429],
                        [192],
                        [13895],
                        [140],
                        [5308],
                        [433],
                        [4310]])
```

We then concatenate the two arrays into a single array using

```
potsdamnorth = [Districts,Inhabitants]
```

which yields

```
[array([['Bornim'],
        ['Nedlitz'],
        ['Bornstedt'],
        ['Sacrow'],
        ['Eiche'],
        ['Grube'],
        ['Golm']], dtype='<U9'),
 array([[ 3429],
        [  192],
        [13895],
        [  140],
        [ 5308],
        [  433],
        [ 4310]])]
```

As we can see, the heterogeneous data are collected in a single array and are displayed as a table. We can access the district names and the number of inhabitants separately by typing

```
print(potsdamnorth[0])
print(potsdamnorth[1])
```

which yields

```
[['Bornim']
 ['Nedlitz']
 ['Bornstedt']
 ['Sacrow']
 ['Eiche']
 ['Grube']
 ['Golm']]
```

```
[[ 3429]
 [  192]
 [13895]
 [  140]
 [ 5308]
 [  433]
 [ 4310]]
```

This data type is very useful when working with experimental data, such as large data arrays with sample IDs that consist of combinations of letters with integers and floating-point numbers.

The third data structure for storing mixed data types in Python is the *tuple*, which uses parentheses instead of the curly brackets used with dicts or the square brackets used with lists. Tuples are also separated by commas. Typing

```
potsdamnorth = (Districts,Inhabitants)
print(potsdamnorth[0])
print(potsdamnorth[1])
```

yields

```
[['Bornim']
 ['Nedlitz']
 ['Bornstedt']
 ['Sacrow']
 ['Eiche']
 ['Grube']
 ['Golm']]

[[ 3429]
 [  192]
 [13895]
 [  140]
 [ 5308]
 [  433]
 [ 4310]]
```

Tuples are faster and require less memory compared with dicts and lists, but they are immutable, whereas lists can be extended and reduced.

2.6   Data Storage and Handling

This section deals with how to store, import, and export data with Python. Many of the data formats typically used in the earth sciences must be converted before being analyzed with Python. Alternatively, the software provides several import routines for reading many binary data formats in the earth sciences, such as those used to store digital elevation models and satellite data. Newer functions in Python are capable of importing mixed data types (e.g., text, integers, and floating-point numbers) from a single file into the workspace.

A computer generally stores data as *binary digits*, or *bits*, for short. A bit is analogous to a two-way switch with two states: $on = 1$ and $off = 0$. These bits are joined together to form larger groups, such as *bytes* (which consist of 8 bits) in order to store more complex types of data. These groups of bits are then used to encode data, such as numbers or characters. Unfortunately, different computer systems and software use different schemes for encoding data. Exchanging binary data between different computer platforms and software can therefore be difficult.

Various formats for exchanging data have been developed in recent decades. The classic example of a data format that can be used with different computer platforms and software is the American Standard Code for Information Interchange (ASCII), which was first published in 1963 by the American Standards Association (ASA). As a 7 bit code, ASCII consists of $2^7 = 128$ characters (codes 0 to 127). Whereas ASCII-1963 lacked lowercase letters, in the ASCII-1967 update, lowercase letters as well as various control characters (e.g., escape and line feed) and symbols (e.g., brackets and mathematical operators) were also included. Since then, a number of variants have appeared in order to facilitate the exchange of text written in languages other than English. Such variants include the expanded ASCII, which contains 255 codes, including the Latin-1 encoding.

The simplest way to exchange data between a certain piece of software and Python is by using the ASCII format. Although some packages (e.g., *Pandas*) provide various functions for reading binary files, such as Microsoft Excel files, most data arrive in the form of ASCII files. Consider a simple data set stored in a table, such as

```
SampleID   Percent C   Percent S
101        0.3657      0.0636
102        0.2208      0.1135
103        0.5353      0.5191
104        0.5009      0.5216
105        0.5415      -999
106        0.501       -999
```

The first row contains the names of the variables, and the columns provide the percentages of carbon and sulfur in each sample. The absurd value of -999 indicates missing data in the data set. Two things must be changed in order to convert this table to Python format. First, Python uses NaN or nan as the representation for *Not2011a-Number*, which can be used to mark missing data or gaps. Second, the hash character (#) should be added at the beginning of the first line. This hash character is used to indicate non-executable text within the body of a program. This text is normally used to include comments in the code, as in the following example:

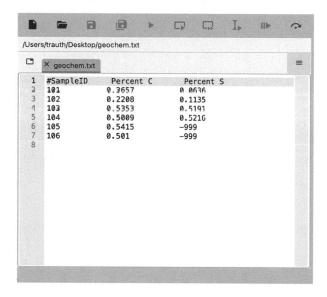

Fig. 2.2 Screenshot of Spyder *Editor* showing the content of the file *geochem.txt*. The first line of text needs to be commented out using the hash character (#) at the beginning of the line, which is followed by the actual data array. The -999 values need to be replaced by NaNs.

```
#SampleID  Percent C    Percent S
101        0.3657       0.0636
102        0.2208       0.1135
103        0.5353       0.5191
104        0.5009       0.5216
105        0.5415       NaN
106        0.501        NaN
```

Python ignores any text that appears after the hash character and continues processing the next line. After editing this table in a text editor, such as the *Spyder Editor*, it can be saved as an ASCII text file *geochem.txt* in the current working directory (Fig. 2.2). We first enter

```
%reset -f
```

after the prompt in the Console. Python can now import the data from this file with the loadtxt() command:

```
import numpy as np
import matplotlib.pyplot as plt
from datetime import datetime

geochem = np.loadtxt('geochem.txt')
```

Python then loads the contents of the file and assigns the array to a variable geo-chem specified by the filename *geochem.txt*. Typing

```
np.whos()
```

yields

```
Name              Shape           Bytes           Type
================================================================

geochem           6 x 3           144             float64

Upper bound on total bytes   =       144
```

The command `savetxt()` now allows geochem to be stored in a text format:

```
np.savetxt('geochem_new.txt',geochem)
```

This ASCII file can be viewed and edited using the Spyder Editor or any other text editor.

Such data files (especially those that are produced by electronic instruments) can appear much more complicated than the example file *geochem.txt*, which has a single header line. In Chaps. 7 and 8, we read some of these complicated and extensive files, which are either binary or text files and usually have long headers that describe their contents. At this point, we now take a look at a text file variant that contains not only one or more header lines, but also unusual data types, such as date and time in a non-decimal format. We use `loadtext()` to perform this task, as shown below:

```
dtype = np.dtype('i4,f8,f8')
C1,C2,C3 = np.loadtxt('geochem.txt',
    dtype=dtype,
    unpack=True,
    skiprows=1)
```

The character string `'i4,f8,f8'` used as input for `dtype()` defines the conversion specifiers enclosed in single quotation marks, where `i4` stands for the 32 bit (or 4 byte) unsigned integers and `f8` stands for 64 bit (or 8 byte) double-precision floating-point numbers. The function `loadtext()` uses `dtype` as an input argument together with `unpack=True` to transpose the returned arrays, and it uses `skip-rows=1` to ignore a single header line while reading the file. The function reads the text from the file and stores it in arrays C1, C2, and C3. We then reshape the arrays to column vectors by typing

```
C1 = np.reshape(C1,(6,1))
C2 = np.reshape(C2,(6,1))
C3 = np.reshape(C3,(6,1))
```

and we concatenate C1, C2, and C3 into a single array C by typing

```
C = np.concatenate((C1,C2,C3),axis=1)
print(C)
```

which yields

```
array([[1.010e+02, 3.657e-01, 6.360e-02],
       [1.020e+02, 2.208e-01, 1.135e-01],
       [1.030e+02, 5.353e-01, 5.191e-01],
       [1.040e+02, 5.009e-01, 5.216e-01],
       [1.050e+02, 5.415e-01,      nan],
       [1.060e+02, 5.010e-01,      nan]])
```

The next examples demonstrate how to read the file *geophys.txt*, which contains a single header line but also the date (in an *MM/DD/YY* format) and time (in an *HH:MM:SS.SS* format). We again use loadtext() to read the file:

```
dtype = np.dtype('i4,f8,f8,f8,U12,U12')
C1,C2,C3,C4,C5,C6 = \
    np.loadtxt('geophys.txt',
        dtype=dtype,
        unpack=True,
        skiprows=1,
        delimiter='\t')
```

We skip the header, read the first column (the sample ID) as a 32 bit unsigned integer (uint32) with specifier i4, read the next three columns (*X*, *Y*, and *Z*) as 64 bit double-precision floating-point numbers (double) with specifier f8, and then read the date and time as character strings with specifier U12. In this code, we use the explicit line break (\) to create a new line after the equal sign. Alternatively, we can instead wrap the code in parentheses by typing

```
dtype = np.dtype('i4,f8,f8,f8,U12,U12')
C1,C2,C3,C4,C5,C6 = (
    np.loadtxt('geophys.txt',
        dtype=dtype,
        unpack=True,
        skiprows=1,
        delimiter='\t')
    )
```

The parentheses used in `loadtext()` are also the reason that we do not need the explicit line breaks inside the function in order to write the input arguments on separate lines. We then convert the date using

```
for i in range(0,5):
    date_time_obj_1 = datetime.strptime(C5[i],'%m/%d/%y')
    print(date_time_obj_1)
```

which yields

```
2013-11-18 00:00:00
2013-11-18 00:00:00
2013-11-18 00:00:00
2013-11-18 00:00:00
2013-11-18 00:00:00
```

in which the date uses `00:00:00` as the default time since no time has been defined and in which the time uses

```
for i in range(0,5):
    date_time_obj_2 = datetime.strptime(C6[i],'%H:%M:%S.%f')
    print(date_time_obj_2)
```

which yields

```
1900-01-01 10:23:09.100000
1900-01-01 10:23:10.200000
1900-01-01 10:23:50.400000
1900-01-01 10:24:05.100000
1900-01-01 10:24:23.300000
```

in which the time uses `1900-01-01` as the default date since no date has been defined.

We can also write data to a formatted text file using `savetxt()`. As an example, we again load the data from *geochem.txt* after adding a hash character (#) to the beginning of the first line and after replacing -999 with `NaN`. Instead of adding the hash character, we can also skip the first line by using `skiprows=1`

```
data = np.loadtxt('geochem.txt',skiprows=1)
```

to load the contents of the text file into a double-precision array `data`. We write the data to a new text file *geochem_formatted.txt* using `savetxt()`:

```
np.savetxt('geochem_formatted.txt',data,
    delimiter='\t',
    fmt='%i %6.4f %6.4f')
```

We then write data to the file using the formatting operators %i for unsigned integers and %6.4f for fixed-point numbers with a field width of six characters and four digits after the decimal point. The control character \t denotes the tabulator as a delimiter. We can view the contents of the file by typing

```
%edit geochem_formatted.txt
```

which opens the file *geochem_formatted.txt*

```
101 0.3657 0.0636
102 0.2208 0.1135
103 0.5353 0.5191
104 0.5009 0.5216
105 0.5415    nan
106 0.5010    nan
```

in the Spyder Editor. The format of the data is as expected.

2.7   Control Flow

Control flow in computer science helps to control the order in which computer code is evaluated. The most important kinds of control flow statements are count-controlled loops (e.g., *for* loops) and conditional statements (e.g., *if* statements). Since we do not deal with the programming capabilities of Python in any depth in this book, the following introduction to the basics of control flow is rather brief and omits certain important aspects of efficient programming. This introduction is limited to the two most important kinds of control flow statements: the aforementioned *for* loops, and *if* statements.

Python for loops execute a series of commands a specified number of times. As an example, we can use such a loop to multiply the elements of an array A by 10, round the result to the nearest integer, and store the result in B by typing

```
%reset -f
```

```
import numpy as np
from numpy.random import default_rng
```

```
np.random.seed(0)
A = np.random.rand(10,1)
B = np.zeros((10,1))
```

```
for i in range(0,10):
    B[i] = np.round(10*A[i])
print(A)
print(B)
```

which yields

```
[[0.5488135 ]
 [0.71518937]
 [0.60276338]
 [0.54488318]
 [0.4236548 ]
 [0.64589411]
 [0.43758721]
 [0.891773  ]
 [0.96366276]
 [0.38344152]]

[[ 5.]
 [ 7.]
 [ 6.]
 [ 5.]
 [ 4.]
 [ 6.]
 [ 4.]
 [ 9.]
 [10.]
 [ 4.]]
```

It is important that the block of code within the for loop, which is B[i] = np. round(10*A[i]) in our example, be indented in order to indicate that it belongs to the loop. *Spaces* should be used instead of *tabulators* (or *tabs*, for short) for indentation (e.g., four spaces for single indentation, as in our example). Please note that we use B = np.zeros((10,1)) to pre-allocate memory prior to using the for loop. We can expand the experiment by using a nested for loop to create a 2D array B. Typing

```
np.random.seed(0)
A = np.random.rand(10,3)
B = np.zeros((10,3))
for i in range(0,10):
    for j in range(0,3):
        B[i,j] = np.round(10*A[i,j])
print(A)
print(B)
```

yields

```
[[0.5488135  0.71518937 0.60276338]
 [0.54488318 0.4236548  0.64589411]
 [0.43758721 0.891773   0.96366276]
 [0.38344152 0.79172504 0.52889492]
 [0.56804456 0.92559664 0.07103606]
 [0.0871293  0.0202184  0.83261985]
 [0.77815675 0.87001215 0.97861834]
 [0.79915856 0.46147936 0.78052918]
 [0.11827443 0.63992102 0.14335329]
 [0.94466892 0.52184832 0.41466194]]

[[ 5.  7.  6.]
 [ 5.  4.  6.]
 [ 4.  9. 10.]
 [ 4.  8.  5.]
 [ 6.  9.  1.]
 [ 1.  0.  8.]
 [ 8.  9. 10.]
 [ 8.  5.  8.]
 [ 1.  6.  1.]
 [ 9.  5.  4.]]
```

In this embedded for loop, the second loop is indented using four spaces, and its associated block of code is again indented, which therefore requires double indentation (i.e., eight spaces).

This book aims to keep all of the recipes independent of the actual dimensions of the data. This is achieved by consistently using shape() and len() to determine the dimension of the arrays instead of using fixed numbers, such as the 10 and 3 in the example above (Sect. 2.4):

```
np.random.seed(0)
A = np.random.rand(10,3)
B = np.zeros(np.shape(A))
for i in range(0,A.shape[0]):
    for j in range(0,A.shape[1]):
        B[i,j] = np.round(10*A[i,j])
print(A)
print(B)
```

The second important statements for controlling the flow of a script (apart from for loops) are if statements, which evaluate an expression and then execute a group of instructions if the expression is true. As an example, we compare the value of two scalars A and B by typing

```
A = 1
B = 2
if A < B:
    print('A is less than B')
```

which yields

```
A is less than B
```

Similar to for loops, if statements require indenting the block of code that is part
of the statement, which is done in our example using four spaces. The script first
evaluates whether A is less than B and, if it is, displays the message A is less
than B in the Console. We can expand the if statement by introducing else,
which provides an alternative statement if the expression is not true. For example,
typing

```
A = 1
B = 2
if A < B:
    print('A is less than B')
else:
    print('A is not less than B')
```

yields

```
A is less than B
```

Alternatively, we can use else together with a second if statement to introduce a
second expression to be evaluated, as shown in the following example:

```
A = 1
B = 2
if A < B:
    print('A is less than B')
else:
    if A >= B:
        print('A is not less than B')
```

These for loops and if statements are extensively used in the subsequent chapters
of this book.

2.8   Scripts and Functions

Python is a powerful programming language. All files that contain Python code use *.py* as an extension and are therefore called *.py files*. These files contain ASCII text and can be edited using a standard text editor. However, the built-in Spyder Editor color-highlights various syntax elements: comments are highlighted in gray, keywords such as *if*, *for*, and *end* are highlighted in blue, and character strings are highlighted in green. This syntax highlighting facilitates Python coding.

Python uses two types of .py files: *scripts* and *functions*. Whereas scripts are a series of commands that operate on data in the workspace, functions are true algorithms with *input variables* (or *input arguments*) and *output variables* (or *output arguments*). The advantages and disadvantages of both types of .py files are now illustrated via an example. We first generate a simple Python script by typing a series of commands in order to calculate the average of the elements of a data array x as follows:

```
%reset -f

import numpy as np

x = np.array([[3,6,2,-3,8]])
print(x)

m,n = x.shape
if m == 1:
    m = n

print(np.sum(x)/m)
```

The first line of the if statement yields the shape of the variable x using the command shape(). In our example, x should be either a column vector – that is, an array with a single column and dimensions (m,1) – or a row vector – that is, an array with a single row and dimensions (1,n). The if statement evaluates a logical expression and executes a group of commands if the expression is true. In the example, the if statement selects either m or n depending on whether m==1 is false or true. Here, the double equal sign (==) makes element-by-element comparisons between the variables (or numbers) to the left and right of the equal signs and returns an array of the same size that is made of elements set to logical 1 where the relationship is true and to logical 0 where it is not true. In our example, m==1 returns 1 if m equals 1 and 0 if m equals any other value. The last line of the if statement computes the average by dividing the sum of elements by m or n. Running the script yields

3.2

After typing

```
%whos
```

we see that the workspace now contains

```
Variable   Type        Data/Info
--------------------------------
m          int         5
n          int         5
np         module      <module 'numpy' from
                       '/Us<...>kages/numpy/__init__.py'>
x          ndarray     1x5: 5 elems, type `int64`, 40 bytes
```

As expected, all variables defined and used in the script appear in the workspace. In our example, these are also the variables m and n. Scripts contain sequences of commands that are applied to variables in the workspace. However, Python functions allow inputs and outputs to be defined. They do not automatically import variables from the workspace. To convert the script above into a function, we need to introduce the following modifications (Fig. 2.3):

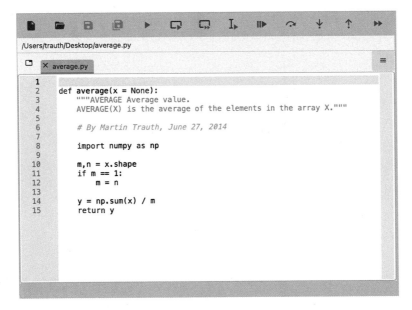

Fig. 2.3 Screenshot of the Spyder *Editor* showing the function average(). The function starts with a line that contains the keyword def, the name of the function average(), and the input variable x. The subsequent lines contain the output for help(average), the copyright and version information, and the actual Python code for computing the average using this function, including the return statement that asks that the output variable y be defined.

```
def average(x = None):
    '''AVERAGE Average value.
    AVERAGE(X) is the average of the elements in the array X.'''

    # By Martin Trauth, June 27, 2014

    import numpy as np

    m,n = x.shape
    if m == 1:
        m = n

    y = np.sum(x) / m
    return y
```

The first line now contains the keyword def (which marks the start of the function header) as well as the function name average() and the input variable (or input argument) x. The lines after the function header contain the output of help(), as indicated by the triple ' signs. Separated by an empty line, line 5 contains the author's name and the version of the .py file. The rest of the file contains the actual operations. The last line now defines the value of the output variable y after the return statement. The workspace can now be cleared before we import average() from our module *average* by typing

```
%reset -f
```

```
import numpy as np
from average import average
```

and using help()

```
help(average)
```

to display the first block of contiguous comment lines, as shown below:

```
Help on function average in module average:
```

```
average(x=None)
    AVERAGE Average value.
    AVERAGE(X) is the average of the elements in the array X.
```

We can now define x and run average() to calculate the mean of the elements in x in order to define result. Typing

```
data = np.array([[3,6,2,-3,8]])

result = average(data)
print(result)
```

again displays

```
3.2
```

whereas typing

```
%whos
```

results in

```
Variable    Type        Data/Info
--------------------------------
average     function    <function average at 0x7fe2a9a02af0>
data        ndarray     1x5: 5 elems, type `int64`, 40 bytes
np          module      <module 'numpy' from '/Us<...>kages/
                        numpy/__init__.py'>
result      float64     3.2
```

which reveals that all variables used in the function do not appear in the work-space. Only the input data and output result as defined by the user are stored in the workspace. The .py files can therefore be applied to data as if they were real functions, whereas scripts contain sequences of commands that are applied to the variables in the workspace.

2.9   Basic Visualization Tools

The *Matplotlib* package provides numerous Python routines for displaying data as graphics. This section introduces the most important graphics functions. The graphics can be modified, printed, and exported to be edited with graphics software other than Python. The simplest function that produces a graph of a variable y versus another variable x is plot(). First, we define two one-dimensional arrays x and y, where y is the sine of x. Array x contains 21 values between 0 and 2π with $\pi/10$ increments, whereas y is the element-by-element sine of x:

```
%reset -f

import numpy as np
import matplotlib.pyplot as plt
```

```
x = np.linspace(0,2*np.pi,21)
y = np.sin(x)
```

These two commands result in two arrays with 21 elements each. Since the two arrays x and y have the same length, we can use plot() to produce a linear 2D graph of y against x, as shown below:

```
plt.plot(x,y)
```

However, the most general way of plotting is to first create a figure window using figure(), to then use plot() to initiate plotting, and to finally display the actual graphic using show(). In particular, if we display subplots or multiple graphics objects, we need show() to see the graphic:

```
plt.figure()
plt.plot(x,y)
plt.show()
```

We may wish to plot two different curves in a single plot, such as the sine and the cosine of x in different colors. The command

```
x = np.linspace(0,2*np.pi,21)
y1 = np.sin(x)
y2 = np.cos(x)

plt.figure()
plt.plot(x,y1,'--',x,y2,'-')
plt.show()
```

creates a dashed blue line that displays the sine of x and a solid orange line that represents the cosine of this array (Fig. 2.4). If we wish to create plots in separate windows, we again use figure() to create a new window that displays the cosine of x after displaying the sine of x:

```
plt.figure()
plt.plot(x,y1,'--')
plt.show()

plt.figure()
plt.plot(x,y2,'-')
plt.show()
```

We can also display the sine of x as a line plot and the cosine of x as a bar plot using

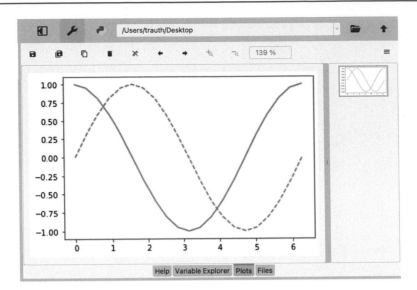

Fig. 2.4 Screenshot of the Spyder *Plots* panel showing two curves in different colors and line types. The graphics can be saved by clicking on the floppy-disk icon, copied to the clipboard by clicking on the copy-to-clipboard icon, and deleted by clicking on the trash icon.

```
plt.figure()
plt.plot(x,y1,'r--')
plt.bar(x,y2,color='b')
plt.show()
```

This command plots y1 versus x as a dashed red line using 'r--', whereas y2 versus x is shown as a group of blue vertical bars. Alternatively, we can plot both graphics in the same figure window but in different plots using subplot(). The syntax subplot(m,n,p) divides the figure window into a grid with m rows and n columns and creates axes in the position specified by p. Here, the first subplot (p=1) is in the first column and first row of the grid, the second subplot (p=2) is in the second column of the first row, and so on:

```
plt.figure()
plt.subplot(2,1,1)
plt.plot(x,y1,'r--')
plt.subplot(2,1,2)
plt.bar(x,y2,color='b')
plt.show()
```

In our example, the figure window is divided into two rows and one column. The 2D linear plot is displayed in the upper half of the figure window, and the bar plot appears in the lower half. Alternatively, we can also use subplots() instead of

subplot() to create the figure window `fig` together with the two axes `ax1` and `ax2`.
If we use subplots(), we do not need figure() and show():

```
fig,(ax1,ax2) = plt.subplots(2,1)
ax1.plot(x,y1,'r--')
ax2.bar(x,y2,color='b')
```

An important modification to the graphics is made by scaling the axes. By default,
Python uses axis limits close to the minima and maxima of the data. Using the
command axis(), however, allows the scale settings to be changed using an array
with the values [xmin,xmax,ymin,ymax], such as

```
plt.figure()
plt.plot(x,y1,'r--')
plt.axis(np.array([0,np.pi,-1,1]))
plt.show()
```

This code sets the limits of the x-axis to 0 and π, whereas the limits of the y-axis
are set to the default values of -1 and $+1$. Important options of axis() are

```
plt.figure()
plt.plot(x,y1,'r--')
plt.axis('square')
plt.show()
```

which makes the x-axis and y-axis the same length, and

```
plt.figure()
plt.plot(x,y1,'r--')
plt.axis('equal')
plt.show()
```

which makes the individual tick mark increments on the x-axis and y-axis the same
length. The function grid() adds a grid to the current plot, whereas title(), xla-
bel(), and ylabel() allow a title to be defined and labels to be applied to the x-
and y-axes:

```
plt.figure()
plt.plot(x,y1,'r--')
plt.title('My first plot')
plt.xlabel('x-axis')
plt.ylabel('y-axis')
plt.grid()
plt.show()
```

The most sophisticated way of plotting using the *Matplotlib* package (similar to the object-oriented syntax of MATLAB) is by typing

```
fig = plt.figure(
    facecolor=(0.9,0.9,0.9))
ax = plt.axes(
    title='My first plot',
    xlabel='x-axis',
    ylabel='y-axis')
ax.plot(x,y1,
    markersize=15,
    marker='s',
    linewidth=2)
ax.grid()
plt.show()
```

which again uses functions such as `figure()`, `plot()`, and `show()` to create and display graphics. However, some of these functions, such as `plot()`, set a number of graphics attributes, such as `markersize`, `marker`, and `linewidth`, in order to customize plots. Furthermore, these functions can also create objects such as `fig` and `ax`, which can later be called by other functions for further modification, such as in the figure window and axes in our example. These are a few examples of how Python functions that are contained within the *Matplotlib* package can be used to create and edit plots. More graphics functions are introduced in the following chapters of this book.

2.10   Generating Code to Recreate Graphics

This section from the MATLAB-based book MATLAB Recipes for Earth Sciences (Trauth 2021a, b) could not be translated to Python.

2.11   Publishing and Sharing MATLAB Code

This section from the MATLAB-based book MATLAB Recipes for Earth Sciences (Trauth 2021a, b) could not be translated to Python.

2.12   Creating Graphical User Interfaces

This section from the MATLAB-based book MATLAB Recipes for Earth Sciences (Trauth 2021a, b) could not be translated to Python (Fig. 2.5).

Fig. 2.5 This figure from the MATLAB-based book MATLAB Recipes for Earth Sciences (Trauth 2021) could not be translated to Python.

Recommended Reading

Matthes E (2019) Python crash course: a hands-on, project-based introduction to programming. No Starch Press.

McKinney W (2017) Python for data analysis, f2. Aufl. O'Reilly Media, Sebastopol

Petrelli M (2021) Introduction to Python in earth science data analysis – from descriptive statistics to machine learning. Springer, Berlin

Trauth MH, Sillmann E (2018) Collecting, processing and presenting geoscientific information, MATLAB and design recipes for earth sciences, 2nd edn. Springer, Berlin

Tenkanen H, Heikinheimo V, Whipp D (2020) Introduction to Python for geographic analysis. https://pythongis.org

Trauth MH (2021a) MATLAB Recipes for Earth Sciences, 5th edn. Springer, Berlin

Trauth MH (2021b) Signal and noise in geosciences, MATLAB Recipes for Data Acquisition inEarth Sciences. Springer, Berlin

Univariate Statistics

3

3.1 Introduction

The statistical properties of a single variable are investigated by means of univariate analysis. Such a variable could, for example, be the organic carbon content of deep-sea sediments, the size of grains in a sandstone layer, or the age of sanidine crystals in volcanic ash. The size of *samples* that we collect from a larger *population* is often limited by financial and logistical constraints. The methods of univariate statistics assist us in drawing from the sample conclusions that apply to the population as a whole. For univariate analysis, we use the *NumPy* (https://numpy. org), *Matplotlib* (https://matplotlib.org), *SciPy* (https://scipy.org), and *scikit-learn* packages (https://scikit-learn.org), which contain all the necessary routines.

We first need to describe the characteristics of the sample using statistical parameters and to compute an *empirical distribution* (*descriptive statistics*) in Sects. 3.2 and 3.3. A brief introduction is provided on the most important statistical parameters (e.g., the *measures of central tendency and dispersion*), followed by Python examples. We then select a *theoretical distribution* that shows similar characteristics to the empirical distribution in Sects. 3.4 and 3.5. A suite of theoretical distributions is introduced, and their potential applications are outlined prior to using Python tools to explore these distributions. We then examine whether the theoretical distribution is a suitable model for the empirical distribution (*hypothesis testing*) before drawing conclusions from the sample that can be applied to the larger population of interest in Sects. 3.6 to 3.12. Section 3.13 introduces methods used to fit distributions to our own data sets. Finally, Sect. 3.14 provides a very brief introduction to the basics of error analysis.

Supplementary Information The online version contains supplementary material available at https://doi.org/10.1007/978-3-031-07719-7_3.

3.2   Empirical Distributions

Let us assume that we have collected a number of measurements x_i from a specific object. The collection of data (or sample) as a subset of the population of interest can be written as a vector or a one-dimensional array x

$$x = (x_1, x_2, \ldots, x_n)$$

that contains a total of n observations. The array x may contain a large number of data points, and understanding its properties may consequently prove difficult. Descriptive statistics are therefore often used to summarize the characteristics of the data. The statistical properties of the data set may be used to define an empirical distribution, which can then be compared with a theoretical distribution.

The most straightforward way of investigating sample characteristics is by displaying the data in graphical form. Plotting all of the data points along a single axis does not reveal a great deal of information about the data set. However, the density of the points along the scale does provide some information about the characteristics of the data. A widely used graphical display of univariate data is the *histogram* (Fig. 3.1). A histogram is a bar plot of a frequency distribution that is organized into intervals, or *classes*. Such histogram plots provide valuable information on the characteristics of the data, such as the *central tendency*, the *dispersion*, and the *general shape* of the distribution. However, quantitative measures provide a more accurate way of describing the data set than does the graphical form. In purely quantitative terms, for example, the *mean* and the *median* define the central tendency of the data set, while the data dispersion is expressed in terms of the *range* and the *standard deviation*.

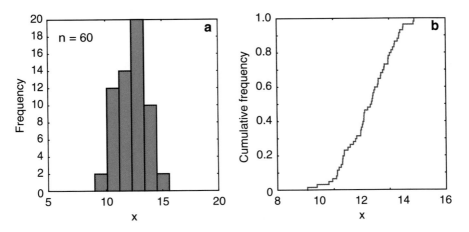

Fig. 3.1 Graphical representation of an empirical frequency distribution. **a** In a *histogram*, the frequencies are organized into *nbin* classes and are plotted as a bar plot. **b** The *cumulative distribution plot* of a frequency distribution displays the totals of all observations less than or equal to a certain value. This plot is normalized to a total number of observations of one.

Measures of Central Tendency

Parameters of central tendency (or location) represent the most important measures for characterizing an empirical distribution (Fig. 3.2). These values help locate the data on a linear scale, and they represent a typical or best value that describes the data. The most popular indicator of central tendency is the *arithmetic mean*, which is the sum of all data points x_i divided by the number of observations n:

$$\bar{x} = \frac{1}{n} \sum_{i=1}^{n} x_i$$

The arithmetic mean of a univariate data set can also be referred to as the mean, or the average. The sample mean is used as an estimate of the population mean μ if the underlying theoretical distribution is a Gaussian (or normal) distribution. However, the arithmetic mean is sensitive to outliers, that is, extreme values that may be very different from the majority of the data, and the *median* is therefore often used as an alternative measure of central tendency. The median is the x-value that lies in the middle of the data set, that is, 50% of all observations lie below the median, and 50% lie above it. The median of a data set sorted in ascending order is defined as

$$\tilde{x} = x_{(n+1)/2}$$

if n is odd and

$$\tilde{x} = \left(x_{(n/2)} + x_{(n/2)+1} \right)/2$$

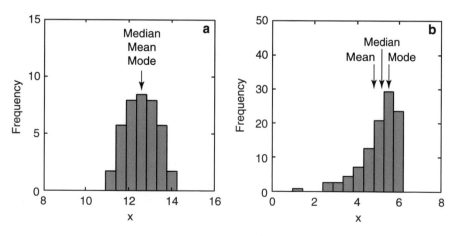

Fig. 3.2 Measures of *central tendency*. **a** In a unimodal symmetric distribution, the mean, the median, and the mode are all identical. **b** In a skewed distribution, the median lies between the mean and the mode. The mean is highly sensitive to outliers, whereas the median and the mode are little influenced by extremely high or low values. The median usually lies in a skewed distribution.

if n is even. Although outliers also affect the median, their absolute values do not influence it. *Quantiles* are a more general way of dividing the data sample into groups that contain equal numbers of observations. For example, three *quartiles* divide the data into four groups, four *quintiles* divide the observations into five groups, and 99 *percentiles* define one hundred groups.

The third important measure of central tendency is the *mode*. The mode is the most frequent x-value or—if the data are grouped into classes—the center of the class with the largest number of observations. The data set has no mode if there are no values that appear more frequently than any of the other values. Frequency distributions with a single mode are called *unimodal*, but there may also be two modes (*bimodal*), three modes (*trimodal*), or four or more modes (*multimodal*) (Fig. 3.3).

The mean, median, and mode are used when several quantities add together to produce a total, whereas the *geometric mean* is often used if these quantities are multiplied. Let us assume that the population of an organism grows by 10% in the first year, by 25% in the second year, and by 60% in the third year. The average rate of growth is not the arithmetic mean since the original number of individuals has increased by a factor (and not a sum) of 1.1 after one year, 1.25 after the second year, and 1.6 after the third year. The average growth of the population is therefore calculated by the geometric mean:

$$\bar{x}_G = (x_1 \cdot x_2 \cdot \ldots \cdot x_n)^{1/n}$$

The average growth of these values is 1.3006, which suggests an approximate per annum growth in the population of 30%. The arithmetic mean would result in an erroneous value of 1.3167, or approximately 32% annual growth. The geometric mean is also a useful measure of central tendency for skewed data, such log-normally distributed data; however, it does not apply if zeros or negative values are present. Finally, the *harmonic mean*

$$\bar{x}_H = n / \left(\frac{1}{x_1} + \frac{1}{x_2} + \cdots + \frac{1}{x_n} \right)$$

is also used to derive a mean value for asymmetric data (e.g., log-normally distributed data), as is the geometric mean, but neither is robust to outliers. The harmonic mean is a better average when the numbers are defined in relation to a particular unit. The commonly quoted example is for averaging velocities. The harmonic mean is also used to calculate the mean of sample sizes.

Measures of Dispersion

Another important property of a distribution is dispersion. Some of the parameters that can be used to quantify dispersion are illustrated in Fig. 3.3. The simplest way to describe the dispersion of a data set is by using the *range*, which is the difference between the highest and lowest values in the data set and is given by

$$\Delta x = x_{max} - x_{min}$$

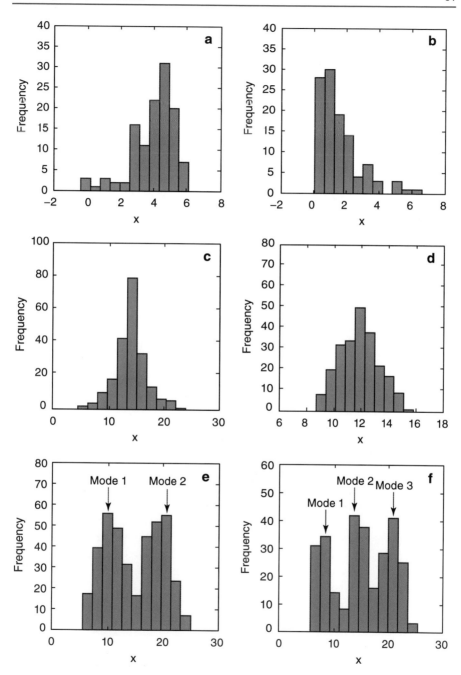

Fig. 3.3 *Dispersion* and *shape* of a distribution. **a–b** Unimodal distributions showing a negative or positive skew. **c–d** Distributions showing high or low kurtosis. **e–f** Bimodal and trimodal distributions showing two or three modes.

Since the range is defined by two extreme data points, it is highly susceptible to outliers and is therefore not a reliable measure of dispersion in most cases. The interquartile range of the data (i.e., the middle 50% of the data) can be used to help overcome this problem.

A more useful measure of dispersion is the *standard deviation*:

$$s = \sqrt{\frac{1}{n-1} \sum_{i=1}^{n} (x_i - \bar{x})^2}$$

The standard deviation is the average deviation of each data point from the mean. The sample standard deviation is used as an estimate of the population standard deviation σ if the underlying theoretical distribution is a Gaussian (or normal) distribution. The *variance* is the third important measure of dispersion and is simply the square of the standard deviation:

$$s^2 = \frac{1}{n-1} \sum_{i=1}^{n} (x_i - \bar{x})^2$$

Although the variance has the disadvantage of not having the same dimensions as the original data, it is used extensively in many applications instead of the standard deviation.

In addition, both *skewness* and *kurtosis* can be used to describe the shape of a frequency distribution (Fig. 3.3). Skewness is a measure of the asymmetry of the tails of a distribution. The classic way of computing the asymmetry of a distribution is by using Karl Pearson's (1857 to 1936) mode skewness,

$$\gamma = \frac{\bar{x} - \tilde{x}}{s}$$

in which the difference between the arithmetic mean and the median is divided by the standard deviation. A negative skew indicates that the distribution is spread out more to the left of the mean value assuming that values increase toward the right along the axis. The sample mean in this case is smaller than the mode. Distributions with positive skewness have large tails that extend toward the right. The skewness of the symmetric normal distribution is zero. Although Pearson's measure is useful, the following formula by Sir Ronald Fisher (1890 to 1962) for calculating the skewness is often used instead:

$$\gamma = \frac{\sum_{i=1}^{n} (x_i - \bar{x})^3 / n}{s^3}$$

The second important measure of the shape of a distribution is *kurtosis*. Again, numerous formulas are available for computing kurtosis, the most popular of which is the following:

$$\omega = \frac{\sum\limits_{i=1}^{n} (x_i - \bar{x})^4/n}{s^4}$$

Kurtosis is a measure of tail weight. A normal distribution has a kurtosis of three, and some definitions of kurtosis therefore subtract three from the term above in order to set the kurtosis of the normal distribution to zero. A low kurtosis suggests that the distribution produces fewer and less extreme outliers than does the normal distribution.

3.3   Examples of Empirical Distributions

As an example, we now analyze the data in the file *organicmatter_one.txt*, which contains the organic carbon content C_{org} of lake sediments in weight percentage (wt%). In order to load the data, we type

```
%reset -f

import numpy as np
import matplotlib.pyplot as plt
from scipy import stats

corg = np.loadtxt('organicmatter_one.txt')
```

The data file contains 60 measurements of C_{org}, which can be displayed by typing

```
plt.figure()
plt.plot(corg,np.zeros((len(corg))),'o')
plt.show()
```

This graph shows some of the characteristics of the data. The organic carbon content of the samples ranges from 9 to 15 wt%, with most of the data clustering between 12 and 13 wt%. Values below 10 wt% and above 14 wt% are rare. While this kind of representation of the data undoubtedly has its advantages, histograms are a much more convenient way of displaying univariate data (Fig. 3.1). Histograms divide the range of the data into nbin equal intervals (also called bins or classes), count the number of observations n in each bin, and display the frequency distribution of observations as a bar plot. The bins are defined either by their edges e or their centers v. There is no fixed rule for the correct number of bins, and the most suitable number depends on the application (e.g., the statistical method) that is used. Ideally, the number of bins n should lie between 5 and 15; it should closely reflect the underlying distribution and should not result in any empty bins (i.e., in classes with no counts). In practice, the square root of the total number of observations `length(corg)` (rounded to the nearest integer using round) is often used as the number of bins. In our example, `nbin` can be calculated using

```
nbin = np.round(np.sqrt(len(corg)))
print(nbin)
```

which yields

```
8.0
```

Using `hist()` from the *Matplotlib* package

```
plt.figure()
plt.hist(corg)
plt.show()
```

displays a histogram with six classes using an automatic binning algorithm. The algorithm returns bins with a uniform width that was chosen to cover the range of elements in `corg` and to reveal the underlying shape of the distribution. Instead, we can also define the number of bins (e.g., 8) by typing

```
plt.figure()
plt.hist(corg,bins=8)
plt.show()
```

Alternatively, we can define the bin edges e defined above in order to display the histogram by typing

```
e = np.arange(9,15+1,1)

plt.figure()
ne = plt.hist(corg,bins=e)
plt.show()

print(ne)
```

which yields the tuple ne, which contains the frequency distribution n and the bin edges e

```
(array([ 2.,  6., 13., 20., 15.,  4.]),
 array([ 9, 10, 11, 12, 13, 14, 15]),
 <BarContainer object of 6 artists>)
```

as well as the histogram plot, which is similar to the one shown in Fig. 3.1a. Python tuples (like dicts and lists) are used to store multiple items in a single variable (see Sect. 2.5). We can also use histogram() from the *NumPy* package to determine n and e without plotting the histogram. We also use the edges e to calculate the bin centers v. Typing

```
n, e = np.histogram(corg,bins=8)
v = np.diff(e[0:2]) * 0.5 + e[0: 1]
print(n)
print(e)
print(v)
```

yields

```
[ 2   2 10   7 14   9 12   4]

[ 9.416751    10.05984113 10.70293125
 11.34602137 11.9891115  12.63220163
 13.27529175 13.91838187 14.561472  ]

[ 9.73829606 10.38138619 11.02447631
 11.66756644 12.31065656 12.95374669
 13.59683681 14.23992694]
```

which lists (from top to bottom) the frequencies n, the bin edges e, and the bin centers v. As an alternative way of plotting the data, the empirical cumulative distribution function can be displayed using the argument cumulative=True (Fig. 3.1b):

```
plt.figure()
plt.hist(corg,cumulative=True,histtype='step')
plt.show()
```

The most important parameters that describe the distribution are the measures of central tendency and the dispersion about the average. The most popular measure of central tendency is the arithmetic mean. Typing

```
print(np.mean(corg))
```

yields

```
12.344804809999998
```

Since this measure is highly susceptible to outliers, we can take the median as an alternative measure of central tendency,

```
print(np.median(corg))
```

which yields

```
12.471171
```

which does not differ by very much in this particular example. However, we will
see later that this difference can be substantial for distributions that are not symmetric. A more general parameter for defining fractions of the data less than or
equal to a certain value is the quantile. Some quantiles have special names, such
as the three quartiles that divide the distribution into four equal parts: 0 to 25%, 25
to 50%, 50 to 75%, and 75 to 100% of the total number of observations. We use
quantile() to compute the three quartiles

```
print(np.quantile(corg,np.array([0.25,0.5,0.75])))
```

which yields

```
[11.43325275  12.471171  13.29648125]
```

Fewer than 25% of the data values are therefore lower than 11.43, 25% lie
between 11.43 and 12.47, another 25% lie between 12.47 and 13.30, and the
remaining 25% are higher than 13.30.

The third parameter in this context is the mode, which is the midpoint of the interval with the highest frequency. Using mode() to identify the most frequent value in
a sample is unlikely to yield a good estimate of the peak in continuous probability
distributions, such as the one in corg. Furthermore, mode() is not suitable for finding
peaks in distributions that have multiple modes. In these cases, it is better to compute
a histogram and calculate the peak of that histogram. We can use where() to locate
the class that has the largest number of observations. Typing

```
print(v[np.where(n==np.max(n))])
```

yields

```
[12.31065656]
```

The index of this element is then used to display the midpoint of the corresponding class v. If there are several elements in n with similar values, this statement
returns several solutions, which suggests that the distribution has several modes.
The median, quartiles, minimum, and maximum of a data set can be summarized
and displayed in a *box and whisker plot*:

```
plt.figure()
plt.boxplot(corg)
plt.show()
```

The boxes have lines at the lower quartile, the median, and the upper quartile values. The whiskers are lines that extend from each end of the boxes in order to show the extent or range of the rest of the data.

The most popular measures of dispersion are the range, variance, and standard deviation. We have already used the range to define the midpoints of the classes. The range is the difference between the highest and lowest values in the data set. Typing

```
print(np.max(corg)-np.min(corg))
```

yields

```
5.144721000000005
```

The variance is the average of the squared deviation of each number from the mean of a data set. Typing

```
print(np.var(corg,ddof=1))
```

yields

```
1.3595162670031486
```

Please note that we must use ddof=1, which stands for *delta degrees of freedom*. By default, ddof is zero, and we must therefore change it to one if we wish to calculate the sample variance with $n-1$ degrees of freedom. The standard deviation is the square root of the variance. Typing

```
print(np.std(corg,ddof=1))
```

yields

```
1.165982961712198
```

again using ddof=1. When using skew() to describe the shape of the distribution, we observe a slightly negative skew. Typing

```
print(stats.skew(corg))
```

yields

```
-0.25291964917461346
```

Finally, the tailedness of the distribution is described by the kurtosis. The result of `kurtosis()` according to both Pearson's (`fisher=False`) and Fisher's (`fisher=True`) formula

```
print(stats.kurtosis(corg,fisher=False))
print(stats.kurtosis(corg,fisher=True))
```

is

```
2.466996016707096
-0.5330039832929039
```

which suggests that our distribution produces fewer and less extreme outliers than does the normal distribution since its Pearson kurtosis is less than three and its Fisher kurtosis is negative.

Most of these functions have corresponding versions for data sets that contain gaps, such as `nanmean()` and `nanstd()`, which treat NaNs as missing values. To illustrate the use of these functions, we introduce a gap to our data set and compute the mean using both `mean()` and `nanmean()` for comparison. Typing

```
corg[25] = np.NaN

print(np.mean(corg))
```

yields

```
nan
```

whereas typing

```
print(np.nanmean(corg))
```

yields

```
12.336529722033896
```

In this example, `mean()` follows the rule that all operations with NaNs result in NaNs, whereas `nanmean()` simply skips the missing value and computes the mean of the remaining data.

As a second example, we now explore a data set characterized by a striking skew. The data represent 120 microprobe analyses on glass shards that were hand-picked from volcanic ash. The volcanic glass was affected by chemical weathering at an initial stage, and the shards therefore exhibit glass hydration and sodium depletion in some sectors. We can study the distribution of sodium (in wt%) in the 120 analyses using the same procedure as above. The data are stored in the file *sodiumcontent_one.txt*:

```
sodium = np.loadtxt('sodiumcontent_one.txt')
```

As a first step, we recommend always displaying the data as a histogram. The square root of 120 suggests 11 classes, and we therefore display the data by typing

```
plt.figure()
plt.hist(corg,bins=11)
plt.show()

n, e = np.histogram(sodium,bins=11)
v = np.diff(e[0:2]) * 0.5 + e[0:-1]
```

Since the distribution has a negative skew, the mean, the median, and the mode differ from one another. Typing

```
print(np.mean(sodium))
print(np.median(sodium))
print(v[np.where(n==np.max(n))])
```

yields

```
5.662788932499999
5.97410535
[6.54074771]
```

The mean of the data is lower than the median, which is in turn lower than the mode. We observe a strong negative skewness, as expected from our data. Typing

```
print(stats.skew(sodium))
```

yields

```
-1.1085610752227795
```

We now introduce an outlier to the data and explore its effect on the statistics of the sodium content. To do so, we use a different data set that is better suited to this example than to the previous data set. The new data set contains higher sodium values of around 17 wt% and is stored in the file *sodiumcontent_two.txt*:

```
clear, close all, clc

sodium = np.loadtxt('sodiumcontent_two.txt')
```

This data set contains only 50 measurements in order to better illustrate the effects
of an outlier. We can use the same script as in the previous example to display the
data in a histogram with seven classes and to compute the number of observations
n in each of the classes v,

```
plt.figure()
plt.hist(sodium,bins=11)
plt.show()

n, e = np.histogram(sodium,bins=11)
v = np.diff(e[0:2]) * 0.5 + e[0:-1]

print(np.mean(sodium))
print(np.median(sodium))
print(v[np.where(n==np.max(n))])
```

which yields

```
16.63794532
16.973911
[17.02385664 17.45143718]
```

The mean of the data is 16.64, the median is 16.94, and there are two modes (i.e.,
two bars of the same height in the histogram) of 17.02 and 17.45, respectively. We
now introduce a single, very low value of 7.0 wt% in addition to the 50 measurements contained in the original data set:

```
sodium = np.append(sodium,7.)

plt.figure()
plt.hist(sodium,bins=11)
plt.show()

n, e = np.histogram(sodium,bins=11)
v = np.diff(e[0:2]) * 0.5 + e[0:-1]
```

The histogram of this data set—which uses a larger number of classes—illustrates the distortion produced in the frequency distribution by this single outlier by
showing several empty classes. The influence of this outlier on the sample statistics is also substantial, as suggested by the result of

```
print(np.mean(sodium))
print(np.median(sodium))
print(v[np.where(n==np.max(n))])
```

which yields

```
16.448966000000002
16.972217
[17.02385664 17.45143718]
```

The most substantial observed change is in the mean of 16.45, which is substantially lower due to the presence of the outlier. This example clearly demonstrates the sensitivity of the mean to outliers. In contrast, the median of 16.97 is relatively unaffected.

3.4   Theoretical Distributions

We have now described the empirical frequency distribution of our sample. A histogram is a convenient way to depict the frequency distribution of the variable x. If we sample the variable sufficiently often and the output ranges are narrow, we obtain a very smooth version of the histogram. An infinite number of measurements $n \rightarrow \infty$ and an infinitely small class width produce the random variable's *probability density function* (PDF). The probability distribution density $f(x)$ defines the probability that the variable has a value equal to x. The integral of $f(x)$ is normalized to unity, that is, the total number of observations is one. The *cumulative distribution function* (CDF) is the sum of the frequencies of a discrete PDF or the integral of a continuous PDF. The cumulative distribution function $F(x)$ is the probability that the variable has a value less than or equal to x.

As a next step, we need to find appropriate theoretical distributions that fit the empirical distributions described in the previous section. The present section therefore introduces the most important theoretical distributions and describes their application.

Uniform Distribution

A *uniform* or *rectangular distribution* is a distribution that has a constant probability (Fig. 3.4). The corresponding probability density function is

$$f(x) = 1/N = const.$$

where the random variable x has any of N possible values. The cumulative distribution function is

$$f(x) = x \cdot 1/N$$

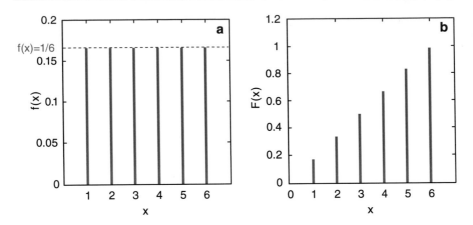

Fig. 3.4 **a** Probability density function $f(x)$ and **b** cumulative distribution function $F(x)$ of a uniform distribution with $N=6$. The 6 discrete values of the variable x have the same probability of 1/6.

The probability density function is normalized to unity,

$$\sum_{-\infty}^{+\infty} f(x)dx = 1$$

that is, the sum of all probabilities is one. The maximum value of the cumulative distribution function is therefore one:

$$F(x)_{max} = 1$$

An example is a rolling die with $N=6$ faces. A discrete variable such as the faces of a die can only take a countable number of values x. The probability of each face is 1/6. The probability density function of this distribution is

$$f(x) = 1/6$$

The corresponding cumulative distribution function is

$$F(x) = x \cdot 1/6$$

where x takes only discrete values (i.e., $x = 1, 2, ..., 6$).

Binomial Distribution

A *binomial distribution* gives the discrete probability of x successes out of N trials, with a probability p of success in any given trial (Fig. 3.5). The probability density function of a binomial distribution is

$$f(x) = \binom{N}{x} p^x (1-p)^{N-x}$$

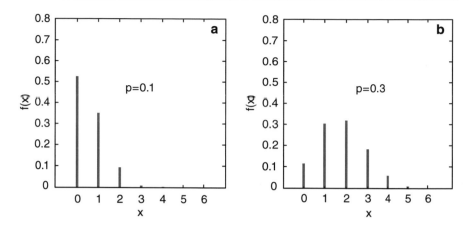

Fig. 3.5 Probability density function $f(x)$ of a binomial distribution, which gives the probability p of x successes out of $N = 6$ trials, with a probability of **a** $p = 0.1$ and **b** $p = 0.3$ of success in any given trial.

The cumulative distribution function is

$$F(x) = \sum_{i=1}^{x} \binom{N}{i} p^i (1-p)^{N-i}$$

where

$$\binom{n}{r} = \frac{n!}{r!(n-r)!}$$

The binomial distribution has two parameters N and p. An example of the application of this distribution can be found when determining the likely outcome of drilling for oil. Let us assume that the probability of drilling success is 0.1, or 10%. The probability of $x = 3$ successful wells out of a total number of $N = 10$ wells is

$$f(3) = \binom{10}{3} 0.1^3 (1 - 0.1)^{10-3} = 0.057 \approx 6\%$$

The probability of exactly 3 successful wells out of 10 trials is therefore approximately 6% in this example.

Poisson Distribution

When the number of trials is $n \to \infty$ and the success probability is $p \to 0$, the binomial distribution approaches a *Poisson distribution* with a single parameter $\lambda = n \cdot p$ (Fig. 3.6) (Poisson 1837). This works well for $n > 100$ and $p < 0.05$ (or 5%).

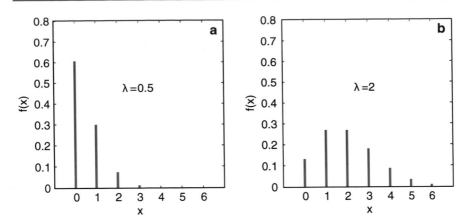

Fig. 3.6 Probability density function $f(x)$ of a Poisson distribution with different values for λ: **a** $\lambda = 0.5$ and **b** $\lambda = 2$.

We therefore use the Poisson distribution for processes characterized by extremely low occurrence, such as earthquakes, volcanic eruptions, storms, and floods. The probability density function is

$$f(x) = \frac{e^{-\lambda}\lambda^x}{x!}$$

and the cumulative distribution function is

$$F(x) = \sum_{i=0}^{x} \frac{e^{-\lambda}\lambda^i}{i!}$$

The single parameter λ describes both the mean and the variance of this distribution.

Normal or Gaussian Distribution

When $p = 0.5$ (symmetric, no skew) and $n \to \infty$, the binomial distribution approaches a *normal* or *Gaussian distribution* defined by the mean μ and the standard deviation σ (Fig. 3.7). The probability density function of a normal distribution is

$$f(x) = \frac{1}{\sigma\sqrt{2\pi}} exp\left(-\frac{1}{2}\left(\frac{x-\mu}{\sigma}\right)^2\right)$$

and the cumulative distribution function is

$$F(x) = \frac{1}{\sigma\sqrt{2\pi}} \int_{-\infty}^{x} exp\left(-\frac{1}{2}\left(\frac{x-\mu}{\sigma}\right)^2\right) dy$$

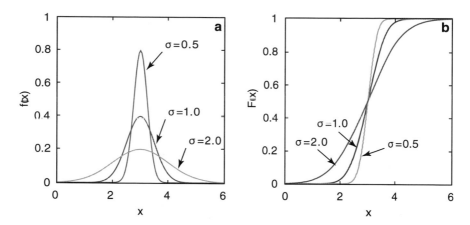

Fig. 3.7 a Probability density function $f(x)$ and **b** cumulative distribution function $F(x)$ of a Gaussian (or normal) distribution with a mean $\mu = 3$ and various values for standard deviation σ.

The normal distribution is therefore used when the mean is both the most frequent and the most likely value. The probability of deviations is equal in either direction and decreases with increasing distance from the mean.

The *standard normal distribution* is a special member of the normal distribution family that has a mean of *zero* and a standard deviation of *one*. We can transform the equation for a normal distribution by substituting $z = (x-\mu)/\sigma$. The probability density function of this distribution is

$$f(x) = \frac{1}{\sqrt{2\pi}} exp\left(-\frac{z^2}{2}\right)$$

This definition of the normal distribution is often called the *z-distribution*.

Logarithmic Normal (or Log-Normal) Distribution

The *logarithmic normal* (or *log-normal*) *distribution* is used when the data have a lower limit, such as with mean annual precipitation or the frequency of earthquakes (Fig. 3.8). In such cases, distributions are usually characterized by major skewness, which is best described by a logarithmic normal distribution. The probability density function of this distribution is

$$f(x) = \frac{1}{\sigma\sqrt{2\pi}x} exp\left(-\frac{1}{2}\left(\frac{lnx - \mu}{\sigma}\right)^2\right)$$

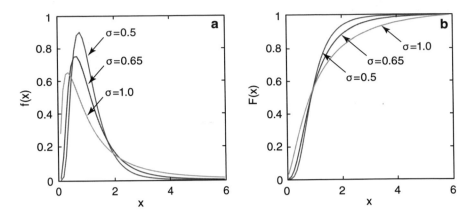

Fig. 3.8 **a** Probability density function $f(x)$ and **b** cumulative distribution function $F(x)$ of a logarithmic normal distribution with a mean $\mu = 0$ and with various values for Φ.

and the cumulative distribution function is

$$F(x) = \frac{1}{\sigma\sqrt{2\pi}} \int_{-\infty}^{x} \frac{1}{y} exp\left(-\frac{1}{2}\left(\frac{lny - \mu}{\sigma}\right)^2\right) dy$$

where $x>0$. The distribution can be described by two parameters: the mean μ and the standard deviation σ. However, the formulas for the mean and the standard deviation are different from those used for normal distributions. In practice, the values of x are logarithmized, the mean and the standard deviation are computed using the formulas for a normal distribution, and the empirical distribution is then compared with a normal distribution.

Student's t-Distribution

Student's t-distribution was first introduced by William S. Gosset (1876 to 1937), who needed a distribution for small samples (Fig. 3.9). Gosset was an employee of the Irish Guinness Brewery and wished to publish his research results anonymously. He thus published his *t*-distribution under the pseudonym *Student* (Student 1908). The probability density function is

$$f(x) = \frac{\Gamma\left(\frac{\Phi+1}{2}\right)}{\Gamma\left(\frac{\Phi}{2}\right)} \frac{1}{\sqrt{\Phi\pi}} \frac{1}{\left(1 + \frac{x^2}{\Phi}\right)^{\frac{\Phi+1}{2}}}$$

where Γ is the Gamma function

$$\Gamma(x) = \lim_{n\to\infty} \frac{n!n^{x-1}}{x(x+1)(x+2)\ldots(x+n-1)}$$

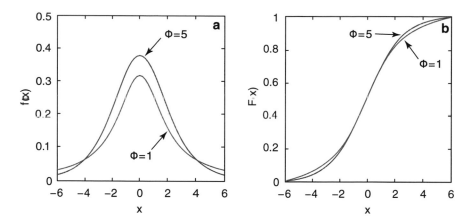

Fig. 3.9   a Probability density function $f(x)$ and **b** cumulative distribution function $F(x)$ of Student's t-distribution with two different values for Φ.

which can be written as

$$\Gamma(x) = \int_0^\infty e^{-y} y^{x-1} dy$$

if $x>0$. The single parameter Φ of the t-distribution is the number of degrees of freedom. In the analysis of univariate data, this distribution has $n{-}1$ degrees of freedom, where n is the sample size. As $\Phi \to \infty$, the t-distribution converges toward the standard normal distribution. Since the t-distribution approaches the normal distribution for $\Phi > 30$, it is rarely used for distribution fitting. However, the t-distribution is used for hypothesis testing using the t-test (Sect. 3.7).

Fisher's *F*-Distribution

The *F-distribution* was named after statistician Sir Ronald Fisher (1890 to 1962). It is used for hypothesis testing via the F-test (Sect. 3.8). The F-distribution has a relatively complex probability density function (Fig. 3.10),

$$f(x) = \frac{\Gamma\left(\frac{\Phi_1+\Phi_2}{2}\right)\left(\frac{\Phi_1}{\Phi_2}\right)^{\frac{\Phi_1}{\Phi_2}}}{\Gamma(\Phi_1/2)\Gamma(\Phi_2/2)} x^{\frac{\Phi_1-2}{2}} \left(1+\frac{\Phi_1}{\Phi_2}x\right)^{-\frac{\Phi_1+\Phi_2}{2}}$$

where $x>0$ and Γ is again the Gamma function. The two parameters Φ_1 and Φ_2 are the numbers of degrees of freedom.

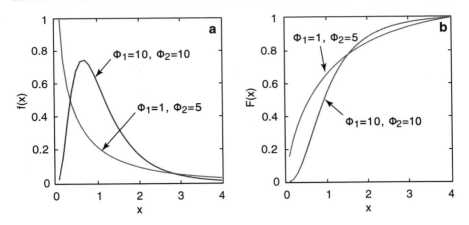

Fig. 3.10 a Probability density function $f(x)$ and **b** cumulative distribution function $F(x)$ of Fisher's F-distribution with different values for Φ_1 and Φ_2.

χ^2- (or Chi-Squared-)Distribution

The χ^2-distribution was introduced by Friedrich Helmert (1876) and Karl Pearson (1900). This distribution has important applications in statistical hypothesis testing (Sect. 3.9). The probability density function of the χ^2-distribution is

$$f(x) = \frac{1}{2^{\Phi/2}\Gamma(\Phi/2)} x^{\frac{\Phi-2}{2}} e^{-\frac{x}{2}}$$

where $x>0$; otherwise, $f(x)=0$. Once again, Γ is the Gamma function, and Φ is the number of degrees of freedom (Fig. 3.11).

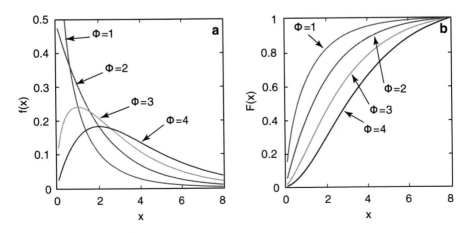

Fig. 3.11 a Probability density function $f(x)$ and **b** cumulative distribution function $F(x)$ of a χ^2-distribution with different values for Φ.

3.5   Examples of Theoretical Distributions

We use `norm.pdf()` and `norm.cdf()` to compute the probability density function and the cumulative distribution function of a Gaussian distribution with `mu=12.3448` and `sigma=1.1660` (evaluated for the values in x) to compare the results with those of our sample data set:

```
%reset -f

import numpy as np
import matplotlib.pyplot as plt
from scipy import stats


mu = 12.3448
sigma = 1.166


x = np.arange(5,20+0.001,0.001)
pdf = stats.norm.pdf(x,mu,sigma)
cdf = stats.norm.cdf(x,mu,sigma)

plt.figure()
plt.plot(x,pdf,x,cdf)
plt.show()
```

We can use these functions to familiarize ourselves with the properties of distributions. This will be important when testing hypotheses in the following sections. The test statistics used in these sections follow the theoretical frequency distributions introduced in the previous sections of this chapter. In particular, the integral (or in the discrete case, the sum) of the theoretical distribution within a certain range $a \leq x \leq b$ is of great importance as it helps in calculating the probability that a measurement falls within this range.

As an example, we can calculate the probability that a measurement falls within the range of $\mu - \sigma \leq x \leq \mu + \sigma$, where μ is the mean and σ is the standard deviation of a Gaussian distribution. Using the PDF of the Gaussian distribution with `mu=12.3448` and `sigma=1.1660`, we find

```
pdf = pdf / np.sum(pdf)
print(np.sum(pdf[np.where((x>mu-sigma) & (x<mu+sigma))]))
```

which yields

```
0.6826894910701773
```

or ~68% after normalizing the PDF to unity. The expression `np.where(x>mu-sigma)` returns the first index of x where x is larger than `mu-sigma`. Simi-

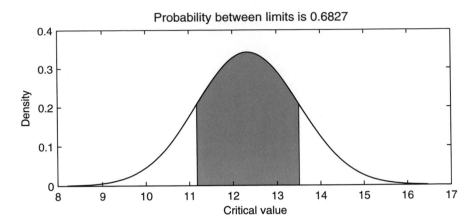

Fig. 3.12 Plot of a standard normal distribution between specified limits. As an example, the shaded area displays the $\mu \pm \sigma$ range of a Gaussian distribution with a mean $\mu = 12.3448$ and a standard deviation $\sigma = 1.1660$.

larly, `np.where(x<mu+sigma)` returns the last index of x where x is smaller than `mu+sigma`. We can also display the PDF together with a shaded region inside the $\mu - \sigma \leq x \leq \mu + \sigma$ range (Fig. 3.12):

```
plt.figure()
plt.plot(x,pdf,color='k')
xint = x[np.where((x>mu-sigma) & (x<mu+sigma))]
pdfint = pdf[np.where((x>mu-sigma) & (x<mu+sigma))]
plt.fill_between(xint,pdfint)
plt.show()
```

Conversely, we can additionally calculate the x-values of the $\mu \pm \sigma$ range of our PDF using the inverse of the cumulative normal distribution function with `norm.ppf()` by typing

```
print(stats.norm.ppf((1-0.6827)/2,mu,sigma))
print(stats.norm.ppf(1-(1-0.6827)/2,mu,sigma))
```

which yields

```
11.178774682265383
13.510825317734616
```

Here, the values for p are calculated from the complement of $\sim 68\%$ (i.e., $\sim 32\%$), which is halved on both tails of the Gaussian distribution, that is, `(1-0.6827)/2` and `1-(1-0.6827)/2`.

The standard deviation σ of the Gaussian distribution is important in defining confidence intervals. In many examples, however, the confidence of one sigma ($\mu \pm 1\sigma$) (or ~68%) that the true value falls within the $\mu \pm 1\sigma$ range is not sufficient, and higher confidence intervals such as two sigma ($\mu \pm 2\sigma$) and three sigma ($\mu \pm 3\sigma$) are therefore also used. We can calculate the corresponding probability that the true value falls within the $\mu \pm 2\sigma$ range and the $\mu \pm 3\sigma$ range by typing

```
print(np.sum(pdf[np.where((x>mu-2*sigma) & (x<mu+2*sigma))]))
print(np.sum(pdf[np.where((x>mu-3*sigma) & (x<mu+3*sigma))]))
```

which yields

```
0.9544997357412237
0.9973002040460477
```

or ~95% and ~99%, respectively. Again, using `norm.ppf()`, we can calculate the upper and lower bounds of the two sigma ($\mu \pm 2\sigma$) range,

```
print(stats.norm.ppf(0.05/2,mu,sigma))
print(stats.norm.ppf(1-0.05/2,mu,sigma))
```

which yields

```
10.059481994026296
14.630118005973703
```

and the three sigma ($\mu \pm 3\sigma$) range,

```
print(stats.norm.ppf(0.01/2,mu,sigma))
print(stats.norm.ppf(1-0.01/2,mu,sigma))
```

which yields

```
9.34138303206198
15.348216967938017
```

3.6   Hypothesis Testing

The following sections deal with methods used to draw conclusions from the statistical sample, which can then be applied to the larger population of interest (*hypothesis testing*) (Sects. 3.6 to 3.12). All hypothesis tests share the same concept and terminology. The *null hypothesis* is an assertion about the population that describes the absence of a statistically significant characteristic or effect, whereas an *alternative hypothesis* is a contrasting assertion. The *p-value* of a hypothesis test is the probability (under the null hypothesis) of observing larger values for the test statistic than those calculated from the sample. The *significance level* α is the threshold of probability that controls the outcome of the tests. If the *p*-value is smaller than α, the null hypothesis can be rejected; the outcome of the test is regarded as *significant* if $p<0.05$ and is regarded as *highly significant* if $p<0.01$.

A hypothesis test can be performed either as a *one-tailed* (one-sided) or as a *two-tailed* (two-sided) test. The term *tail* stems from the concept of tailing off the data to the far left or far right of a probability density function, for instance, as in the standard normal distribution used in the Mann–Whitney and Ansari–Bradley tests (Sects. 3.11 and 3.12). As an example, the Mann–Whitney test compares the medians of two data sets. The one-tailed Mann–Whitney test is used to test against the alternative hypothesis that the median of the first sample is either smaller or larger than the median of the second sample at a significance level of 5% (or 0.05). The two-tailed Mann–Whitney test is used when the medians are not equal at a 5% significance level, that is, when it makes no difference which of the medians is larger. In this case, the significance level is halved, that is, 2.5% is used instead of 5%.

We can display the standard normal distribution for the one-tailed test by typing

```
%reset -f

import numpy as np
import matplotlib.pyplot as plt
from scipy import stats

x = np.arange(-4,4,0.001)
pdf = stats.norm.pdf(x,0.,1.)

plt.figure()
plt.plot(x,pdf,color='k')
xint = x[np.where(x>stats.norm.ppf(0.95,0.,1.))]
pdfint = pdf[np.where(x>stats.norm.ppf(0.95,0.,1.))]
plt.fill_between(xint,pdfint)
plt.show()
```

which yields a plot with one blue tail to the right in which the 5% area is shaded (Fig. 3.13a). Similarly, we can display the standard normal distribution for the two-tailed test by typing

```
x = np.arange(-4,4,0.001)
pdf = stats.norm.pdf(x,0.,1.)
```

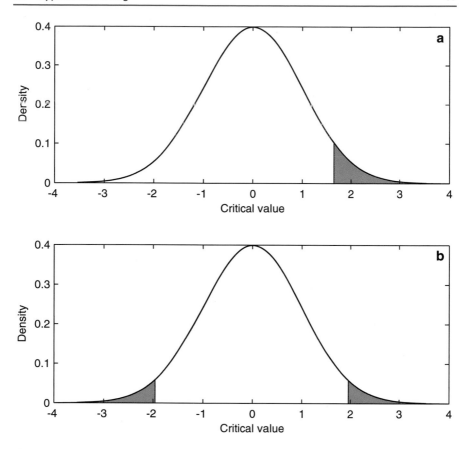

Fig. 3.13 Plot of a standard normal distribution with **a** one or **b** two critical regions, which are shown as shaded areas that contain a total of 5% of the area under the curve.

```
plt.figure()
plt.plot(x,pdf,color='k')
xint1 = x[np.where(x<stats.norm.ppf(0.05/2,0.,1.))]
xint2 = x[np.where(x>stats.norm.ppf(1-0.05/2,0.,1.))]
pdfint1 = pdf[np.where(x<stats.norm.ppf(0.05/2,0.,1.))]
pdfint2 = pdf[np.where(x>stats.norm.ppf(1-0.05/2,0.,1.))]
plt.fill_between(xint1,pdfint1,color=(0.3,0.5,0.8))
plt.fill_between(xint2,pdfint2,color=(0.3,0.5,0.8))
plt.show()
```

which yields a plot with two blue tails (one to the left and one to the right) in which the 2.5% areas are shaded (Fig. 3.13b).

Note that we cannot prove the null hypothesis: In other words *not guilty* is not the same as *innocent*. In practice, we design hypotheses based on our data, test them, and then continue to work with those we could not show to be false.

The inherent possibility of proving a hypothesis to be false is therefore an important requirement of our hypotheses.

The next sections introduce the most important hypothesis tests for earth science applications: the two-sample t-test for comparing the means of two data sets, the two-sample F-test for comparing the variances of two data sets, and both the χ^2-test and the Kolmogorov–Smirnov test for comparing distributions (Sects. 3.7 to 3.10). The Mann–Whitney and Ansari–Bradley tests are alternatives to the t-test and F-test for comparing the medians and dispersions, respectively, of two data sets without requiring a normality assumption for the underlying population (Sects. 3.11 and 3.12).

3.7   The t-Test

Student's t-test was created by William S. Gosset (Student 1908) and compares the means of two distributions. The *one-sample t-test* is used to test the hypothesis that the mean of a Gaussian-distributed population has a value specified in the null hypothesis. The *two-sample t-test* is employed to test the hypothesis that the means of two Gaussian distributions are identical. In the following text, the two-sample t-test is introduced to demonstrate hypothesis testing. Let us assume that two independent sets of n_a and n_b measurements have been carried out on the same object—for instance, measurements on two sets of rock samples taken from two separate outcrops. The t-test can be used to determine whether both samples come from the same population—for example, the same lithologic unit (null hypothesis)—or from two different populations (alternative hypothesis). Both sample distributions must be Gaussian, and the variances for the two sets of measurements should be similar. The appropriate test statistic for the difference between the two means is then

$$\hat{t} = \frac{|\bar{a} - \bar{b}|}{\sqrt{\frac{n_b + n_b}{n_a n_b} \cdot \frac{(n_a - 1) \cdot s_a^2 + (n_b - 1) \cdot s_b^2}{n_a + n_b - 2}}}$$

where n_a and n_b are the sample sizes and s_a^2 and s_b^2 are the variances of the two samples a and b, respectively. The null hypothesis can be rejected if the measured t-value is higher than the critical t-value, which depends on the number of degrees of freedom $\Phi = n_a + n_b - 2$ and the significance level α. The one-tailed test is used to test against the alternative hypothesis that the mean of the first sample is either smaller or larger than the mean of the second sample at a significance level of 5% (or 0.05). The one-tailed test would require modifying the equation above by replacing the absolute value of the difference between the means with the actual difference between the means. The two-tailed t-test is used when the means are not equal at a 5% significance level, that is, when it makes no difference which of the means is larger. In this case, the significance level is halved, that is, 2.5% is used to compute the critical t-value.

We can now load two example data sets from two independent series of measurements. The first example shows the performance of the two-sample *t*-test on two distributions with means of 25.5 and 25.3 and standard deviations of 1.3 and 1.5, respectively:

```
%reset -f
```

```
import numpy as np
import matplotlib.pyplot as plt
from scipy import stats

corg = np.loadtxt('organicmatter_two.txt')
corg1 = corg[0:,0]
corg2 = corg[0:,1]
```

We compare two data sets corg1 and corg2. We first plot both histograms in a single graph:

```
plt.figure()
plt.hist(corg1)
plt.hist(corg2)
plt.show()
```

We then compute the sample sizes, the means, and the standard deviations:

```
na = len(corg1)
nb = len(corg2)
ma = np.mean(corg1)
mb = np.mean(corg2)
sa = np.std(corg1,ddof=1)
sb = np.std(corg2,ddof=1)
```

We next calculate the *t*-value using the translation of the equation for the *t*-test statistic to Python code,

```
tcalc = np.abs((ma-mb))/np.sqrt(((na+nb)/(na*nb)) *
    (((na-1)*sa**2+(nb-1)*sb**2)/(na+nb-2)))
print(tcalc)
```

which yields

```
0.7279110523246475
```

We can now compare the calculated `tcalc` value of 0.7279 with the critical `tcrit` value. This can be accomplished using `t.ppf()`, which yields the inverse of the *t*-distribution function with `na-nb-2` degrees of freedom at the 5% significance level. This is a two-sample *t*-test, that is, the means are not equal. Computing the two-tailed critical `tcrit` value by entering `1-0.05/2` yields the upper (positive) `tcrit` value, which we compare with the absolute value of the difference between the means. Typing

```
tcrit = stats.t.ppf(1-0.05/2,na+nb-2)
```

yields

```
1.9802722492407059
```

Since the `tcalc` value calculated from the data is smaller than the critical `tcrit` value, we cannot reject the null hypothesis without another cause. We therefore conclude that the two means are identical at a 5% significance level. Alternatively, we can apply `ttest_ind()` to the two independent samples `corg1` and `corg2` at an `alpha=0.05` (i.e., a 5% significance level). The command

```
statistic,pvalue = stats.ttest_ind(corg1,corg2)
print(statistic)
print(pvalue)
```

yields

```
0.7279110523246475
0.46810997986836755
```

The result means that we cannot reject the null hypothesis without another cause at a 5% significance level. The *p*-value of 0.4681 (or ~47%, which is much greater than the significance level of 0.05, or 5%) suggests that the chances of observing more extreme *t*-values than the values in this example from similar experiments would be 4,681 in 10,000.

The second synthetic example shows the performance of the two-sample *t*-test in an example with very different means of 24.3 and 25.5 and with standard deviations again of 1.3 and 1.5, respectively:

```
corg = np.loadtxt('organicmatter_three.txt')
corg1 = corg[0:,0]
corg2 = corg[0:,1]
```

As before, we compare the two data sets `corg1` and `corg2`. We plot both histograms in a single graph:

```
plt.figure()
plt.hist(corg1)
plt.hist(corg2)
plt.figure()
```

We then compute the sample sizes, the means, and the standard deviations:

```
na = len(corg1)
nb = len(corg2)
ma = np.mean(corg1)
mb = np.mean(corg2)
sa = np.std(corg1,ddof=1)
sb = np.std(corg2,ddof=1)
```

We next calculate the *t*-value using the translation of the equation for the *t*-test statistic to Python code,

```
tcalc = np.abs((ma-mb))/np.sqrt(((na+nb)/(na*nb)) *
    (((na-1)*sa**2+(nb-1)*sb**2)/(na+nb-2)))
print(tcalc)
```

which yields

```
4.7364005886749485
```

We can now compare the calculated `tcalc` value of 4.7364 with the critical `tcrit` value. Again, this can be accomplished using `t.ppf()` at a 5% significance level. The function `t.ppf()` yields the inverse of the *t*-distribution function with `na-nb-2` degrees of freedom at the 5% significance level. This is again a two-sample *t*-test, that is, the means are not equal. Computing the two-tailed critical `tcrit` value by entering `1-0.05/2` yields the upper (positive) `tcrit` value, which we compare with the absolute value of the difference between the means. Typing

```
tcrit = stats.t.ppf(1-0.05/2,na+nb-2)
print(tcrit)
```

yields

```
1.9802722492407059
```

Since the `tcalc` value calculated from the data is now larger than the critical `tcrit` value, we can reject the null hypothesis and conclude that the means are not identical at a 5% significance level. Alternatively, we can apply `ttest_ind()` to the two independent samples of `corg1` and `corg2` at an `alpha=0.05` (or a 5%) significance level. The command

```
statistic,pvalue = stats.ttest_ind(corg1,corg2)
print(statistic)
print(pvalue)
```

yields

```
4.7364005886749485
6.113768768578045e-06
```

The result suggests that we can reject the null hypothesis. The *p*-value is extremely low and very close to zero.

3.8   The F-Test

The two-sample *F-test* discussed by Snedecor and Cochran (1989) compares the variances s_a^2 and s_b^2 of two distributions in which $s_a^2 > s_b^2$. An example can be found when comparing the natural heterogeneity of two samples based on replicated measurements. The sample sizes n_a and n_b should be above 30. Both the sample and the population distribution must be Gaussian. The appropriate test statistic with which to compare the variances is then

$$\widehat{F} = \frac{s_a^2}{s_b^2}$$

The two variances are significantly different (i.e., we can reject the null hypothesis without another cause) if the measured *F*-value is higher than the critical *F*-value, which in turn depends on the number of degrees of freedom $\Phi_a = n_a - 1$ and $\Phi_b = n_b - 1$, respectively, and the significance level α. The one-sample *F*-test, in contrast, virtually performs a χ^2-test of the hypothesis that the data come from a normal distribution with a specific variance (see Sect. 3.9). We first apply the two-sample *F*-test to two distributions with very similar standard deviations of 1.2550 and 1.2097:

```
%reset -f

import numpy as np
import matplotlib.pyplot as plt
from scipy import stats

corg = np.loadtxt('organicmatter_four.txt')
corg1 = corg[0:,0]
corg2 = corg[0:,1]
```

The *F*-value is the quotient of the larger variance divided by the smaller variance.
We can now compute the standard deviations by typing

```
s1 = np.std(corg1,ddof=1)
s2 = np.std(corg2,ddof=1)
print(s1)
print(s2)
```

which yields

```
1.254958845144188
1.2096912224649008
```

The *F*-distribution has two parameters df1 and df2, which represent the number of
observations in each of the distributions minus one, where

```
df1 = len(corg1)-1
df2 = len(corg2)-1
print(df1)
print(df2)
```

yields

```
59
59
```

We then sort the standard deviations by their absolute values

```
if s1 > s2:
    slarger = s1
    ssmaller = s2
else:
    slarger = s2
    ssmaller = s1
print(slarger)
print(ssmaller)
```

and get

```
1.254958845144188
1.2096912224649008
```

We can now compare the calculated *F*-value with the critical *F*-value using
f.ppf() at a significance level of 0.05 (or 5%). The function f.ppf() returns the
inverse of the *F*-distribution function with df1 and df2 degrees of freedom at the

5% significance level. This is a two-tailed test, and we must therefore divide the
p-value of 0.05 by two. Typing

```
Fcalc = slarger**2/ssmaller**2
Fcrit = stats.f.ppf(1-0.05/2,df1,df2)
print(Fcalc)
print(Fcrit)
```

yields

```
1.0762419323639238
1.674131963667053
```

Since the F-value calculated from the data is smaller than the critical F-value, we
cannot reject the null hypothesis without another cause. We therefore conclude
that the variances are identical at a 5% significance level. Alternatively, we can
create a function ftest() and apply it to the two independent samples corg1 and
corg2 at an alpha$=$0.05 (or a 5%) significance level:

```
def ftest(x,y):
    s1 = np.std(x,ddof=1)
    s2 = np.std(y,ddof=1)
    df1 = len(x)-1
    df2 = len(y)-1
    if s1 > s2:
        sl = s1
        ss = s2
    else:
        sl = s2
        ss = s1
    f = sl**2/ss**2
    p = 2*(1-stats.f.cdf(f,df1,df2))
    return f, p
```

The command

```
statistic,pvalue = ftest(corg1,corg2)
print(statistic)
print(pvalue)
```

yields

```
1.0762419323639238
0.7787390940882846
```

The result means that we cannot reject the null hypothesis without another cause at a 5% significance level. The *p*-value of 0.7787 (or ~ 78%, which is much greater than the significance level) means that the chances of observing more extreme values of *F* than the value in this example from similar experiments would be 7,787 in 10,000.

We now apply this test to two distributions with very different standard deviations of 1.8799 and 1.2939, respectively:

```
corg = np.loadtxt('organicmatter_five.txt')
corg1 = corg[0:,0]
corg2 = corg[0:,1]
```

We again compare the calculated `Fcalc` value with the critical `Fcrit` value at a 5% significance level using `f.ppf()` in order to compute `Fcrit`

```
s1 = np.std(corg1,ddof=1)
s2 = np.std(corg2,ddof=1)

df1 = len(corg1)-1
df2 = len(corg2)-1

if s1 > s2:
    slarger = s1
    ssmaller = s2
else:
    slarger = s2
    ssmaller = s1

Fcalc = slarger**2/ssmaller**2
Fcrit = stats.f.ppf(1-0.05/2,df1,df2)
print(Fcalc)
print(Fcrit)
```

and get

```
3.496686940871093
1.674131963667053
```

Since the `Fcalc` value calculated from the data is now larger than the critical `Fcrit` value, we can reject the null hypothesis. The variances are therefore different at a 5% significance level.

Alternatively, we can apply `ftest()` by performing a two-sample *F*-test on the two independent samples `corg1` and `corg2` at an `alpha=0.05` (or 5%) significance level. Typing

```
statistic,pvalue = ftest(corg1,corg2)
print(statistic)
print(pvalue)
```

yields

```
3.496686940871093
3.415254814731483e-06
```

The result suggests that we can reject the null hypothesis. The p-value is extremely low and very close to zero, which suggests that the null hypothesis is highly unlikely.

3.9   The χ^2-Test

Karl Pearson's χ^2-*test* involves comparing distributions, thereby allowing two distributions to be tested for derivation from the same population (Pearson 1900). This test is independent of the distribution that is used and can therefore be used to test the hypothesis that the observations were drawn from a specific theoretical distribution.

Let us assume that we have a data set that consists of multiple chemical measurements from a sedimentary unit. We could use the χ^2-test to test the null hypothesis that these measurements can be described by a Gaussian distribution with a typical central value and a random dispersion around it. The n data are grouped into K classes, where n should be above 30. The frequencies within the classes O_k should not be lower than four and should certainly never be zero. The appropriate test statistic is then

$$\widehat{\chi}^2 = \sum_{k=1}^{K} \frac{(O_k - E_k)^2}{E_k}$$

where E_k are the frequencies expected from the theoretical distribution (Fig. 3.14). The null hypothesis can be rejected if the measured χ^2-value is higher than the critical χ^2-value, which depends on the number of degrees of freedom $\Phi = K{-}Z$, where K is the number of classes and Z is the number of parameters that describe the theoretical distribution plus the number of variables (e.g., $Z = 2 + 1$ for the mean and the variance from a Gaussian distribution of a data set for a single variable and $Z = 1 + 1$ for a Poisson distribution for a single variable).

As an example, we can test the hypothesis that our organic carbon measurements contained in *organicmatter_one.txt* follow a Gaussian distribution. We must first load the data into the workspace and compute the frequency distribution n_obs for the data measurements using eight classes:

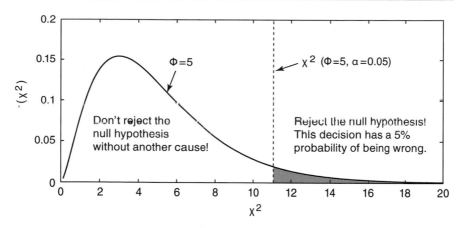

Fig. 3.14 Principles of a χ^2-test. The alternative hypothesis that the two distributions are different can be rejected if the measured χ^2 is lower than the critical χ^2. The χ^2 depends on $\Phi = K-Z$, where K is the number of classes and Z is the number of parameters that describe the theoretical distribution plus the number of variables. In the example, the critical $\chi^2(\Phi = 5, \alpha = 0.05)$ is 11.0705. Since the measured $\chi^2 = 5.7602$ is below the critical χ^2, we cannot reject the null hypothesis. In our example, we can conclude that the sample distribution is not significantly different from a Gaussian distribution.

```
%reset -f

import numpy as np
import matplotlib.pyplot as plt
from scipy import stats

corg = np.loadtxt('organicmatter_one.txt')

n_obs,e = np.histogram(corg,bins=8)
v = np.diff(e[0:2])*0.5+e[0:-1]
```

We then use `norm.pdf()` to create the expected frequency distribution n_exp with the mean and standard deviation of the data in corg:

```
n_exp = stats.norm.pdf(v,np.mean(corg),np.std(corg,ddof=1))
```

The data need to be scaled so that they are similar to the original data set:

```
n_exp = n_exp/np.sum(n_exp)
n_exp = np.sum(n_obs)*n_exp
```

The first command normalizes the observed frequencies n_exp to a total of one. The second command scales the expected frequencies n_exp to the sum of n_obs. We can now display both histograms for comparison:

```
plt.figure()
plt.subplot(1,2,1)
plt.bar(v,n_obs,color='r')
plt.subplot(1,2,2)
plt.bar(v,n_exp,color='b')
plt.show()
```

An alternative way of plotting the data in corg is to use a normal probability plot:

```
stats.probplot(corg,dist='norm',plot=plt)
```

The function probplot() plots the data in corg with blue circles superimposed by a red line, which represents the normal distribution. If the data in corg are indeed normally distributed, they will all fall on the red line. In our example, the data seem to agree well with the line, except for the tails of the normal distribution.

Visually inspecting these plots reveals that the empirical distribution is similar to the theoretical distribution. However, we advise using a more quantitative approach to test the hypothesis of similarity. The χ^2-test explores the squared differences between the observed and expected frequencies. The quantity chi2calc is the sum of the squared differences divided by the expected frequencies. Typing

```
chi2calc = np.sum((n_obs-n_exp)**2/n_exp)
print(chi2calc)
```

yields

```
7.285412350787187
```

The critical chi2crit value can be calculated using the chi2inv value. The χ^2-test requires the number of degrees of freedom Φ. In our example, we test the hypothesis that the data have a Gaussian distribution, that is, we estimate the two parameters μ and σ. The number of degrees of freedom is $\Phi = 8-(2+1) = 5$. We can now test our hypothesis at a 5% significance level. The function chi2.ppf() computes the inverse of the χ^2 CDF with parameters specified by Φ for the corresponding probabilities. Typing

```
chi2crit = stats.chi2.ppf(1-0.05,5)
print(chi2crit)
```

yields

```
11.070497693516351
```

Since the critical `chi2crit` value of 11.0705 is well above the measured `chi2calc` value of 7.2854, we cannot reject the null hypothesis without another cause. We can therefore conclude that our data follow a Gaussian distribution. Alternatively, we can apply `chisqaure()` to the sample. The command

```
chisq,p = stats.chisquare(n_obs,n_exp,ddof=0,axis=0)
print(chisq)
print(p)
```

yields

```
7.285412350787187
0.39978002411496866
```

The result means that we cannot reject the null hypothesis without another cause at a 5% significance level. The *p*-value of 0.3998 (or ~40%, which is much greater than the significance level) means that the chances of observing either the same result or a more extreme result from similar experiments in which the null hypothesis is true would be 3,998 in 10,000.

3.10   The Kolmogorov–Smirnov Test

The *Kolmogorov–Smirnov* (or *K–S*) *test* (introduced by Andrei N. Kolmogorov and Nikolai W. Smirnov) is similar to the χ^2-test in that it also involves comparing distributions and thereby allows two distributions to be tested for derivation from the same population (Kolmogorov 1933; Smirnov 1939). This test is independent of the type of distribution that is used and can therefore be used to test the hypothesis that the observations were drawn from a specific theoretical distribution.

Let us again assume that we have a data set that consists of multiple chemical measurements from a sedimentary unit. We can use the K–S test to test the null hypothesis that these measurements can be described by a Gaussian distribution with a typical central value (the mean) and a random dispersion around the mean (the standard deviation). The appropriate test statistic is then

$$KS = max|F_n(x) - F(x)|$$

where $F_n(x)$ is the cumulative frequency distribution of the n measurements and $F(x)$ is the cumulative frequency distribution expected from the theoretical distribution. The null hypothesis can be rejected if the measured KS value is higher than the critical KS value.

As an example, we can again test the hypothesis that our organic carbon measurements contained in *organicmatter_one.txt* follow a Gaussian distribution. We must first load and standardize the data in order to have a mean of zero and a standard deviation of one:

```
%reset -f
import numpy as np
import matplotlib.pyplot as plt
from scipy import stats

corg = np.loadtxt('organicmatter_one.txt')
corg = (corg-np.mean(corg))/np.std(corg,ddof=1)
```

We then compute the cumulative frequency distribution cn_obs of corg:

```
corg = np.sort(corg)
cn_obs = np.arange(1,len(corg)+1)/len(corg)
```

We next use norm.cdf() to create the cumulative frequency distribution that is expected from the theoretical distribution cn_exp with a mean of zero and a standard deviation of one:

```
cn_exp = stats.norm.cdf(corg)
```

The test statistic is the maximum difference between the two cumulative frequency distributions cn_obs and cn_exp,

```
kscalc = max(cn_obs-cn_exp)
print(kscalc)
```

which yields

```
0.07567749152482395
```

We can compare the two cumulative frequency distributions in a plot by typing

```
plt.plot(corg,cn_obs)
plt.plot(corg,cn_exp)
plt.plot([corg[np.where(cn_obs-cn_exp==max(cn_obs-cn_exp))],
    corg[np.where(cn_obs-cn_exp==max(cn_obs-cn_exp))]],
    [cn_exp[np.where(cn_obs-cn_exp==max(cn_obs-cn_exp))],
    cn_obs[np.where(cn_obs-cn_exp==max(cn_obs-cn_exp))]],
    color='k',linestyle=':')
```

The vertical black line marks the location of the maximum difference between the two cumulative frequency distributions. The critical kscalc values are solutions of an nth order polynomial, which can be found in Table 3.1 (O'Connor and Kleyner 2012). For sample sizes larger than 40 and a significance level of 0.05 (or 5%), we calculate

Table 3.1 Critical values of KS for the Kolmogorov–Smirnov test (O'Connor and Kleyner 2012).

n	Level of Significance α			
	0.10	0.05	0.02	0.01
1	0.95000	0.97500	0.99000	0.99500
2	0.77639	0.84189	0.90000	0.92929
3	0.63604	0.70760	0.78456	0.82900
4	0.56522	0.62394	0.68887	0.73424
5	0.50945	0.56328	0.62718	0.66853
6	0.46799	0.51926	0.57741	0.61661
7	0.43607	0.48342	0.53844	0.57581
8	0.40962	0.45427	0.50654	0.54179
9	0.38746	0.43001	0.47960	0.51332
10	0.36866	0.40925	0.45662	0.48893
11	0.35242	0.39122	0.43670	0.46770
12	0.33815	0.37543	0.41918	0.44905
13	0.32549	0.36143	0.40362	0.43247
14	0.31417	0.34890	0.38970	0.41762
15	0.30397	0.33760	0.37713	0.40420
16	0.29472	0.32733	0.36571	0.39201
17	0.28627	0.31796	0.35528	0.38086
18	0.27851	0.30936	0.34569	0.37062
19	0.27136	0.30143	0.33685	0.36117
20	0.26473	0.29408	0.32866	0.35241
21	0.25858	0.28724	0.32104	0.34427
22	0.25283	0.28087	0.31394	0.33666
23	0.24746	0.27490	0.30728	0.32954
24	0.24242	0.26931	0.30104	0.32286
25	0.23768	0.26404	0.29516	0.31657
26	0.23320	0.25907	0.28962	0.31064
27	0.22898	0.25438	0.28438	0.30502
28	0.22497	0.24993	0.27942	0.29971
29	0.22117	0.24571	0.27471	0.29466
30	0.21756	0.24170	0.27023	0.28987
31	0.21412	0.23788	0.26596	0.28530
32	0.21085	0.23424	0.26189	0.28094
33	0.20771	0.23076	0.25801	0.27677
34	0.20472	0.22743	0.25429	0.27279
35	0.20185	0.22425	0.26073	0.26897
36	0.19910	0.22119	0.24732	0.26532
37	0.19646	0.21826	0.24404	0.26180
38	0.19392	0.21544	0.24089	0.25843
39	0.19148	0.21273	0.23786	0.25518
40	0.18913	0.21012	0.23494	0.25205
>40	$1.22/n^{0.5}$	$1.36/n^{0.5}$	$1.51/n^{0.5}$	$1.63/n^{0.5}$

```
kscrit = 1.36/len(corg)**0.5
print(kscrit)
```

which yields

```
0.17557524502806957
```

Since the critical `kscrit` value of 0.1756 is well above the measured `kscalc` value of 0.0757, we cannot reject the null hypothesis without another cause. We can therefore conclude that our data follow a Gaussian distribution.

Alternatively, we can apply `kstest()` to the sample. Typing

```
statistic,pvalue = stats.kstest(corg,'norm')
print(statistic)
print(pvalue)
```

yields

```
0.07567749152482395
0.8562157307523406
```

The result means that we cannot reject the null hypothesis without another cause at a 5% significance level. The p-value of 0.8562 (or ~86%, which is much greater than the significance level) means that the chances of observing either the same result or a more extreme result from similar experiments in which the null hypothesis is true would be 8,562 in 10,000. The output `0.0757` corresponds to `kscalc` in our experiment without using `kstest`.

3.11 Mann–Whitney Test

The *Mann–Whitney test* (also known as the *Wilcoxon rank-sum test*) was introduced by Henry B. Mann and Donald R. Whitney (1947) and can be used to determine whether two samples come from the same population (e.g., the same lithologic unit) (null hypothesis) or from two different populations (alternative hypothesis). In contrast to the t-test, which compares the means of Gaussian-distributed data, the Mann–Whitney test compares the medians without requiring a normality assumption for the underlying population (i.e., it is a non-parametric hypothesis test).

The test requires that the samples have similar dispersions. We first combine both sets of measurements (samples 1 and 2) and arrange them together in ascending order. We then sum the ranks of samples 1 and 2, where the sum of all ranks is $R = n(n + 1)/2$, with n as the total number of measurements. The published literature is full of different versions of how to calculate the test statistic. Here, we use the version that can be found in Hedderich and Sachs (2012, pp. 484),

$$U_1 = n_1 n_2 + \frac{n_1(n_1 + 1)}{2} - R_1$$

$$U_2 = n_1 n_2 + \frac{n_2(n_2 + 1)}{2} - R_2$$

where n_1 and n_2 are the sizes of samples 1 and 2, respectively, and R_1 and R_2 are the sums of the ranks of samples 1 and 2, respectively. The required test statistic U is the smaller of the two variables U_1 and U_2, which we compare with a critical U value that depends on the sample sizes n_1 and n_2, respectively, and on the significance level α. Alternatively, we can use the U value to calculate

$$\widehat{Z} = \frac{\left|U - \frac{n_1 n_1}{2}\right|}{\sqrt{\frac{n_1 n_2 (n_1 + n_2 + 1)}{12}}}$$

if $n_1 \geq 8$ and $n_2 \geq 8$ (Mann and Whitney 1947; Hedderich and Sachs 2012, page 486). The null hypothesis can be rejected if the absolute measured z-value is greater than the absolute critical z-value, which depends on the significance level α (Sect. 3.4).

In practice, data sets often contain tied values, that is, some of the values in sample 1 and/or sample 2 are identical. In this case, the average ranks of the tied values are used instead of the true ranks. That means that the equation for the z-value must be corrected for tied values,

$$\widehat{Z} = \frac{\left|U - \frac{n_1 n_1}{2}\right|}{\sqrt{\left(\frac{n_1 n_2}{S(S-1)}\right) \cdot \left(\frac{S^3 - S}{12} - \sum_{i=1}^{r} \frac{t_i^3 - t_i}{12}\right)}}$$

where $S = n_1 + n_2$, r is the number of tied values and t_i is the number of occurrences of the ith tied value. Again, the null hypothesis can be rejected if the absolute measured z-value is greater than the absolute critical z-value, which depends on the significance level α.

The *SciPy* package provides ranksums() for performing a Mann–Whitney test to determine whether two samples data1 and data2 that each consist of eight measurements with some tied values come from the same population (null hypothesis) or from two different populations (alternative hypothesis). We define two samples that each consist of eight measurements:

```
%reset -f

import numpy as np
from scipy import stats

data1 = np.array([5,5,8,9,13,13,13,15])
data2 = np.array([3,3,4,5,5,8,10,16])
```

We use ranksums() to perform a Mann–Whitney test on the same samples,

```
statistic,pvalue = stats.ranksums(data1,data2)
print(statistic)
print(pvalue)
```

which yields

```
1.6278255976825613
0.10356187117653398
```

The result means that we cannot reject the null hypothesis at a 5% significance level. The p-value of 0.1071 (or ~ 11%, which is larger than the significance level) means that the chances of observing either the same result or a more extreme result from similar experiments in which the null hypothesis is true would be 11 in 100.

We can also use this script to test whether the two samples in *organicmatter_ two.mat* come from the same lithological unit (null hypothesis) or from two different units (alternative hypothesis) without requiring a normality assumption for the underlying population. We load the data using

```
corg = np.loadtxt('organicmatter_two.txt')
corg1 = corg[0:,0]
corg2 = corg[0:,1]
```

We use ranksums() to perform a Mann–Whitney test on the same samples,

```
statistic,pvalue = stats.ranksums(corg1,corg2)
print(statistic)
print(pvalue)
```

which yields

```
0.5406097975139224
0.5887765633694566
```

The result h=0 means that we cannot reject the null hypothesis at a 5% significance level without another cause. The p-value of 0.5888 (or ~ 59%, which is much larger than the significance level) means that the chances of observing either the same result or a more extreme result from similar experiments in which the null hypothesis is true would be 5,888 in 10,000.

3.12   The Ansari–Bradley Test

The *Ansari–Bradley test* was introduced by Abdur R. Ansari and Ralph A. Bradley (1960) and can be used to determine whether two samples come from the same distribution (null hypothesis) or whether they come from distributions with the same median and shape, but different dispersions (e.g., the variances) (alternative hypothesis). In contrast to the *F*-test, which compares the dispersions of normally distributed data, the Ansari–Bradley test compares dispersions without requiring a normality assumption for the underlying population, that is, it is a non-parametric hypothesis test.

The test requires that the samples have similar medians, which can be achieved by subtracting the medians from the samples. The test combines both sets of measurements (samples 1 and 2) and arranges them together in ascending order. There are different ways to calculate the test statistic A_n. Here, we use the method given by Hedderich and Sachs (2012, pp. 463),

$$A_n = \sum_{i=1}^{n} \left(\frac{n+1}{2} - \left| i - \frac{n+1}{2} \right| \right) V_i$$

where the value of the indicator function V_i is 1 for values from sample 1 and 0 for values from sample 2. The test statistic is therefore equal to the sum of the absolute values of the deviations from the mean value $(n+1)/2$ (Hedderich and Sachs 2012). The data are first concatenated and sorted (as in the Mann–Whitney test introduced in Sect. 3.11). Subsequently, the smallest and largest values are assigned rank 1, the second-smallest and second-largest values are assigned rank 2, and so forth. The smaller the A_n, the larger the dispersion of the values between the two samples 1 and 2. Again, the ranking of the data may also be corrected for tied values, as was previously carried out in the Mann–Whitney test. For $n \leq 20$, we can find the critical values for the text statistic A_n in Table 3.1 in the open-access article by Ansari and Bradley (1960). For larger values of n, we use the standard normal distribution,

$$Z = \frac{A_n - \mu_{A_n}}{\sqrt{\sigma_{A_n}^2}}$$

with

$$\mu_{A_n} = \begin{cases} \frac{1}{4}n_1(n+2), & \text{if } n \text{ is even} \\ \frac{1}{4}n_1(n+1)^2/n, & \text{if } n \text{ is odd} \end{cases}$$

$$\sigma_{A_n} = \begin{cases} n_1 n_2 (n^2 - 4)/(48(n-1)), & \text{if } n \text{ is even} \\ n_1 n_2 (n+1)(n^2 + 3)/(48n^2), & \text{if } n \text{ is odd} \end{cases}$$

which is again from Hedderich and Sachs (2012, page 463). The null hypothesis can be rejected if the absolute measured z-value is higher than the absolute critical z-value, which depends on the significance level α.

Our first example uses the Ansari–Bradley test to determine whether two samples data1 and data2 (each consisting of eight measurements) come from the same distribution (null hypothesis) or from distributions with the same median and shape but different dispersions (alternative hypothesis). We define two samples using

```
%reset -f

import numpy as np
from numpy.random import default_rng
from scipy import stats

data1 = np.array([7,14,22,36,40,48,49,52])
data2 = np.array([3,5,6,10,17,18,20,39])
```

The *SciPy* package provides ansari() for performing the Ansari–Bradley test by typing

```
statistic,pvalue = stats.ansari(data1,data2)
print(statistic)
print(pvalue)
```

which yields

```
33.0
0.6060606060606061
```

The result means that we cannot reject the null hypothesis without another cause at a 5% significance level. The p-value of 0.6061 (or ~61%, which is greater than the significance level) means that the chances of observing either the same result or a more extreme result from similar experiments in which the null hypothesis is true would be 6,061 in 10,000.

The second example demonstrates the handling of tied values in a data set with a large sample size (>50 measurements). We create such a data set of 100 measurements by using a random number generator:

```
rng = default_rng()
data1 = 3.4 + rng.standard_normal(100)
data2 = 4.3 + rng.standard_normal(100)
```

We then replace some values in data1 and data2 in order to introduce replicate (or tied) values:

```
data1[49:55] = 2.5
data2[24:28] = 2.5
```

Using

```
statistic,pvalue = stats.ansari(data1,data2)
print(statistic)
print(pvalue)
```

yields

```
4918.0
0.5187725818910889
```

This result means that we cannot reject the null hypothesis without another cause at a 5% significance level. The p-value of 0.5188 (or ~ 52%, which is greater than the significance level) means that the chances of observing either the same result or a more extreme result from similar experiments in which the null hypothesis is true would be 5,188 in 10,000.

We can use the same script to test whether the two samples in *organicmatter_ four.mat* come from the same distribution (null hypothesis) or from distributions with the same median and shape but different dispersions (alternative hypothesis). We clear the workspace and load the data:

```
corg = np.loadtxt('organicmatter_four.txt')
corg1 = corg[0:,0]
corg2 = corg[0:,1]
```

We use ansari() to perform the Ansari–Bradley test by typing

```
statistic,pvalue = stats.ansari(corg1,corg2)
print(statistic)
print(pvalue)
```

which yields

```
1861.0
0.7448408885112794
```

The result means that we cannot reject the null hypothesis without another cause at a 5% significance level. The *p*-value of 0.7448 (or ~74%, which is greater than the significance level) means that the chances of observing either the same result or a more extreme result from similar experiments in which the null hypothesis is true would be 7,448 in 10,000.

3.13   Distribution Fitting

In Sect. 3.9 we computed the mean and the standard deviation of a sample and designed a normal distribution based on these two parameters. We then used the χ^2-test to test the hypothesis that our data indeed follow a Gaussian (or normal) distribution. Distribution-fitting functions that are contained in the *scikit-learn* package provide powerful tools for estimating distributions directly from the data.

The function for fitting normal distributions to the data is norm.fit(). To demonstrate the use of this function, we first generate 100 synthetic Gaussian-distributed sets of values with a mean of 6.4 and a standard deviation of 1.4:

```
%reset -f

import numpy as np
import numpy.ma as ma
import matplotlib.pyplot as plt
from numpy.random import default_rng
from scipy import stats
from sklearn.mixture import GaussianMixture

rng = np.random.default_rng(0)
data = 6.4 + 1.4*rng.standard_normal(100)
```

We then define the midpoints v of nine histogram intervals with edges e, display the results, and calculate the frequency distribution n:

```
plt.figure()
plt.hist(data)
plt.show()

n,e = np.histogram(data,bins=8)
v = np.diff(e[0:2])*0.5+e[0:-1]
```

The function norm.fit() yields estimates of the mean muhat, and the standard deviation sigmahat of the normal distribution for the observations in data

```
mu,std = stats.norm.fit(data)
print(mu)
print(std)
```

yields

```
6.513535370887004
1.3469826455752003
```

These values for the mean and the standard deviation are similar to those that we
defined initially. We can now calculate the frequency distribution of the normal
distribution with the mean muhat and the standard deviation sigmahat. Then, we
can scale the resulting probability density function y to the same total number of
observations in data and plot the result:

```
res = 1/20
x = np.arange(2,12,res)
pdf = stats.norm.pdf(x,mu,std)
pdf = len(data)*(1/res)*pdf/sum(pdf)

plt.figure()
plt.bar(v,n)
plt.plot(x,pdf,color='r')
plt.show()
```

In the earth sciences, we often encounter mixed distributions. Examples include
multimodal grain size distributions (Sect. 8.10), multiple preferred paleocurrent
directions (Sect. 10.6), and multimodal chemical ages in monazite that reflect
multiple episodes of deformation and metamorphism in a mountain belt. Fitting
Gaussian mixture distributions to the data is done to determine the means and variances of the individual distributions that combine to produce the mixed distributions. The methods described in this section help in determining the episodes of
deformation in a mountain range or in separating the different paleocurrent directions caused by tidal flow in an ocean basin.

As a synthetic example of Gaussian mixture distributions we generate two sets
of 100 random numbers ya and yb with means of 6.4 and 13.3 and standard deviations of 1.4 and 1.8, respectively. We then vertically concatenate the series using
hstack() and store the 200 data values in the variable data:

```
rng = np.random.default_rng(0)
ya = 6.4 + 1.4*rng.standard_normal(100)
yb = 13.3 + 1.8*rng.standard_normal(100)

data = ma.hstack([ya,yb])
```

Plotting the histogram reveals a bimodal distribution. We can also determine the frequency distribution n using histogram():

```
e = np.arange(-0.5,30.5+1,1)
n,e = np.histogram(data,e)
v = np.diff(e[0:2])*0.5+e[0:-1]

plt.figure()
plt.bar(v,n)
plt.show()
```

We use GaussianMixture() to fit a Gaussian mixture distribution with k components to the data. The function fits the model by maximum likelihood using the *Expectation–Maximization (EM) algorithm*. The EM algorithm, which was introduced by Arthur Dempster, Nan Laird, and Donald Rubin (1977), is an iterative method that alternates between performing an expectation step and a maximization step. The expectation step computes an expectation of the logarithmic likelihood with respect to the current estimate of the distribution. The maximization step computes the parameters that maximize the expected logarithmic likelihood computed in the expectation step. The function GaussianMixture() creates an object gm of the *GaussianMixture* class. We can now determine the Gaussian mixture distribution with two components in a single dimension. Typing

```
gm = GaussianMixture(n_components=2,
    random_state=0).fit(data.reshape(-1, 1))
print(gm.weights_)
print(gm.means_)
```

yields

```
[0.47194769 0.52805231]
[[13.41685425]
 [ 6.6834305 ]]
```

Thus, we obtain the means and relative mixing proportions of both distributions. In our example, both normal distributions with means of 13.4169 and 6.6834, respectively, contribute ~50% (~0.47 and ~0.53, respectively) to the mixture distribution. The object gm contains several layers of information, including the mean gm.means_ and the covariances gm.covariances_, which we use to calculate the probability density function y of the mixed distribution:

```
x = np.arange(0,20+1/30,1/30)
y1 = np.ravel(stats.norm.pdf(x,gm.means_[0],
    (gm.covariances_[0])**0.5))
y2 = np.ravel(stats.norm.pdf(x,gm.means_[1],
    (gm.covariances_[1])**0.5))
```

The object gm also contains information on the relative mixing proportions of the two distributions in the layer gm.weights_. We can use this information to scale y1 and y2 to the correct proportions relative to each other:

```
y1 = gm.weights_[0]*y1/np.trapz(y1,x)
y2 = gm.weights_[1]*y2/np.trapz(y2,x)
```

We can now superimpose the two scaled probability density functions y1 and y2. The integral of the original data is determined by using trapz() to perform trapezoidal numerical integration:

```
y = y1 + y2
y = np.trapz(n,v)*y/np.trapz(y,x)
```

Finally, we can plot the probability density function y on the bar plot of the original histogram of data:

```
plt.figure()
plt.bar(v,n)
plt.plot(x,y,color='r')
plt.show()
```

We can then see that the Gaussian mixture distribution more or less matches the histogram of the data (Fig. 3.15).

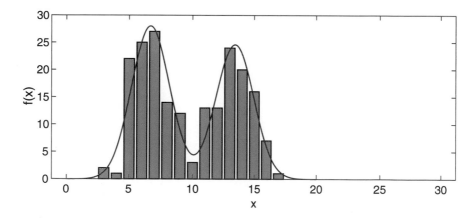

Fig. 3.15 Fitting Gaussian mixture distributions. As a synthetic example of Gaussian mixture distributions, we generate two sets of 100 random numbers with means of 6.4 and 13.3 and standard deviations of 1.4 and 1.8, respectively. The *Expectation–Maximization (EM) algorithm* is used to fit a Gaussian mixture distribution (solid line) with two components to the data (bars).

3.14   Error Analysis

The previous sections on univariate statistics provided a basis for investigating and evaluating measurement errors. These errors cannot be avoided, but they are of the utmost importance when it comes to reaching statistical conclusions. Unfortunately, thorough error analysis still occurs—albeit rarely—in many areas of the earth sciences. This is surprising—if not alarming—because a statement that does not include associated uncertainty is of no value to the recipients of the information. If, for example, the gold contained within a rock is given as 6 g per ton without any confidence limits, an astute investor will hold back on making any commitment. If confidence limits are provided, then the investor will be happy to invest provided that these limits do not exceed ± 2 g per ton. A confidence interval of ± 8 g per ton would mean that the actual gold content could be more than twice as high or could be zero and would therefore a bad investment.

Let us look briefly at the principles of error calculation without becoming too deeply immersed in the subject. A deeper investigation is not necessary because excellent books are already available on the topic (albeit without Python examples), including *An Introduction to Error Analysis* by John R. Taylor (1997). First of all, it is important to understand that there are two different types of measurement errors (not only in geosciences): *random errors* and *systematic errors* (Fig. 3.16). If we try to determine the gold content of a rock using a measuring device, we will only achieve finite measurement precision. This becomes clear when the gold content of the same sample is determined several times in a row: The results are never identical and usually scatter around a value due to technical

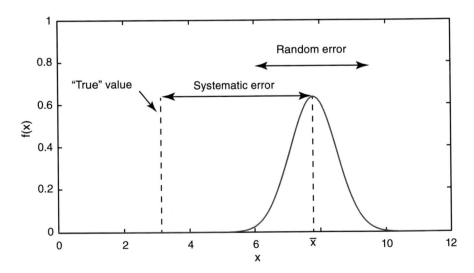

Fig. 3.16 Illustration of the difference between *random error* and *systematic error*. While random errors can be reduced by increasing the sample size, this does have any effect on systematic errors.

inaccuracies resulting from imperfections in the measuring instrument. However, external influences (e.g., temperature fluctuations and vibrations) can also be a factor, as can errors in recording the measured values (e.g., reading errors). In the ideal case, all these disturbances are random such that the distribution of n measured values x_i can be given as

$$x_i = (x_1, x_2, \ldots, x_n)$$

which follows a Gaussian distribution (Scct. 3.4). In this case, we can use the sample mean to obtain a best estimate of the gold concentration in the rock:

$$\bar{x} = \frac{1}{n} \sum_{i=1}^{n} x_i$$

The sample standard deviation s_r describes the *random error* and therefore defines the reliability of a single measurement (Fig. 3.16):

$$s_r = \sqrt{\frac{1}{n-1} \sum_{i=1}^{n} (x_i - \bar{x})^2}$$

The standard deviation is the mean square deviation of the individual measurements x_i from the sample mean. Since ± 2 g per ton is the standard deviation, we know that a single measurement must fall within the 6 ± 2 g per ton interval with a ~68% probability (see Sect. 3.5). Together with the mean, the standard deviation describes our set of measurements. However, both the mean and the standard deviation of our repeat measurements are only of finite precision depending on the number of measurements. Thus, the Gaussian distribution that is calculated from our measured values and displayed as a histogram is only an approximation of the Gaussian distribution that describes the actual gold content and the scatter due to measuring errors.

If we know the standard deviation s_r of the single measurement, we can calculate the standard error of the mean depending on the number of measurements N:

$$s_{\bar{y}} = s / \sqrt{n}$$

The standard error of the mean provides an estimate of the error in the sample mean with respect to the population (or *true*) mean. As we can see, the error of the mean decreases with the square root of the sample size n. Due to the square-root relationship, the error first decreases rapidly with increasing n and then decreases very slowly. Therefore, we must greatly increase the sample size to see any effect. For example, the results of repeated measurements of gold content might be given as 6 ± 1.2 g per ton as the sample mean and the best estimate for the population (*true*) mean and as ± 1.2 g per ton as the standard error of the mean for n measurements. Assuming a normal population distribution, we then know that the true mean has a ~68% probability of falling within the 6 ± 1.2 g per ton interval (see Sect. 3.5).

In contrast to random errors that can be determined by evaluating repeated measurements, *systematic errors* (s_s) are much more difficult to detect and quantify. They are usually consistent downward or upward deviations in measured values caused by an incorrectly calibrated instrument. The equation

$$s_s = \mu - \overline{x}$$

expresses systematic errors as the difference between the (usually unknown) population mean (or *true* value) μ and the sample mean of the measurements (Fig. 3.16).

As an example, when I was a student in a mapping course, a professor asked us perform measurements with his geological compass. Each student did their best, but the professor commented that each reading was *wrong*. He was obviously bored with the beginner's course and had thus simply adjusted the scale on his compass by five degrees in order to challenge the students. None of us dared to question the accuracy of the professor's compass, and all of our readings were therefore wrong because they included a 5-degree systematic error (apart from any random errors). As there was no possibility to detect this error on the device itself, repeated measurements with another device would have allowed the systematic error to be detected and eliminated. If we know the random and systematic errors of a measurement, we can indicate the *precision* and *accuracy* of our data. Precise measurements are consistent about the mean, but the *true* value may be different from the mean value of the measurements due to systematic errors that limit accuracy. Increasing the sample size n increases the precision, but not the accuracy.

We now turn to a simple Python experiment in order to gain a better understanding of measurement uncertainties. After clearing the workspace and resetting the random number generator, we generate measurement data by randomly sampling a population with a mean of 3.4 and a standard deviation of 1.2. This could mean that the true gold content of a rock is 3.4 g per ton, but this value cannot be determined more precisely than ± 1.2 g per ton due to measurement errors. As a result of time and budgetary limitations, let us assume that we can only take 12 measurements, which are stored in data,

```
%reset -f

import numpy as np
import matplotlib.pyplot as plt
from numpy.random import default_rng
from scipy import stats

samplesize = 12
sqrtsamplesize = np.sqrt(samplesize)
rng = np.random.default_rng(0)
data = 3.4 + 1.5*rng.standard_normal(samplesize)
print(data)
```

which yields

```
[3.58859533 3.20184271 4.36063398 3.55735018
 2.59649594 3.94239258 5.35600007 4.82062144
 2.34439715 1.50186779 2.46508831 3.46198897]
```

We display the frequency distribution of the measured values in a histogram and calculate the mean and the standard deviation by typing

```
plt.figure()
plt.hist(data)
plt.show()

print(np.mean(data))
print(np.std(data,ddof=1))
print(np.std(data,ddof=1)/sqrtsamplesize)
```

which yields

```
3.4331062031703756
1.1032297537289668
```

The error in the mean, which depends on the number of measurements (in this case, $n = 12$), is obtained by

```
print(np.std(data,ddof=1)/sqrtsamplesize)
```

which yields

```
0.3184749976467118
```

When we specify the result, a number of rules apply regarding how many digits to specify. First, the measured value and the measurement uncertainty must have the same number of digits. Second, the number of digits after rounding is defined by the measurement uncertainty. Third, we need to search for the first non-zero digit in the measurement uncertainty starting from the left. If this digit is greater than 2, then this digit is the rounding digit; if this digit is 1 or 2, then the next digit to the right is the rounding digit. Fourth, the measurement uncertainty is always rounded up. In our example, the first digit of the measurement uncertainty is indeed 2, and we therefore use the first digit to the right of the decimal point (i.e., 7) as the rounding digit. The result of our measurements is therefore 4.4 ± 2.8 g per ton, which suggests that the true value falls within this interval with a ~ 68% probability. As we can easily see from our synthetic example, this value deviates considerably from the original value of 3.4 ± 1.5 (which we defined at the beginning) due to the small sample size of $n = 12$. However, the *true* value of 3.4 still falls within the interval of 4.4 ± 2.8 g per ton.

Increasing the sample size, however, dramatically improves the result since the error of the mean depends on the number of measurements,

```
samplesize = 1000
sqrtsamplesize = np.sqrt(samplesize)
rng = np.random.default_rng(0)
data = 3.4 + 1.5*rng.standard_normal(samplesize)

plt.figure()
plt.hist(data)
plt.show()

print(np.mean(data))
print(np.std(data,ddof=1))
```

which yields

```
3.3279575848555196
1.4658625581392208
```

The error of the mean for $n = 1{,}000$ measurements is

```
print(np.std(data,ddof=1)/sqrtsamplesize)
```

which yields

```
0.0463546442048093
```

As expected, the value is much smaller than the corresponding value for $n = 12$. This result suggests that the true value lies in the interval 3.35 ± 0.05 g per ton. Again, the *true* value of 3.4 still falls within the interval of 3.35 ± 0.05 g per ton.

Next, we examine a case for which our value y depends on j measured variables $x_1, x_2, \ldots$ and their j uncertainties $s_1, s_2, \ldots$, which we use to calculate y and its uncertainty s. We can use Gauss's law of error propagation to calculate s if the uncertainties $s_1, s_2, \ldots$ are not independent of one another,

$$s = \sqrt{\sum_{j=1}^{m} \left(\frac{\partial y}{\partial x_j} \cdot s_j \right)^2 + 2 \sum_{j=1}^{m-1} \sum_{k=i+1}^{m} \left(\frac{\partial y}{\partial x_j} \right) \left(\frac{\partial y}{\partial x_k} \right) \cdot s(x_j x_k)}$$

where $\partial y/\partial x_j$ is the partial derivative of y with respect to the jth variable x_j, s_j is the standard deviation of this variable, and $s(x_j x_k)$ is the covariance of the uncertainties of variables x_j and x_k (e.g., Taylor 1997, page 212). If the variables are not dependent on one another, the covariance term is zero, and the equation is reduced to

$$ s = \sqrt{ \sum_{j=1}^{m} \left(\frac{\partial y}{\partial x_j} \cdot s_j \right)^2 } $$

Python is a great tool with which to simulate error propagation. As an example, we can use the software to estimate the uncertainty in the difference y between the measurements of two quantities x_1 and x_2 based on their individual uncertainties s_1 and s_2. The experiment reveals the difference in uncertainty when the measurements are dependent on each other and when they are independent. We define a large sample size of 100,000,000 measurements in order to yield very precise values for the errors:

```
samplesize = 100000000
```

To calculate both the difference y between x_1 and x_2 and the error in this difference, we use two Gaussian distributions with mean values of 6.3 and 3.4 and with standard deviations of 1.2 and 1.5, respectively. We calculate the difference between the means of the two distributions based on the assumption that the two distributions are dependent on each other,

$$ \bar{y} = \bar{x_1} - \bar{x_2} $$

which yields

$$ \bar{y} = 6.3 - 3.4 = 2.9 $$

with the real variables x_1 and x_2 having standard deviations of s_1 and s_2, respectively, and a covariance of $s(x_1 x_2)$. If we solve the equation for Gauss's law of error propagation from above for our example of a difference between the two means, we then obtain

$$ s_f = \sqrt{ s_1^2 + s_2^2 - 2 \cdot (x_1 x_2) } $$

which yields

$$ s_f = \sqrt{ 1.5^2 + 1.2^2 - 2 \cdot 1.5 \cdot 1.2 } = 0.3 $$

which is the error in the difference between the means where the two measurements are perfectly dependent on one another. In another Python experiment, we use the same set of random numbers r to add uncertainty to the data,

```
rng = np.random.default_rng(0)
r = rng.standard_normal(samplesize)
age1 = 3.4 + 1.2*r
age2 = 6.3 + 1.5*r
dage = age2-age1

plt.figure()
plt.hist(data,bins=30)
plt.show()

print(np.mean(dage))
print(np.std(dage,ddof=1))
```

which yields

```
2.9000297901909815
0.30000440854326643
```

for the difference in the means and its error. We now run the same experiment again but with the two measurements independent of each other. The covariance term in

$$s_f = \sqrt{s_1^2 + s_2^2 - 2 \cdot s(x_1 x_2)}$$

is then zero, which yields

$$s_f = \sqrt{1.5^2 + 1.2^2 - 0} \approx 1.92$$

which is the error in the difference between two completely independent measurements. In another Python experiment, we use different sets of random numbers to add uncertainty to the data,

```
rng = np.random.default_rng(0)
age1 = 3.4 + 1.2*rng.standard_normal(samplesize)
age2 = 6.3 + 1.5*rng.standard_normal(samplesize)
dage = age2-age1

plt.figure()
plt.hist(data,bins=30)
plt.show()

print(np.mean(dage))
print(np.std(dage,ddof=1))
```

which yields

```
2.899813389856444
1.9209998186394632
```

for the approximate difference between the means and the error of the difference. The small deviations from the exact values that are used to create the data are due to the limited sample size. As we can see from the example, either we add the same Gaussian random number to the mean, or we take different, independent random numbers to simulate the error in the difference y between dependent and independent measurement errors.

Recommended Reading

Ansari AR, Bradley RA (1960) Rank-Sum Tests for Dispersion. Annals of Mathematical Statistics, 31:1174–1189. [Open access]

Dempster AP, Laird NM, Rubin DB (1977) Maximum Likelihood from Incomplete Data via the EM Algorithm. Journal of the Royal Statistical Society, Series B (Methodological) 39(1):1–38

Fisher RA (1935) The Design of Experiments. Oliver and Boyd, Edinburgh

Grabe M (2010) Generalized Gaussian Error Calculus. Springer, Berlin Heidelberg New York

Helmert FR (1876) Über die Wahrscheinlichkeit der Potenzsummen der Beobachtungsfehler und über einige damit im Zusammenhang stehende Fragen. Zeitschrift Für Mathematik Und Physik 21:192–218

Kolmogorov AN (1933) On the Empirical Determination of a Distribution Function. Italian Giornale Dell'istituto Italiano Degli Attuari 4:83–91

Mann HB, Whitney DR (1947) On a Test of Whether one of Two Random Variables is Stochastically Larger than the Other. Ann Math Stat 18:50–60

Miller LH (1956) Table of Percentage Points of Kolmogorov Statistics. J Am Stat Assoc 51:111–121

O'Connor PDT, Kleyner A (2012) Practical Reliability Engineering, 5th ed. John Wiley & Sons, New York

Pearson ES (1990) Student—A Statistical Biography of William Sealy Gosset. In: Plackett RL, with the assistance of Barnard GA, Oxford University Press, Oxford

Pearson K (1900) On the criterion that a given system of deviations from the probable in the case of a correlated system of variables is such that it can be reasonably supposed to have arisen from random sampling. Phil Mag 50:157–175

Poisson SD (1837) Recherches sur la Probabilité des Jugements en Matière Criminelle et en Matière Civile, Précédées des Règles Générales du Calcul des Probabilités. Bachelier, Imprimeur-Libraire pour les Mathematiques, Paris

Popper K (1959) The Logic of Scientific Discovery. Hutchinson & Co., London

Hedderich J, Sachs L (2012) Angewandte Statistik: Methodensammlung mit R, 14th edn. Springer, Berlin Heidelberg New York

Smirnov NV (1939) On the Estimation of the Discrepancy between Empirical Curves of Distribution for Two Independent Samples. Bulletin of Moscow 2:3–16

Spiegel MR (2011) Schaum's Outline of Statistics, 4nd Ed. Schaum's Outlines, McGraw-Hill Professional, New York

Student, (1908) On the Probable Error of the Mean. Biometrika 6:1–25

Taylor JR (1997) An Introduction to Error Analysis: The Study of Uncertainties in Physical Measurements, 2nd edn. University Science Books, Sausalito, California

Wilcoxon F (1945) Individual Comparisons by Ranking Methods. Biometrics Bulletin 1:80–83

Bivariate Statistics

4

4.1 Introduction

Bivariate analysis is used to obtain a better understanding of the relationship between two variables x and y, such as the length and width of a fossil, the sodium and potassium content of volcanic glass, or the organic matter content along a sediment core. When the two variables are measured on the same object, x is usually identified as the *independent variable,* and y as the *dependent variable*. If both variables have been generated in an experiment, the variable manipulated by the experimenter is described as the independent variable. In some cases, neither variable is manipulated, and neither is independent.

Bivariate statistics are used to describe the strength of a relationship between two variables either with a single measure—such as Pearson's correlation coefficient for linear relationships—or with an equation obtained by regression analysis (Fig. 4.1). The equation that describes the relationship between variables x and y can be used to predict the y-response from any arbitrary x within the range of the original data values used for the regression analysis. This is of particular importance if one of the two variables is difficult to measure. In such a case, the relationship between the two variables is first determined via regression analysis on a small training set of data. The regression equation can then be used to calculate the second variable.

This chapter first introduces correlation coefficients in Sect. 4.2 and then explains the widely used methods of linear and nonlinear regression analysis in Sect. 4.3 and Sects. 4.9 to 4.11. A selection of additional methods that are used to assess uncertainties in regression analysis are also explained in Sects. 4.4 to 4.8. All methods are illustrated by means of synthetic examples since such examples

Supplementary Information The online version contains supplementary material available at https://doi.org/10.1007/978-3-031-07719-7_4.

M. H. Trauth, *Python Recipes for Earth Sciences*, Springer Textbooks in Earth Sciences, Geography and Environment, https://doi.org/10.1007/978-3-031-07719-7_4

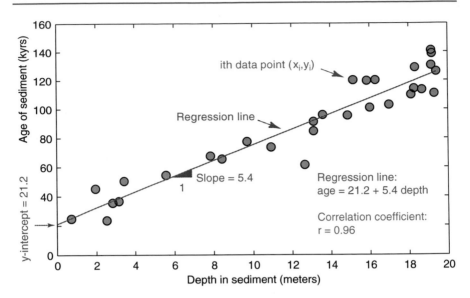

Fig. 4.1 Display of a bivariate data set. The thirty data points represent the *age* of a sediment (in kiloyears before present) at a certain *depth* (in meters) below the sediment–water interface. The combined distribution of the two variables suggests a linear relationship between *age* and *depth*, i.e., the rate of increase in the sediment age is constant with depth. A Pearson's correlation coefficient (explained in the text) of $r = 0.96$ supports a strong linear interdependency between the two variables. Linear regression yields the equation $age = 21.2 + 5.4 \cdot depth$, which indicates an increase in sediment age of 5.4 kyrs per meter of sediment depth (the slope of the regression line).

provide an excellent means of assessing the final outcome. For bivariate analysis, we use the *NumPy* (https://numpy.org), *Matplotlib* (https://matplotlib.org), and *SciPy* packages (https://scipy.org), which contain all the necessary routines.

4.2   Correlation Coefficients

Correlation coefficients are often used in the early stages of bivariate analysis. They provide only a very rough estimate of a rectilinear trend in a bivariate data set. Unfortunately, the literature is full of examples in which the importance of correlation coefficients is overestimated or in which outliers in the data set lead to an overestimation of the population correlation coefficient.

The most popular correlation coefficient is *Pearson's linear product-moment correlation coefficient* ρ (Pearson 1895) (Fig. 4.2). We estimate a population's correlation coefficient ρ from the n sample data, that is, we compute the sample correlation coefficient r, which is defined as

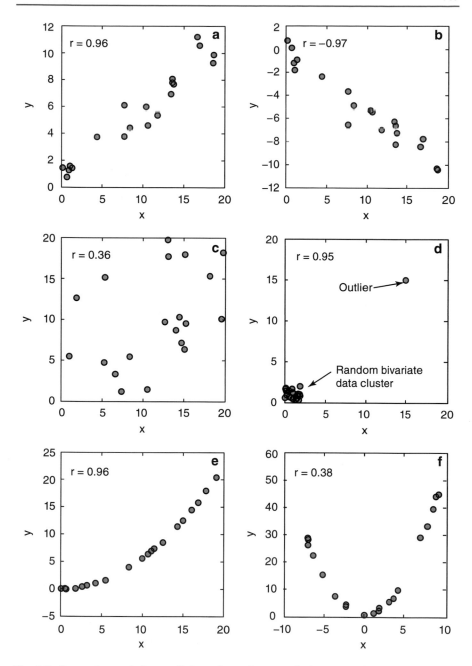

Fig. 4.2 Pearson's correlation coefficient r for various sample data sets. **a–b** Positive and negative linear correlation, **c** random scatter with no linear correlation, **d** an outlier causing a misleading value of r, **e** a curvilinear relationship causing a high r since the curve is close to a straight line, and **f** a curvilinear relationship that is clearly not described by r. Modified from Swan and Sandilands (1995).

$$r_{xy} = \frac{\sum_{i=1}^{n} (x_i - \bar{x})(y_i - \bar{y})}{(n-1)s_x s_y}$$

The numerator of Pearson's correlation coefficient is known as the *corrected sum of products* of the bivariate data set. Dividing the numerator by $(n-1)$ yields the *covariance*,

$$cov_{xy} = \frac{1}{(n-1)} \sum_{i=1}^{n} (x_i - \bar{x})(y_i - \bar{y})$$

which is the summed product of deviations in the data x_i and y_i from the sample means $\bar{x}$ and $\bar{y}$ divided by $(n-1)$. Covariance is a widely used measure in bivariate statistics, although it has the disadvantage of being dependent on the dimensions of the data. Dividing the covariance by the univariate standard deviations s_x and s_y removes this dependency and leads to Pearson's correlation coefficient. The prerequisite for the validity of the correlation coefficient is that both x and y be normally distributed.

One way to test the significance of Pearson's correlation coefficient is to determine the probability of an r-value for a random sample from a population with a $\rho = 0$. The significance of the correlation coefficient can be estimated using a t-statistic:

$$t = r\sqrt{\frac{n-2}{1-r^2}}$$

The correlation coefficient is significant if the calculated t is greater than the critical t ($n-2$ degrees of freedom, $\alpha = 0.05$). This test is also only valid if both variables are Gaussian distributed.

Pearson's correlation coefficient is very sensitive to disturbances in the bivariate data set. Several alternatives therefore exist to Pearson's correlation coefficient, such as *Spearman's rank correlation coefficient*, which was proposed by English psychologist Charles Spearman (1863–1945). Spearman's coefficient can be used to measure statistical dependence between two variables without requiring a normality assumption for the underlying population (i.e., it is a non-parametric measure of correlation) (Spearman 1904, 1910). Furthermore, since Spearman's coefficient uses the ranks of the values in x and y rather than the actual numerical values, it can be used to find correlations in nonlinear data and even in non-numerical data, such as fossil names or rock types in stratigraphic sequences. After replacing the numerical values in x and y with their ranks (whereby multiple values in x and y are replaced with their respective average ranks), the sample Spearman's rank correlation coefficient is defined as

$$r_{xy} = \frac{6 \sum\limits_{i=1}^{n} d_i^2}{n(n^2 - 1)}$$

where d_i is the difference between the ranks of the two variables. Since this correlation coefficient is based on ranks rather than on numerical values, it is less sensitive to outliers than is Pearson's correlation coefficient.

Another alternative to Pearson's correlation coefficient is *Kendall's tau rank correlation coefficient*, which was proposed by British statistician Maurice Kendall (1907–1983). This coefficient is also a non-parametric measure of correlation, similar to Spearman's rank correlation coefficient (Kendall 1938). Kendall's tau rank correlation coefficient compares the ranks of the numerical values in x and y, which means that there is a total of $0.5\ n(n-1)$ pairs to compare. Pairs of observations (x_i, y_i) and (x_j, y_j) are said to be concordant if the ranks for both observations are the same, and they are said to be discordant if they are not. The sample Kendall's tau rank correlation coefficient is defined as

$$\tau = \frac{(P - Q)}{\frac{1}{2}n(n - 1)}$$

where P is the number of concordant pairs and Q is the number of discordant pairs. Kendall's correlation coefficient typically has a lower value than does Spearman's correlation coefficient.

The following example illustrates the use of the correlation coefficients and highlights potential pitfalls when using these measures of linear trends. The example also illustrates the resampling methods that can be used to explore the confidence level of the estimate for ρ. The synthetic data consist of two variables: the age of a sediment in kiloyears (kyrs) before present (BP) and the depth below the sediment–water interface in meters. An advantage of using synthetic data sets is that we are able to fully understand the linear model behind the data.

The data are represented as two columns contained in the file *agedepth_1.txt*. These data were generated using a series of thirty random levels below the sediment surface. The linear relationship age=5.6*meters+20 was used to compute noise-free values for the variable age. This is the equation of a straight line with a slope of 5.6 and an intercept with the y-axis of 20. Some Gaussian noise with a mean of zero and a standard deviation of ten has been added to the age data:

```
%reset -f
```

```
import numpy as np
import numpy.ma as ma
from scipy import stats
import matplotlib.pyplot as plt
```

```
from scipy.stats import bootstrap
from numpy.random import default_rng

rng(0)
np.random.seed(0)
meters = 20*np.random.rand(30)
age = 5.6*meters + 20
rng = default_rng(0)
age = age + 10*rng.standard_normal(np.shape(meters))

plt.figure()
plt.plot(meters,age,
    marker='o',
    linestyle='none')
plt.show()

agedepth = ma.hstack([meters,age])
```

Instead, a synthetic bivariate data set can also be loaded from the file *agedepth_1. txt*:

```
agedepth = np.loadtxt('agedepth_1.txt')
```

We then define two new variables meters and age and generate a scatter plot of the data:

```
meters = agedepth[0:,0]
age = agedepth[0:,1]

plt.figure()
plt.plot(meters,age,
    marker='o',
    linestyle='none')
plt.show()
```

We observe a strong linear trend in the plot, which suggests some interdependency between the two variables meters and age. This trend can be described by Pearson's correlation coefficient r, where $r = 1$ indicates a perfect positive correlation (i.e., age increases with meters), $r = 0$ suggests no correlation, and $r = -1$ indicates a perfect negative correlation. We use corrcoef() to compute Pearson's correlation coefficient,

```
r = np.corrcoef(meters,age)
print(r)
```

which results in the following output:

```
[[1.        0.956335]
 [0.956335 1.       ]]
```

The function corrcoef() calculates an array (or matrix) of Pearson's correlation coefficients for all possible combinations of the two variables age and meters. The value of $r = 0.9563$ suggests that the two variables age and meters are strongly dependent on each other.

However, Pearson's correlation coefficient is highly sensitive to outliers, as illustrated by the following example. Let us generate a normally distributed cluster of thirty data points with a mean of zero and a standard deviation of one. In order to obtain identical data values, we reset the random number generator by using 0 as the seed:

```
rng = default_rng(0)
x = rng.standard_normal((30))
y = rng.standard_normal((30))

plt.figure()
plt.plot(x,y,
    marker='o',
    linestyle='none')
plt.show()
```

As expected, the correlation coefficient for these random data is very low. Typing

```
r = np.corrcoef(x,y)
print(r)
```

yields

```
[[ 1.         -0.33210663]
 [-0.33210663  1.        ]]
```

We now introduce a single outlier to the data set in the form of an exceptionally high (x,y) value in which x=y. The correlation coefficient for the bivariate data set, including the outlier (x,y)=(5,5), is much larger than before. Typing

```
x = np.append(x,5)
y = np.append(y,5)

plt.figure()
plt.plot(x,y,
    marker='o',
    linestyle='none')
```

```
plt.show()

r = np.corrcoef(x,y)
print(r)
```

yields

```
[[1.          0.3415448]
 [0.3415448 1.          ]]
```

Increasing the absolute (x,y) values for this outlier results in a dramatic increase in the correlation coefficient. Typing

```
x[30] = 10
y[30] = 10

plt.figure()
plt.plot(x,y,
    marker='o',
    linestyle='none')
plt.show()

r = np.corrcoef(x,y)
print(r)
```

yields

```
[[1.          0.74130704]
 [0.74130704 1.          ]]
```

which reaches a value close to $r=1$ if the outlier has a value of $(x,y)=(20,20)$. Typing

```
x[30] = 20
y[30] = 20

plt.figure()
plt.plot(x,y,
    marker='o',
    linestyle='none')
plt.show()

r = np.corrcoef(x,y)
print(r)
```

yields

```
[[1.          0.92488156]
 [0.92488156 1.          ]]
```

We can compare the sensitivity of Pearson's correlation coefficient with that of Spearman's correlation coefficient and Kendall's correlation coefficient using pearsonr(), spearmanr(), and kendalltau() from the *stats* module of the *SciPy* package. In contrast to corrcoef(), these functions do not calculate correlation matrices that we can use later (e.g., in Chap. 9) when calculating correlations within multivariate data sets. We type

```
rp,rppvalue = stats.pearsonr(x,y)
rs,rspvalue = stats.spearmanr(x,y)
rk,rkpvalue = stats.kendalltau(x,y)
print(rp)
print(rs)
print(rk)
```

which yields

```
0.9248815580753484
-0.17298387096774195
-0.12258064516129034
```

We observe that the alternative measures of correlation result in reasonable values, in contrast to the overestimated value for Pearson's correlation coefficient, which mistakenly suggests a strong interdependency between the variables. Although outliers are easy to identify in a bivariate scatter, erroneous values can be easily overlooked in large multivariate data sets (Chap. 9).

Various methods exist for calculating the significance of Pearson's correlation coefficient. The function pearsonr() also includes the possibility of evaluating the quality of the result. The *p*-value is the probability of obtaining a correlation that is at least as large as the observed value when the true correlation is zero. If the *p*-value is small, then the correlation coefficient *r* is significant. Typing

```
rp,rppvalue = stats.pearsonr(x,y)
print(rp,rppvalue)
```

yields

```
0.9248815580753484 1.0489538147952255e-13
```

In our example, the *p*-value rppvalue is close to zero, which suggests that the correlation coefficient is significant. We can conclude from this experiment that this particular significance test fails to detect correlations that can be attributed to an

outlier. We therefore next try an alternative *t*-test statistic to determine the significance of the correlation between *x* and *y*. According to this test, we can reject the
null hypothesis, that is, that there is no correlation if the calculated *t* is larger than
the critical *t* ($n-2$ degrees of freedom, $\alpha = 0.05$). Typing

```
tcalc = rp * ((len(x)-2)/(1-rp**2))**0.5
tcrit = stats.t.ppf(1-0.05,len(x)-2)
print(tcalc)
print(tcrit)
```

yields

```
13.09814600608973
1.6991270265334972
```

This result indeed indicates that we can reject the null hypothesis, and there is
therefore no correlation.

As alternatives to detecting outliers, *resampling schemes* and *surrogates*
(e.g., the *bootstrap* or *jackknife* methods) represent powerful tools for assessing
the statistical significance of the results. These techniques are particularly useful
when scanning large multivariate data sets for outliers (see Chap. 9). Resampling
schemes repeatedly resample the original data set of *n* data points either by choosing $n-1$ subsamples *n* times (the jackknife) or by choosing an arbitrary set of subsamples with *n* data points *with replacement* (the bootstrap). The statistics of these
subsamples provide better information on the characteristics of the population than
do the statistical parameters (mean, standard deviation, correlation coefficients)
computed from the full data set. This code allows for resampling our bivariate data
set, including the outlier $(x,y)=(20,20)$:

```
ind = np.linspace(0,30,31,
    dtype=int)
rhos1000 = np.zeros((1000,4))
for i in range(1000):
    indices = np.random.choice(ind,
        size=31,
        replace=True)
    r = np.reshape(np.corrcoef(x[indices],
        y[indices]),(1,4))
    rhos1000[i,:] = r
```

This code first resamples the data one thousand times; it then calculates the correlation coefficient for each new subsample and stores the result in the variable rhos1000. Since corrcoef() delivers a 2-by-2 matrix (as mentioned above),
rhos1000 has dimensions 1,000-by-4 (i.e., 1,000 values for each element of the

2-by-2 matrix). Plotting the histogram of the 1,000 values for the second element, that is, the correlation coefficient of (x,y), illustrates the dispersion of this measure with respect to the presence or absence of the outlier. Since the distribution of rhos1000 contains many empty classes, we use a large number of bins:

```
plt.figure()
plt.hist(rhos1000[:,1],bins=30)
plt.show()
```

The histogram shows a strong peak close to $r = 1$ and a cluster of correlation coefficients at around $r = 0$ that follow the normal distribution (Fig. 4.3). The interpretation of this histogram is relatively straightforward. When the subsample contains the outlier, the correlation coefficient is close to one, but subsamples without the outlier yield a very low correlation coefficient (close to zero), which suggests the absence of any strong interdependence between the two variables x and y.

Bootstrapping therefore provides a simple yet powerful tool for either accepting or rejecting our first estimate of the correlation coefficient for the population. Applying the procedure above to the synthetic sediment data yields a clear unimodal Gaussian distribution for the correlation coefficients of the subsamples. Typing

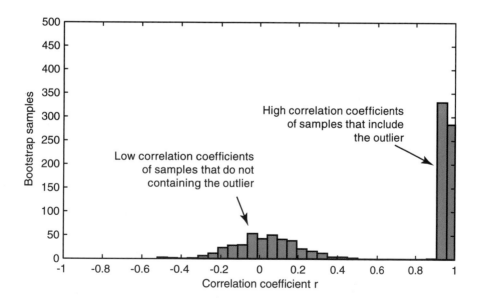

Fig. 4.3 Bootstrap result for Pearson's correlation coefficient r from 1,000 subsamples. The histogram shows a roughly normally distributed cluster of correlation coefficients at around $r = 0$, which suggests that these subsamples do not include the outlier. However, the strong peak close to $r = 1$ suggests that an outlier with high values for the two variables x and y is present in the corresponding subsamples.

```
agedepth = np.loadtxt('agedepth_1.txt')

meters = agedepth[0:,0]
age = agedepth[0:,1]

rp, rppvalue = stats.pearsonr(meters,age)
print(rp,rppvalue)
```

yields

```
0.9563350017670429 1.7163469052071033e-16
```

before we bootstrap the correlation coefficient using

```
ind = np.linspace(0,19,30,dtype=int)
rhos1000 = np.zeros((1000,4))
for i in range(1000):
    indices = np.random.choice(ind,
        size=30,
        replace=True)
    r = np.reshape(np.corrcoef(meters[indices],
        age[indices]),(1,4))
    rhos1000[i,:] = r

plt.figure()
plt.hist(rhos1000[:,1],bins=30)
plt.show()
```

Most of the values for rhos1000 fall within the interval [0.92, 0.98]. Since the correlation coefficients for the resampled data sets (in our example) have an approximately normal distribution, we can use their mean as a good estimate for the true correlation coefficient. Typing

```
print(np.mean(rhos1000))
```

yields

```
0.9672350831170088
```

This value is similar to our first result of $r = 0.9563$, but now we have confidence in the validity of this result. In our example, however, the distribution of the bootstrap estimates of the correlations from the age-depth data is quite skewed because the upper limit is fixed at one. Nevertheless, the bootstrap method is a valuable tool for assessing the reliability of Pearson's correlation coefficient for bivariate analysis.

4.3   Classical Linear Regression Analysis

Linear regression offers another way of describing the relationship between the two variables x and y. While Pearson's correlation coefficient provides only a rough measure of a linear trend, linear models obtained via regression analysis allow for arbitrary y-values to be predicted for any given value of x within the data range. Statistically testing the significance of the linear model provides some insights into the accuracy of these predictions.

Classical regression assumes that y responds to x and that the entire dispersion in the data set is contained within the y-value (Fig. 4.4 and 4.5). That means that x is then the independent variable (also known as the predictor variable, or the regressor). The values of x are defined by the experimenter and are often regarded as being free of errors. An example is the location x within a sediment core from which the variable y has been measured. The dependent variable y contains errors because its magnitude cannot be determined accurately. Linear regression minimizes the deviations Δy between the data points xy and the value y predicted by the best-fit line $y = b_0 + b_1 x$ using a least-squares criterion. The basic equation for a general linear model is

$$y = b_0 + b_1 x$$

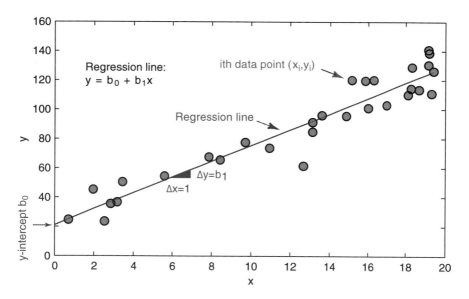

Fig. 4.4 Linear regression. Whereas classical regression minimizes the Δy deviations, reduced major axis regression minimizes the triangular area $0.5 \cdot (\Delta x \Delta y)$ between the data points and the regression line, where Δx and Δy are the distances between the predicted and the true x- and y-values, respectively. The intercept of the line with the y-axis is b_0, and the slope is b_1. These two parameters define the equation of the regression line.

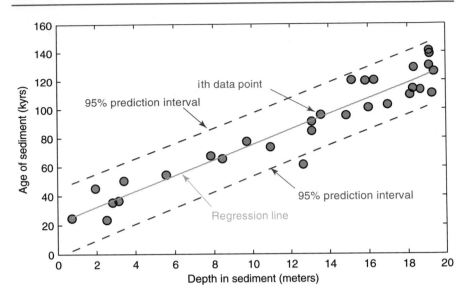

Fig. 4.5 The result of linear regression. The plot shows the original data points (circles), the regression line (solid line), and the error bounds (dashed lines) of the regression. Note that the 95% prediction intervals are actually curved even though they almost appear to be straight lines in the example.

where b_0 and b_1 are the regression coefficients. The value of b_0 is the intercept with the y-axis, and b_1 is the slope of the line. The squared sum of the Δy deviations to be minimized is

$$\sum_{i=1}^{n} (\Delta y_i)^2 = \sum_{i=1}^{n} (y_i - (b_0 + b_1 x_i))^2$$

Partially differentiating the right-hand term in the equation and setting it to zero yields a simple equation for the regression coefficient b_1:

$$b_1 = \frac{\sum_{i=1}^{n} (x_i - \bar{x})(y_i - \bar{y})}{\sum_{i=1}^{n} (x_i - \bar{x})^2}$$

The regression line passes through the data centroid defined by the sample means, and we can therefore compute the other regression coefficient b_0

$$b_0 = \bar{y} - b_1 \bar{x}$$

using the univariate sample means and the slope b_1 computed earlier.

As an example, let us again load the synthetic age-depth data from the file *agedepth_1.txt*. We can define two new variables meters and age and generate a scatter plot of the data:

```
%reset -f

import numpy as np
import matplotlib.pyplot as plt

agedepth = np.loadtxt('agedepth_1.txt')

meters = agedepth[0:,0]
age = agedepth[0:,1]
```

A linear trend in the bivariate scatter plot together with a correlation coefficient greater than $r = 0.9$ suggests a strong linear dependence between meters and age. In geological terms, this implies that the sedimentation rate was constant through time. We now try to fit a linear model to the data that will help us predict the age of the sediment at levels for which we have no age data. The function polyfit() computes the coefficients of a polynomial $p(x)$ of a specific degree that fits the data y in a least-squares sense. In our example, we fit a first-degree (linear) polynomial to the data. Typing

```
p = np.polyfit(meters,age,1)
print(p)
```

yields

```
[ 5.36669965 21.7607019 ]
```

where p is a row vector that contains the polynomial coefficients in descending powers. Since we are working with synthetic data, we know the values for the slope and the intercept with the y-axis. The estimated slope (5.3667) and the intercept with the y-axis (21.7607) are in good agreement with the true values of 5.6 and 20, respectively. Both the data and the fitted line can be plotted on the same graph:

```
plt.figure()
plt.plot(meters,age,
    marker='o',
    linestyle='none')
plt.plot(meters,p[0]*meters+p[1],
    color=(0.8,0.5,0.3))
plt.show()
```

Instead of using the equation for the regression line, we can also use polyy1d() to calculate the y-values:

```
pval = np.poly1d(p)

plt.figure()
plt.plot(meters,age,
    marker='o',
    linestyle='none')
plt.plot(meters,pval(meters),
    color=(0.8,0.5,0.3))
plt.show()
```

The coefficients p(x) and the equation obtained by linear regression can now be used to predict y-values for any given x-value. However, we can only do this within the depth interval for which the linear model was fitted, that is, between 0 and 20 m. As an example, the age of the sediment at a depth of 17 m is given by

```
print(pval(17))
```

which yields

```
112.99459593875632
```

This result suggests that the sediment at a depth of 17 m has an age of ~113 kyrs.

4.4   Analyzing the Residuals

When we compare how much the predicted values vary from the actual or observed values, we are performing an analysis of the residuals. The statistics of the residuals provide valuable information on the quality of a model fitted to the data. For instance, a trend in the residuals suggests that the model does not fully describe the data. In such cases, a more complex model (e.g., a polynomial of a higher degree) should be fitted to the data. Residuals are ideally purely random, that is, they are Gaussian distributed with a mean of zero. We therefore test the hypothesis that our residuals are Gaussian distributed by visually inspecting the histogram and by employing a χ^2-test, as introduced in Chap. 3:

```
%reset -f

import numpy as np
import matplotlib.pyplot as plt
from scipy import stats

agedepth = np.loadtxt('agedepth_1.txt')

meters = agedepth[0:,0]
```

```
age = agedepth[0:,1]

p = np.polyfit(meters,age,1)
print(p)

pval = np.poly1d(p)
res = age - pval(meters)
```

Since plotting the residuals does not reveal any obvious pattern of behavior, no model that is more complex than a straight line should be fitted to the data:

```
plt.figure()
plt.stem(meters,res)
plt.show()
```

An alternative way of plotting the residuals is with stem plot using stem():

```
plt.figure()
plt.subplot(2,1,1)
plt.plot(meters,age,
    marker='o',
    linestyle='none')
plt.plot(meters,pval(meters),
    color='r')
plt.subplot(2,1,2)
plt.stem(meters,res)
plt.show()
```

To explore the distribution of the residuals, we can choose six classes and display the corresponding frequencies:

```
plt.figure()
plt.hist(res,bins=6)
plt.show()
```

The χ^2-test can be used to test the hypothesis that the residuals follow a Gaussian distribution (Sect. 3.9). We use chisquare() to perform the χ^2-test,

```
nres,e = np.histogram(res,bins=6)
chisq,p = stats.chisquare(nres)
print(chisq)
print(p)
```

which yields

```
11.200000000000003
0.04755564396470476
```

The result means that we cannot reject the null hypothesis without another cause at a 5% significance level. However, the quality of the result is not very good because the sample size of 30 measurements is very small.

4.5   Bootstrap Estimates of the Regression Coefficients

In this section, we use the *bootstrap* method to obtain a better estimate of the regression coefficients. As an example, we use `random.choice()` with 1,000 samples (Fig. 4.6):

```
%reset -f

import numpy as np
import matplotlib.pyplot as plt
from numpy.random import default_rng

agedepth = np.loadtxt('agedepth_1.txt')

meters = agedepth[0:,0]
age = agedepth[0:,1]

p = np.polyfit(meters,age,1)
```

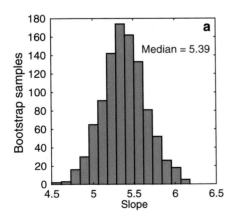 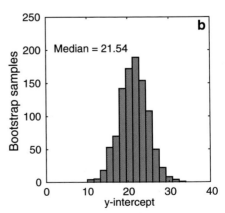

Fig. 4.6 Histogram of **a** the first (slope of the regression line) and **b** the second (y-axis intercept of the line) regression coefficient as estimated via bootstrap resampling. The first coefficient is well constrained, but the second coefficient shows a broad scatter.

```
ind = np.linspace(0,19,30,dtype=int)
p1000 = np.zeros((1000,2))
for i in range(1000):
    indices = np.random.choice(ind,
        size=30,
        replace=True)
    p = np.polyfit(meters[indices],
        age[indices],1)
    p1000[i,:] = p
```

The statistic of the first coefficient, that is, the slope of the regression line, is

```
plt.figure()
plt.hist(p1000[:,0],bins=30)
plt.show()

print(np.median(p1000[:,0]))
```

which yields

```
5.120606217265465
```

We use the median instead of the mean since we cannot expect the bootstrap results to be Gaussian distributed. In contrast, the statistic of the second coefficient shows a striking dispersion. Typing

```
plt.figure()
plt.hist(p1000[:,1],bins=30)
plt.show()

print(np.median(p1000[:,1]))
```

yields

```
24.22153549031753
```

The true values used to simulate our data set were 5.6 for the slope and 20 for the intercept with the y-axis, whereas the corresponding coefficients calculated using polyfit() were 5.3667 and 21.7607, respectively (Sect. 4.3).

4.6   Jackknife Estimates of the Regression Coefficients

The *jackknife* method is a resampling technique that is similar to the bootstrap
method. From a sample with n data points, n subsamples with $n-1$ data points are
taken. The parameters of interest (e.g., the regression coefficients) are calculated
for each of the subsamples. The mean and dispersion of the coefficients are then
computed. The disadvantage of this method is the limited number of n subsamples: A jackknife estimate of the regression coefficients is therefore less precise
than a bootstrap estimate.

The relevant code for the jackknife is easy to generate:

```
%reset -f

import numpy as np
import matplotlib.pyplot as plt
from numpy.random import default_rng

agedepth = np.loadtxt('agedepth_1.txt')

meters = agedepth[0:,0]
age = agedepth[0:,1]

p = np.polyfit(meters,age,1)

p30 = np.zeros((30,2))
for i in range(30):
    j_meters = np.copy(meters)
    j_age = np.copy(age)
    j_meters = np.delete(j_meters,i,0)
    j_age = np.delete(j_age,i,0)
    p = np.polyfit(j_meters,j_age,1)
    p30[i,:] = p
```

The jackknife for subsamples with $n-1=29$ data points can be obtained using a
simple for loop. The ith data point within each iteration is deleted, and regression
coefficients are calculated for the remaining data points. The mean of the i subsamples provides an improved estimate of the regression coefficients. As with the
bootstrap result, the slope of the regression line (first coefficient) is well defined,
whereas the intercept with the y-axis (second coefficient) has a large uncertainty,

```
print(np.median(p30[:,0]))
```

which yields

```
5.3663093479597155
```

compared with 5.1206 calculated by the bootstrap method and

```
print(np.median(p30[:,1]))
```

which yields

```
21.79637039186352
```

compared with 24.2215 from the bootstrap method (Sect. 4.5). As before, the true
values are 5.6 and 20. The histograms of the jackknife results from 30 subsamples
(Fig. 4.7).

```
plt.figure()
plt.subplot(1,2,1)
plt.hist(p30[:,0])
plt.subplot(1,2,2)
plt.hist(p30[:,1])
plt.show()
```

do not display such clear distributions for the coefficients as do the histograms of
the bootstrap estimates. We have therefore seen that resampling using either the
jackknife or the bootstrap method is a simple and valuable way to test the quality
of regression models. The next section introduces an alternative approach for quality estimation that is much more commonly used than the resampling methods.

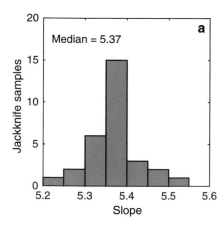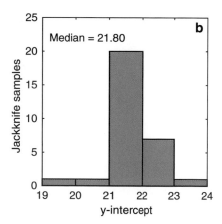

Fig. 4.7 Histogram of **a** the first (slope of the regression line) and **b** the second (y-axis intercept
of the line) regression coefficient as estimated via jackknife resampling. Note that the parameters
are not as well defined as those for bootstrapping.

4.7   Cross-Validation

A third method for testing the quality of the result of a regression analysis involves *cross-validation*. With cross-validation, the regression line is computed by using $n-1$ data points. The nth data point is predicted, and the discrepancy between the prediction and the actual value is computed. Subsequently, the mean of the discrepancies between the actual and predicted values is determined.

In this example, the cross-validation is computed for $n = 30$ data points. The resulting 30 regression lines, which are each computed using $n-1 = 29$ data points, display some dispersion in their slopes and *y*-axis intercepts:

```
%reset -f

import numpy as np
import matplotlib.pyplot as plt

agedepth = np.loadtxt('agedepth_1.txt')

meters = agedepth[0:,0]
age = agedepth[0:,1]

p = np.polyfit(meters,age,1)
print(p)

p_age = np.zeros((30,1))
p_error = np.zeros((30,1))

plt.figure()
plt.plot(meters,age,
    marker='o',
    linestyle='none').
for i in range(0,30):
    j_meters = meters;
    j_age = age;
    j_meters = np.delete(j_meters,i)
    j_age = np.delete(j_age,i)
    p = np.polyfit(j_meters,j_age,1)
    pval = np.poly1d(p)
    plt.plot(meters,pval(meters),
        color=(0.8,0.5,0.3),
        linewidth=0.1)
    p_age =pval(meters)
    p_error[i,0:] = p_age[i] - age[i]
plt.show()
```

In the ideal case, the prediction error should be Gaussian distributed with a mean of zero,

```
print(np.mean(p_error))
```

which yields

```
-0.048670587295191345
```

The standard deviation is the mean of the deviations of the true data points from the predicted straight line,

```
print(np.std(p_error,ddof=1))
```

which yields

```
10.975717618676232
```

Cross-validation provides valuable information on the goodness-of-fit of the regression result and can also be used for quality control in other fields, such as in temporal and spatial prediction (Chaps. 5 and 7).

4.8   Reduced Major Axis Regression

In some examples, neither variable is manipulated, and both can therefore be considered to be independent. In such cases, several methods are available for computing a best-fit line that minimizes the distance from both x and y. As an example, the method of *reduced major axis* (RMA) minimizes the triangular area $0.5 \cdot (\Delta x \Delta y)$ between the data points and the regression line, where Δx and Δy are the distances between the predicted values and the true values of x and y (Fig. 4.4). Although this optimization appears to be complex, it can be shown that the first regression coefficient b_1 (the slope) is simply the ratio of the standard deviations of y and x:

$$b_1 = s_y/s_x$$

As with classical regression, the regression line passes through the data centroid, which is defined by the sample mean. We can therefore compute the second regression coefficient b_0 (i.e., the y-intercept)

$$b_0 = \bar{y} - b_1 \bar{x}$$

using the univariate sample means and the slope b_1 computed earlier. Let us again load the age-depth data from the file *agedepth_1.txt* and define two variables, meters and age. It is assumed that both the variables contain errors and that the scatter of the data can be explained by dispersions of meters and age:

```
%reset -f

import numpy as np
import matplotlib.pyplot as plt

agedepth = np.loadtxt('agedepth_1.txt')

meters = agedepth[0:,0]
age = agedepth[0:,1]
```

The formula above is used for computing the slope of the regression line b_1,

```
p = np.zeros((2,1))
p[0] = np.std(age,ddof=1)/np.std(meters,ddof=1)
print(p)
```

which yields

```
[[5.6117361]
 [0.       ]]
```

The second coefficient b_0 (i.e., the y-axis intercept) can therefore be computed by

```
p[1] = np.mean(age) - p[0]*np.mean(meters)
```

which yields

```
[[ 5.6117361 ]
 [18.70366978]]
```

The regression line can be plotted by

```
plt.figure()
plt.plot(meters,age,
    marker='o',
    linestyle='none')
plt.plot(meters,p[0]*meters+p[1],
    color=(0.8,0.5,0.3))
plt.show()
```

This regression line differs slightly from the line obtained from classical regression. Note that the regression line from RMA is *not* the bisector of the lines produced by the *xy* and *yx* classical linear regression analyses, that is, the lines produced using either *x* or *y* as the independent variable while computing the regression lines.

4.9   Curvilinear Regression

It is apparent from our previous analysis that a linear regression model provides a good way of describing the scaling properties of the data. However, we may wish to check whether the data could be equally well described by a polynomial fit of a higher degree, such as a second-degree polynomial:

$$y = b_0 + b_1 x + b_2 x^2$$

To clear the workspace and reload the original data, we type

```
%reset -f

import numpy as np
import numpy.ma as ma
import matplotlib.pyplot as plt
from numpy.random import default_rng

agedepth = np.loadtxt('agedepth_1.txt')

meters = agedepth[0:,0]
age = agedepth[0:,1]
```

A second-degree polynomial can then be fitted by using `polyfit()`,

```
p = np.polyfit(meters,age,2)
print(p)
```

which yields

```
[ 0.05890582  4.10867582 26.0381405 ]
```

The first coefficient is close to zero, that is, it has little influence on predictions. The second and third coefficients are similar to those obtained by linear regression. Plotting the data yields a curve that resembles a straight line:

```
pval = np.poly1d(p)

plt.figure()
plt.plot(meters,age,
    marker='o',
    linestyle='none')
plt.plot(meters,pval(meters),
    color=(0.8,0.5,0.3))
plt.show()
```

We now use another synthetic data set that we generate using a quadratic relationship between meters and age:

```
np.random.seed(0)
meters = 20*np.random.rand(30)
age = 1.6*meters**2 -1.1*meters + 50
rng = default_rng(0)
age = age + 40*rng.standard_normal(np.shape(meters))

plt.figure()
plt.plot(meters,age,
    marker='o',
    linestyle='none')
plt.show()

agedepth = ma.hstack([meters,age])
```

A synthetic bivariate data set can be also loaded from the file *agedepth_2.txt*:

```
agedepth = np.loadtxt('agedepth_2.txt')

meters = agedepth[0:,0]
age = agedepth[0:,1]

plt.figure()
plt.plot(meters,age,
    marker='o',
    linestyle='none')
plt.show()
```

Fitting a second-order polynomial produces a convincing regression result,

```
p = np.polyfit(meters,age,2)
print(p)
```

which yields

```
[ 1.83562317 -7.0652954  74.15255975]
```

As shown above, the true values of the three coefficients are $+1.8$, -7.1, and $+74.2$,
which means that there are some discrepancies between the true values and the
coefficients that were estimated using `polyfit()`. Let us plot the result (Fig. 4.8):

```
pval = np.poly1d(p)
```

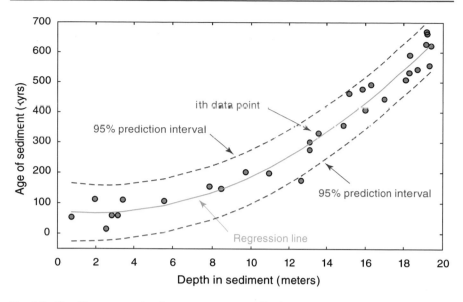

Fig. 4.8 Curvilinear regression from measurements of barium content. The plot shows the original data points (circles), the regression line for a polynomial of degree $n = 2$ (solid line), and the prediction error (dashed lines) of the regression.

```
plt.figure()
plt.plot(meters,age,
    marker='o',
    linestyle='none')
plt.plot(meters,pval(meters),
    color=(0.8,0.5,0.3))
plt.show()
```

The plot shows that the quadratic model for these data is a good one. The quality of the result could again be tested by exploring the residuals, by employing resampling schemes, or by using cross-validation. Combining regression analysis with one of these methods is a powerful tool in bivariate data analysis, whereas Pearson's correlation coefficient should be used only as a preliminary test for linear relationships.

4.10 Nonlinear and Weighted Regression

Many bivariate data sets in the earth sciences follow a more complex trend than a simple linear or curvilinear trend. Classic examples of nonlinear trends include the exponential decay of radionuclides and the exponential growth of algae populations. In such cases, Python provides various tools for fitting nonlinear models to the data. An easy-to-use routine for fitting such models is nonlinear regression using `curve_fit()`. To demonstrate the use of `curve_fit()`, we generate a bivariate data set in which one variable is exponentially correlated with a second variable. We first generate evenly spaced values between 0.5 and 3 at intervals of 0.1 and add some Gaussian noise with a standard deviation of 0.2 in order to make the data unevenly spaced. The resulting 26 data points are sorted and stored in `data1`:

```
%reset -f

import numpy as np
import matplotlib.pyplot as plt
from scipy.optimize import curve_fit
from numpy.random import default_rng

rng = default_rng(0)
data1 = np.linspace(0.5,3.,26)
data1 = data1 + 0.2*rng.standard_normal(np.shape(data1))
data1 = np.sort(data1,axis=0)
```

We can then compute the second variable `data2`, which is the exponent of the first variable `data1` multiplied by 0.2 and increased by 3. We again add Gaussian noise—this time with a standard deviation of 0.5—to the data and display the result:

```
data2 = 3 + 0.2*np.exp(data1)
data2 = data2 + 0.5*rng.standard_normal(np.shape(data2))

plt.figure()
plt.plot(data1,data2,
    marker='o',
    linestyle='none')
plt.show()
```

Nonlinear regression is used to estimate the two coefficients of the exponential function, that is, the multiplier 0.2 and the summand 3. The function `curve_fit()` returns a vector `popt` of coefficient estimates for the nonlinear regression of the responses in `data2` to the predictors in `data1` using the model specified by `func()`. The second output `pcov` is the estimated covariance of `popt`. The diagonals of `popt` provide the variance for the parameter estimate `popt`. To compute errors of one standard deviation on the parameters, we can use `perr = np.sqrt(np.diag(pcov))`. We can design a function handle model `func()` that represents an exponential function with an input variable `x` and with `a` and `b` representing the two coefficients of the exponential function,

```
def func(x,a,b):

    return a*np.exp(x)+b

popt,pcov = curve_fit(func,data1,data2)
print(popt)
```

which yields

```
[0.22560247 2.94436195]
```

We can now use the resulting coefficients popt[0] and popt[1] to compare the results with the original data in a graphic:

```
plt.figure()
plt.plot(data1,data2,
    marker='o',
    linestyle='none')
plt.plot(data1,popt[0]*np.exp(data1)+popt[1])
plt.show()
```

As we can see from the output of popt and the graphics, the fitted red curve describes the data reasonably well. We can now also use curve_fit() to perform a weighted regression. Let us assume that we know the one-sigma errors of the values in data3:

```
data3 = abs(rng.standard_normal(np.shape(data2)))

plt.figure()
plt.plot(data1,data2,
    marker='o',
    linestyle='none')
plt.errorbar(data1,data2,data3,
    color= [0.3,0.5,0.8],
    linestyle='none')
plt.show()
```

To make a weighted fit, we define the model function func() and then use curve_fit() with the optional input sigma for the errors,

```
popt2,pcov2 = curve_fit(func,data1,data2,
    sigma=data3)
yfit2 = func(data1,*popt2)
print(popt2)
```

which yields

```
[0.25687643 2.53249989]
```

As before, curve_fit() computes weighted parameter estimates popt. We again compare the results with the original data:

```
plt.figure()
plt.plot(data1,data2,
    marker='o',
    linestyle='none')
plt.errorbar(data1,data2,data3,
    color=[0.3,0.5,0.8],
    linestyle='none')
plt.plot(data1,
    popt2[0]*np.exp(data1)+popt2[1],
    linestyle='-',
    color=[0.8,0.5,0.3])
plt.legend(['Data',
    'Unweighted',
    'Weighted'])
plt.show()
```

Comparing the coefficients p and the red curves from the weighted regression with the previous results from the unweighted regression reveals slightly different results (Fig. 4.9):

```
plt.figure()
plt.plot(data1,data2,
    marker='o',
    linestyle='none').
plt.errorbar(data1,data2,data3,
    color=[0.3,0.5,0.8],
    linestyle='none')
plt.plot(data1,
    popt[0]*np.exp(data1)+popt[1],
    linestyle=':',
    color=[0.8,0.5,0.3])
plt.plot(data1,
    popt2[0]*np.exp(data1)+popt2[1],
    linestyle='-',
    color=[0.8,0.5,0.3])
plt.legend(['Data',
    'Unweighted',
```

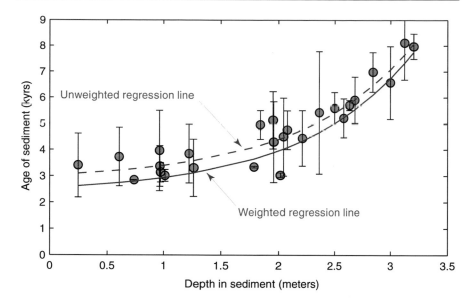

Fig. 4.9 Comparison of unweighted (dashed line) and weighted (solid line) regression results from synthetic data. The plot shows the original data points (circles), the error bars for all data points, and the regression line for an exponential model function. In the unweighted regression, the fitted curve is moved toward the first two data points with a large error, while in the weighted regression, it is moved toward the third data point with a small error.

```
        'Weighted'])
   plt.show()
```

As an example, in the unweighted regression, the fitted curve is moved toward the first two data points with a large error, while in the weighted experiment, it is moved toward the third data point with a small error.

4.11   Classical Linear Regression of Log-Transformed Data

A common error in regression analysis occurs when bivariate data with an exponential relationship are log-transformed, a best-fit line is then calculated using a least-squares linear regression, and the result is finally transformed back. Below is a simple Python example that demonstrates a better fitting of a curve to bivariate data with an exponential relationship.

Log-transforming the y-values has two important consequences that can influence the result: First, if the relationship between the data is of the type $y = a_0 + a_1 \cdot e^x$, then logarithmizing y-values in the form $\log(y)$ does not provide a complete linearization of the data because the parameters a_0 and a_1 are not taken into account. Second, classical regression assumes that y responds to x and that the entire dispersion in the data set is contained within the y-value (see Sect. 4.3).

This means that x is the independent variable defined by the experimenter and can be regarded as being free of errors. Linear regression minimizes the deviations Δy between the data points xy and the value y predicted by the best-fit line $y = b_0 + b_1 x$ using a least-squares criterion. The classic linear regression method makes two assumptions about the data: (1) that there is a linear relationship between x and y and (2) that the unknown errors around the means have a normal (Gaussian) distribution with a similar variance for all data points. Logarithmizing the y-values violates this assumption because errors then have a log-normal distribution and the regression therefore places less weight on the larger y-values.

To see the difference in the results, we first create a synthetic data set. The y-values, which are stored in data2, have an exponential relationship with the x-values, which are stored in data1. After computing the x- and y-values, we add Gaussian noise to the y-values in data2:

```
%reset -f

import numpy as np
import matplotlib.pyplot as plt
from scipy.optimize import curve_fit
from numpy.random import default_rng

rng = default_rng(10)
data1 = np.linspace(0.5,3.,26)
data1 = data1 + 0.2*rng.standard_normal(np.shape(data1))
data1 = np.sort(data1,axis=0)
data2 = 3 + 0.2*np.exp(data1)
data2 = data2 + 0.5*rng.standard_normal(np.shape(data2))
```

Below is the linear fit of the logarithmized data using polyfit():

```
p = np.polyfit(data1,np.log(data2),1)
```

We then use curve_fit() to calculate the nonlinear fit without transforming the data (see Sect. 4.10):

```
def func(x,a,b):
    return a*np.exp(x)+b

popt,pcov = curve_fit(func,data1,data2)
print(popt)
```

We can also calculate the true (noise-free) line using the exponential equation from above before adding noise:

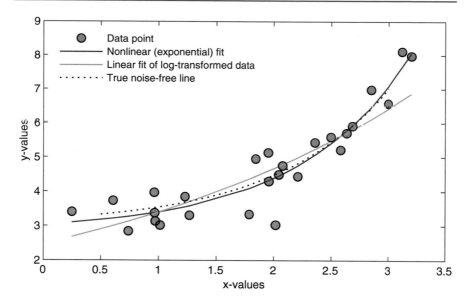

Fig. 4.10 Comparison of nonlinear regression (orange line) and classical linear regression of log-transformed data (yellow line) with an exponential relationship. The yellow curve, which is calculated via linear regression of the log-transformed data, has a much lower curvature than the nonlinear fit. The dotted black line is the noise-free curve. The yellow line is statistically incorrect because it is the result of using the wrong method.

```
trueline = np.zeros((251,2))
trueline[:,0] = np.linspace(0.5,3.,251)
trueline[:,1] = 3 + 0.2*np.exp(trueline[:,0])
```

Displaying the data clearly shows the difference:

```
plt.figure()
plt.plot(data1,data2,
    marker='o',
    linestyle='none')
plt.plot(data1,
    popt[0]*np.exp(data1)+popt[1],
    color=(0.8477,0.3242,0.0977))
plt.plot(data1,
    np.exp(data1*p[0]+p[1]),
    color=(0.9258,0.6914,0.1250))
plt.plot(trueline[:,0],trueline[:,1],
    linestyle=':',
    color='k').
plt.show()
```

The yellow curve calculated via linear regression of the log-transformed data has a much lower curvature than does the nonlinear fit (Fig. 4.10). The dotted black line is the noise-free curve. Of course, the result of the nonlinear regression is not a perfect match with the true line due to the noise in the data used with curve_fit(). However, the yellow line is statistically incorrect because it is the result of using of the wrong method.

Recommended Reading

Kendall M (1938) A new measure of rank correlation. Biometrika 30:81–89
Pearson K (1895) Notes on regression and inheritance in the case of two parents. Proc R Soc Lond 58:240–242
Spearman C (1904) The proof and measurement of association between two things. Am J Psychol 15:72–101
Spearman C (1910) Correlation calculated from faulty data. Br J Psychol 3:271–295

Further Reading

Albarède F (2002) Introduction to geochemical modeling. Cambridge University Press, Cambridge
Davis JC (2002) Statistics and data analysis in geology, 3rd edn. Wiley, New York
Draper NR, Smith, H (1998) Applied regression analysis. Wiley series in probability and statistics. Wiley, New York
Efron B (1982) the jackknife, the bootstrap, and other resampling plans. Society of Industrial and Applied Mathematics CBMS-NSF Monographs 38
Fisher RA (1922) The goodness of fit of regression formulae, and the distribution of regression coefficients. J Roy Stat Soc 85:597–612
MacTavish JN, Malone PG, Wells TL (1968) RMAR; a reduced major axis regression program designed for paleontologic data. J Paleontol 42(4):1076–1078
Pearson K (1894–1898) mathematical contributions to the theory of evolution, Part I to IV. Philosophical Transactions of the Royal Society 185–191

Time Series Analysis

5

5.1   Introduction

Time series analysis is used to investigate the temporal behavior of a variable $x(t)$. Examples include investigations into long-term records of mountain uplift, sea-level fluctuations, orbitally induced insolation variations (and their influence on the ice-age cycles), millennium-scale variations in the atmosphere–ocean system, the effect of the El Niño/Southern Oscillation on tropical rainfall and sedimentation (Fig. 5.1), and tidal influences on noble gas emissions from bore holes. The temporal pattern of a sequence of events can be random, clustered, cyclic, or chaotic. Time series analysis provides various tools with which to detect these temporal patterns. Understanding the underlying processes that produced the observed data allows us to predict future values of the variable. For time series analysis, we use the *NumPy* (https://numpy.org), *Matplotlib* (https://matplotlib.org), and *SciPy* packages (https://scipy.org), which contain all the necessary routines.

Section 5.2 discusses signals in general and contains a technical description of how to generate synthetic signals for time series analysis. The use of spectral analysis to detect cyclicities in a single time series (auto-spectral analysis) and to determine the relationship between two time series as a function of frequency (cross-spectral analysis) is then demonstrated in Sects. 5.3 and 5.4. Since most

Supplementary Information The online version contains supplementary material available at https://doi.org/10.1007/978-3-031-07719-7_5.

Sections 5.9–5.10 can be found in the MATLAB version of this book but could not be translated to Python. However, they are listed here with corresponding section numbers in order to create identical chapter numbering (and thereby also identical figure numbering and computer scripts) between the two versions of the book. The reader can thus use both books side by side to trace the translation of the scripts from MATLAB to Python (and back).

© The Author(s), under exclusive license to Springer Nature Switzerland AG 2022   151
M. H. Trauth, *Python Recipes for Earth Sciences*, Springer
Textbooks in Earth Sciences, Geography and Environment,
https://doi.org/10.1007/978-3-031-07719-7_5

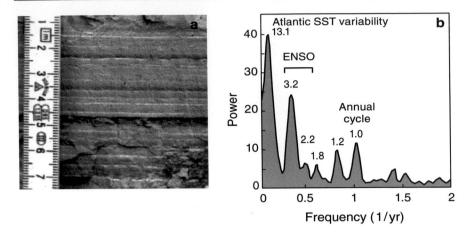

Fig. 5.1 a Photograph of ca. 30 kyr old varved sediments from a lake in the Andes of northwestern Argentina. The distribution of the source rocks and the inter-annual precipitation pattern in the area suggest that the reddish-brown layers reflect the cyclic recurrences of enhanced precipitation that involves increased erosion and sediment input into the lake. **b** The power spectrum of a red-color intensity transect across 70 varves is dominated by major peaks at frequencies of ca. 0.076, 0.313, 0.455, and 1.0 yrs^{-1}. Similar to today's cyclicities, these cyclicities suggest a strong influence of tropical Atlantic sea surface temperature (SST) variability, of the El Niño/Southern Oscillation (ENSO), and of the annual cycle that occurred 30 kyrs ago (Trauth et al. 2003).

time series in the earth sciences have uneven time intervals, various interpolation techniques and subsequent methods of spectral analysis are required, and these are introduced in Sect. 5.5. The use of evolutionary power spectra to map changes in cyclicity through time is demonstrated in Sect. 5.6. An alternative technique for analyzing unevenly spaced data is explained in Sect. 5.7. Section 5.8 introduces the very popular wavelet power spectrum, which is able to map temporal variations in the spectra in a similar way to the method demonstrated in Sect. 5.6. Section 5.9 then introduces methods for detecting and removing abrupt transitions in central tendency and dispersion within time series. Section 5.10 presents methods used to align stratigraphic sequences. The chapter then closes in Sect. 5.11 with an overview of nonlinear techniques, with special attention given to recurrence plots.

5.2   Generating Signals

A time series is an ordered sequence (and not a series in a mathematically rigorous sense) of n values of a variable $x(t)$ at certain times t_k:

$$x(t_k) = x(t_1), x(t_2), \ldots, x(t_n)$$

If the time interval between any two successive observations $x(t_k)$ and $x(t_{k+1})$ is constant, the time series is said to be equally spaced, and the sampling interval is

$$\Delta t = t_{k+1} - t_k$$

The sampling frequency f_s is the inverse of the sampling interval Δt. We generally try to sample at regular time intervals or constant sampling frequencies, but in many earth science examples, this is not possible. As an example, imagine deep-sea sediments sampled at five-centimeter intervals along a sediment core. Radiometric age determinations at certain levels in the sediment core have revealed substantial fluctuations in sedimentation rates. Despite the even spacing along the sediment core of the samples, these samples are not equally spaced on the time axis. Hence, the quantity

$$\Delta t = T/n$$

(where T is the full length of the time series and n is the number of data points) represents only an average sampling interval. In general, a time series $x(t_k)$ can be represented as the linear sum of a periodic component $x_p(t_k)$, a trend $x_{tr}(t_k)$, and random noise $x_{ns}(t_k)$:

$$x(t_k) = x_p(t_k) + x_{tr}(t_k) + x_{ns}(t_k)$$

The decomposition of a climate signal into periodic, trend, and noise components (shown above) corresponds to the definition of climate in terms of trend and variability, which was developed at the end of the 19th century (Brückner 1890; Hann 1901; Köppen 1931). Mudelsee (2014) added an extremal component $x_e(t_k)$ both to facilitate the analysis and to recognize the growing importance of climate extremes as documented, for example, in the Fourth Assessment Report by the IPCC (Solomon et al. 2007). The long-term component $x_{tr}(t_k)$ is a linear or higher-degree trend that can be removed by fitting a polynomial of a certain degree and subtracting the values of this polynomial from the data (see Chap. 4). Methods for removing noise $x_{ns}(t_k)$ are described in Chap. 6. The periodic (or, in a mathematically less rigorous sense, cyclic) component $x_p(t_k)$ can be approximated by a linear combination of sine (or cosine) waves that have different amplitudes A_i, frequencies f_i, and phase angles ψ_i:

$$x_p(t_k) = \sum_i A_i \cdot sin(2\pi f_i t_k - \psi_i)$$

The phase angle ψ helps in detecting temporal shifts between signals of the same frequency. Two signals x and y with the same period are out of phase unless the difference between ψ_x and ψ_y is equal to zero (Fig. 5.2).

The frequency f of a periodic signal is the inverse of the period τ. The *Nyquist frequency* f_{nyq} is half the sampling frequency f_s and represents the maximum frequency that the data can produce. This frequency, which was defined by American mathematician Claude Elwood Shannon and named after Swedish-American electrical engineer Harry Nyquist, prevents the phenomenon of *aliasing* in time-discrete systems. Aliasing means that ambiguities occur in digital signals if the sampling frequency is too low. As an example, audio compact disks (CDs) are sampled at frequencies of 44,100 Hz (Hz, where 1 Hz = 1 cycle per second), but the corresponding Nyquist frequency is 22,050 Hz, which is the highest frequency a CD player can theoretically produce. The performance limitations of anti-alias filters used by

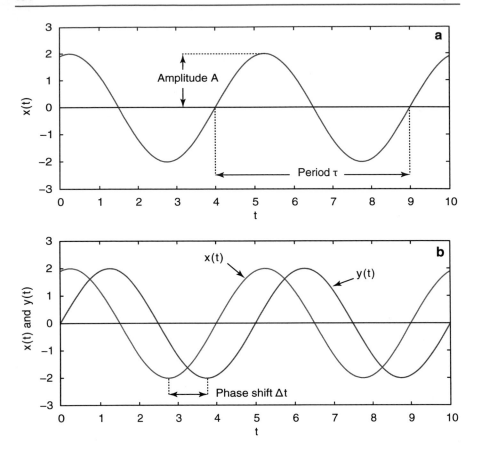

Fig. 5.2   a Periodic signal x, which is a function of time t defined by the amplitude A and the period τ, which is the inverse of the frequency f. **b** Two signals x and y with the same period are out of phase if the difference between ψ_x and ψ_y is not equal to zero.

CD players further reduce the frequency band and result in a cutoff frequency of around 20,050 Hz, which is the true upper frequency limit of a CD player.

We can now generate synthetic signals in order to illustrate the use of time series analysis tools. When using synthetic data, we know in advance what features the time series contains (e.g., periodic or random components), and we can introduce a linear trend and/or gaps in the time series. The user will encounter plenty of examples of the possible effects of varying the parameter settings as well as potential artifacts and errors that can result from applying spectral analysis tools. We begin with simple data and then apply the methods to more complex time series. The first example illustrates how to generate a basic synthetic data series that is characteristic of earth science data. First, we create a time axis t that runs from 1 to 1,000 in steps of one unit (i.e., the sampling frequency is also one). We then generate a simple periodic signal y that consists of a sine wave with a period of five and an amplitude of two:

```
%reset -f

import numpy as np
import matplotlib.pyplot as plt

t = np.linspace(1,1000,1000)
x = 2*np.sin(2*np.pi*t/5)

plt.figure()
plt.plot(t,x)
plt.axis(np.array([0,200,-4,4]))
plt.show()
```

The period of $\tau = 5$ corresponds to a frequency of $f = 1/5 = 0.2$. However, natural data series are more complex than a simple periodic signal. The slightly more complicated signal can be generated by superimposing several periodic components with different periods. As an example, we compute such a signal by adding three sine waves with periods $\tau_1 = 50$ ($f_1 = 0.02$), $\tau_2 = 15$ ($f_2 \approx 0.07$), and $\tau_3 = 5$ ($f_3 = 0.2$). The corresponding amplitudes are $A_1 = 2$, $A_2 = 1$, and $A_3 = 0.5$:

```
t = np.linspace(1,1000,1000)
x = 2*np.sin(2*np.pi*t/50) + \
    np.sin(2*np.pi*t/15) + \
    0.5*np.sin(2*np.pi*t/5)

plt.figure()
plt.plot(t,x)
plt.axis(np.array([0,200,-4,4]))
plt.show()
```

By restricting the t-axis to the interval [0, 200], only one fifth of the original data series is displayed (Fig. 5.3a). However, we recommended that long data series be generated (as in the example) in order to avoid edge effects when applying spectral analysis tools for the first time.

In contrast to our synthetic time series, real data also contain various disturbances, such as random noise and first- or higher-order trends. In order to reproduce the effects of noise, a random number generator can be used to compute Gaussian noise with a mean of zero and a standard deviation of one. The seed of the algorithm should be set to zero using random.seed(0). One thousand random numbers are then generated using random.randn():

```
np.random.seed(0)
ns = np.random.randn(1000)
```

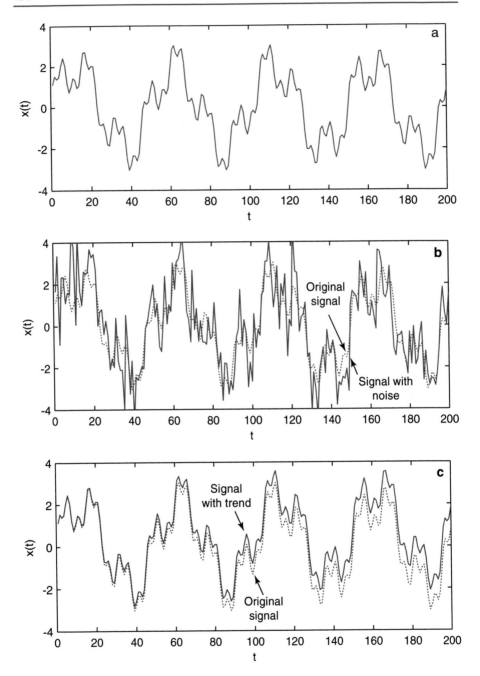

Fig. 5.3 **a** Synthetic signal with periodicities of $\tau_1 = 50$, $\tau_2 = 15$, and $\tau_3 = 5$, each with different amplitudes, and **b** the same signal superimposed with Gaussian noise. **c** The time series shows a striking linear trend.

We add this noise to the original data, that is, we generate a signal that contains additive noise (Fig. 5.3b). Displaying the data illustrates the effect of noise on a periodic signal. Since no record is entirely free of noise in reality, it is important to become familiar with the influence of noise on power spectra:

```
xns = x + ns

plt.figure()
plt.plot(t,x,'b-',t,xns,'r-')
plt.axis(np.array([0,200,-4,4]))
plt.show()
```

Signal processing methods are often applied to remove a major part of the noise, although many filtering methods make arbitrary assumptions concerning the signal-to-noise ratio (Chap. 6). Moreover, filtering introduces artifacts and statistical dependencies to the data, which may have a profound influence on the resulting power spectra.

Finally, we introduce a linear long-term trend to the data by adding a straight line with a slope of 0.005 and an intercept with the y-axis of zero (Fig. 5.3c). Such trends are common in the earth sciences. As an example, consider the glacial–interglacial cycles observed in marine oxygen-isotope records superimposed on a long-term cooling trend over the last six million years:

```
xt = x + 0.005 * t

plt.figure()
plt.plot(t,x,'b-',t,xt,'r-')
plt.axis(np.array([0,200,-4,4]))
plt.show()
```

In reality, more complex trends exist, such as higher-order trends and trends that are characterized by variations in gradient. In practice, it is recommended that such trends be eliminated by fitting polynomials to the data and subtracting the corresponding values. Our synthetic time series now contains many characteristics of a typical earth science data set and can illustrate the use of the spectral analysis tools that are introduced in the next section.

5.3 Auto-Spectral and Cross-Spectral Analysis

Auto-spectral analysis aims to describe the distribution of the variance contained in a single signal $x(t)$ as a function of frequency or wavelength. A simple way of describing the variance in a signal over a time lag k is by means of the autocovariance. The autocovariance cov_{xx} of the signal $x(t)$ with n data points sampled at constant time intervals Δt is

$$cov_{xx}(k) = \frac{1}{n-k-1} \sum_{t=1}^{n-k} (x_i - \bar{x})(x_{i+k} - \bar{x})$$

The autocovariance series clearly depends on the amplitude of $x(t)$. Normalizing the covariance by the variance s^2 of $x(t)$ yields the autocorrelation sequence. Autocorrelation involves correlating a series of data with itself as a function of a time lag k:

$$corr_{xx}(k) = \frac{cov_{xx}(k)}{cov_{xx}(0)} = \frac{cov_{xx}(k)}{s_x^2}$$

A popular method used to compute power spectra in the earth sciences is that introduced by Blackman and Tukey (1958). The *Blackman–Tukey method* uses the complex Fourier transform $X_{xx}(f)$ of the autocorrelation sequence $corr_{xx}(k)$,

$$X_{xx}(f) = \sum_{k=0}^{m} corr_{xx}(k) e^{i2\pi fk/f_s}$$

where m is the maximum lag and f_s is the sampling frequency. The Blackman–Tukey auto-spectrum is the absolute value of the Fourier transform of the autocorrelation series. In some fields, the *power spectral density* is used as an alternative way of describing the auto-spectrum. The Blackman–Tukey power spectral density *PSD* is calculated by

$$PSD = \frac{X_{xx}^*(f)X_{xx}(f)}{f_s} = \frac{|X_{xx}(f)|^2}{f_s}$$

where $X_{xx}^*(f)$ is the conjugate complex of the Fourier transform of the autocorrelation series $X_{xx}(f)$ and f_s is the sampling frequency. The actual computation of the power spectrum can only be performed at a finite number of different frequencies by employing a *Fast Fourier Transform* (FFT). The FFT is a method of computing a discrete Fourier transform with reduced execution time. Most FFT algorithms divide the transform into two portions each of size $n/2$ at each step of the transformation. The transform is therefore limited to blocks with dimensions equal to a power of two. In practice, the spectrum is computed using a number of frequencies that is close to the number of data points in the original signal $x(t)$.

The discrete Fourier transform is an approximation of the continuous Fourier transform. The continuous Fourier transform assumes an infinite signal, but discrete real data are limited at both ends, that is, the signal amplitude is zero beyond either end of the time series. In the time domain, a finite signal corresponds to an infinite signal multiplied by a rectangular window that has a value of one within the limits of the signal and a value of zero elsewhere. In the frequency domain, multiplying the time series by this window is equivalent to a convolution of the power spectrum of the signal with the spectrum of the rectangular window (see Sect. 6.4 for a definition of convolution). However, the spectrum of the window is a $\sin(x)/x$ function, which has a main lobe and numerous side lobes on either side of the main peak. As a result, all maxima in a power spectrum *leak*, that is, they lose power on either side of the peaks (Fig. 5.4).

A popular way of overcoming the problem of *spectral leakage* is by *windowing*, in which the sequence of data is simply multiplied by a smooth bell-shaped curve with positive values. Several window shapes are available, such as *Bartlett* (triangular), *Hamming* (cosinusoidal), and *Hanning* (slightly different cosinusoidal) (Fig. 5.4). The use of these windows slightly modifies the equation for the Blackman–Tukey auto-spectrum to

$$X_{xx}(f) = \sum_{k=0}^{m} corr_{xx}(k)w(k)e^{i2\pi fk/f_s}$$

where $w(k)$ is the windowing function. The Blackman–Tukey method therefore performs auto-spectral analysis in three steps: 1) calculating the autocorrelation sequence $corr_{xx}(k)$, 2) windowing, and finally, 3) computing the discrete Fourier transform. Python allows power spectral analysis to be performed with a number of modifications to the method described above. One useful modification is the *Welch method* (Welch 1967) (Fig. 5.5), which involves dividing the time series into overlapping segments, computing the power spectrum for each segment, and then averaging the power spectra. The advantage of averaging the spectra is obvious: It simply improves the signal-to-noise ratio of a spectrum. The disadvantage is the loss of resolution in the spectra.

Cross-spectral analysis correlates two time series in the frequency domain. The cross-covariance is a measure of the variance between two signals over a time lag

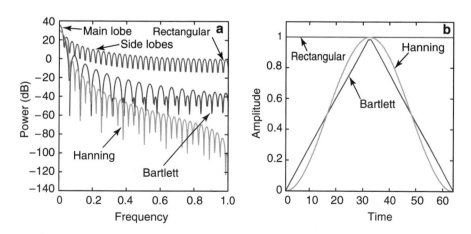

Fig. 5.4 Spectral leakage. **a** The amplitudes of the side lobes relative to that of the main lobe are reduced by multiplying the corresponding time series by **b** a smooth bell-shaped window function. A number of different windows—each with their own advantages and disadvantages—are available for use instead of the default rectangular window. These windows include *Bartlett* (triangular) and *Hanning* (cosinusoidal) windows.

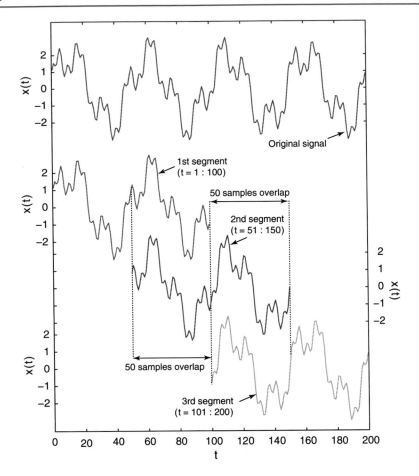

Fig. 5.5 Principle of Welch's power spectral analysis. The time series is first divided into overlapping segments; the power spectrum for each segment is then computed, and all spectra are averaged in order to improve the signal-to-noise ratio of the power spectrum.

k. The cross-covariance cov_{xy} of two signals $x(t)$ and $y(t)$ with n data points sampled at constant time intervals Δt is

$$cov_{xy}(k) = \frac{1}{n-k-1} \sum_{t=1}^{n-k} (x_i - \bar{x})(y_{i+k} - \bar{y})$$

The cross-covariance series again depends on the amplitudes of $x(t)$ and $y(t)$. Normalizing the covariance by the standard deviations of $x(t)$ and $y(t)$ yields the cross-correlation sequence:

$$corr_{xy}(k) = \frac{cov_{xy}(k)}{cov_{xy}(0)} = \frac{cov_{xy}(k)}{s_x s_y}$$

The Blackman–Tukey method uses the complex Fourier transform $X_{xy}(f)$ of the cross-correlation sequence $corr_{xy}(k)$,

$$X_{xy}(f) = \sum_{k=0}^{m} corr_{xy}(k)e^{i2\pi fk/f_s}$$

where m is the maximum lag and f_s is the sampling frequency. The absolute value of the complex Fourier transform $X_{xy}(f)$ is the cross-spectrum, while the angle of $X_{xy}(f)$ represents the phase spectrum. The phase difference is important in calculating leads and lags between two signals, which is a measure often used to propose causalities between two processes that have been documented by the signals. The correlation between two spectra can be calculated by means of the coherence:

$$C_{xy} = \frac{\left|X_{xy}(f)\right|^2}{X_{xx}(f)X_{yy}(f)}$$

The coherence is a real number between 0 and 1, where 0 indicates no correlation and 1 indicates maximum correlation between $x(t)$ and $y(t)$ at frequency f. A certain degree of coherence is an important precondition for computing phase shifts between two signals.

5.4   Examples of Auto-Spectral and Cross-Spectral Analysis

The *SciPy* package provides numerous methods for spectrally analyzing time series. As an example, `periodogram()` computes the power spectral density Pxx of a time series x(t) using the periodogram method. This method was invented by Arthur Schuster in 1898 and calculates the power spectrum by performing a Fourier transform directly on a sequence without requiring prior calculation of the autocorrelation sequence. At the time of its introduction in 1958, indirectly computing the power spectrum via an autocorrelation sequence was faster than directly calculating the Fourier transformation for the full data series x(t). After the introduction both of the Fast Fourier Transform (FFT) by Cooley and Tukey (1965) and of subsequent faster computer hardware, the higher computing speed of the Blackman–Tukey approach compared with the periodogram method became relatively unimportant.

For this next example, we again use the synthetic time series x, xns, and xt generated in Sect. 5.2 as the input data:

```
%reset -f

import numpy as np
import matplotlib.pyplot as plt
from scipy import signal
```

```
t = np.linspace(1,1000,1000)
t = np.transpose(t)
x = 2*np.sin(2*np.pi*t/50) + \
    np.sin(2*np.pi*t/15) + \
    0.5*np.sin(2*np.pi*t/5)

np.random.seed(0)
ns = np.random.randn(1000)
xns = x + ns

xt = x + 0.005*t
```

We then compute the periodogram by calculating the Fourier transform of the sequence x. The fastest possible Fourier transform using `fft.rfft()` computes the Fourier transform for `nfft` frequencies, where `nfft` is the next power of two that is closest to the number of data points n in the original signal x. Since the length of the data series is n=1000, the Fourier transform is computed for `nfft=1024` frequencies, while the signal is padded with `nfft-n=24` zeros:

```
Xxx = np.fft.rfft(x,1024)
```

If x is real (as in our example), then Xxx is symmetric, with the first `(1+nfft/2)` values in Xxx being unique and all other values being symmetrically redundant. The power spectral density is defined as `Pxx2=np.abs(Xxx)**2/Fs`, where Fs is the sampling frequency. The function `periodogram()` also scales the power spectral density by the length of the data series, that is, it divides by Fs=1 and `len(x)=1000`.

```
Pxx2 = np.abs(Xxx)**2/1000
```

We now drop the redundant part of the power spectrum and use only the first `(1+nfft/2)` points. We also multiply the power spectral density by two in order to keep the same energy as in the symmetric spectrum except for the first data point:

```
Pxx = np.hstack((Pxx2[0],2*Pxx2[1:512]))
```

The corresponding frequency axis runs from 0 to Fs/2 in `Fs/(nfft-1)` steps, where Fs/2 is the Nyquist frequency. Since Fs=1 in our example, the frequency axis is

```
f = np.arange(0,1/2,1/1024)
```

We then plot the power spectral density Pxx in the Nyquist frequency range from 0 to Fs/2, which ranges from 0 to 1/2 in our example. The Nyquist frequency range corresponds to the first 512 (or nfft/2) data points. We can plot the power spectral density over the frequency by typing

```
plt.figure()
plt.plot(f,Pxx)
plt.grid
plt.show()
```

The graphical output shows that there are three major peaks at the positions of
the original frequencies of the three sine waves (i.e., 1/50, 1/15, and 1/5). Alternatively, we can also plot the power spectral density over the period by typing

```
plt.figure()
plt.plot(1.0/f,Pxx)
plt.axis(np.array([0,100,0,1000]))
plt.grid
plt.show()
```

and we observe the three periods of 50, 15, and 5, as expected. Since the values on
the x-axis of this plot are not evenly spaced (in contrast to those on the frequency
axis), we find that the long periods are poorly resolved, and a broad peak appears
at a period of 50 in the graphics. The code for the power spectral density can be
rewritten to make it independent of the sampling frequency,

```
Fs = 1

t = np.arange(1/Fs,1000/Fs+1/Fs,1/Fs)
t = np.transpose(t)
x = 2*np.sin(2*np.pi*t/50) + \
    np.sin(2*np.pi*t/15) + \
    0.5*np.sin(2*np.pi*t/5)

nfft = 2**np.ceil(np.log2(t.size))
Xxx = np.fft.rfft(x,int(nfft))

Pxx2 = np.abs(Xxx)**2/Fs/len(x)
Pxx = np.hstack((Pxx2[0],2*Pxx2[1:512]))
f = np.arange(0,Fs/2,Fs/(nfft-1))

plt.figure()
plt.plot(f,Pxx)
plt.grid
plt.axis(np.array([0,0.5,0,np.amax(Pxx)]))
plt.show()
```

where `nfft=2**np.ceil(np.log2(t.size))` computes the next power of two
closest to the length of the time series x(t). This code allows the sampling frequency to be modified and the differences in the results to be explored. We can
now compare the results with those obtained using periodogram():

```
f,Pxx = signal.periodogram(x,fs=1,nfft=1024)
```

This function allows the windowing of the signals with various window shapes to overcome spectral leakage. However, we use the default rectangular window to compare the results with the experiment outlined above. The power spectrum Pxx is computed using an FFT of length nfft=1024, which is the next power of two closest to the length of the series x and which is padded with zeros to make up the number of data points equal to the value of nfft. A sampling frequency fs of one is used within the function in order to obtain the correct frequency scaling for the *f*-axis. We display the results by typing

```
plt.figure()
plt.plot(f,Pxx)
plt.grid
plt.xlabel('Frequency')
plt.ylabel('Power')
plt.show()
```

or alternatively,

```
plt.figure()
plt.plot(1.0/f,Pxx)
plt.axis(np.array([0,100,0,1000]))
plt.grid
plt.xlabel('Period')
plt.ylabel('Power')
plt.show()
```

The graphical output is almost identical to the first plot and again shows that there are three major peaks at the positions of the original frequencies (or periods) of the three sine waves. The same procedure can also be applied to the noisy data:

```
f,Pxx = signal.periodogram(xns,fs=1,nfft=1024)

plt.figure()

plt.plot(f,Pxx)
plt.grid
plt.xlabel('Frequency')
plt.ylabel('Power')
plt.show()
```

Let us now increase the noise level by introducing Gaussian noise with a mean of zero and a standard deviation of five:

```
np.random.seed(0)
nx = int(x.shape[0])
n = 5*np.random.randn(nx)
xns = x + n

f,Pxx = signal.periodogram(xns,fs=1,nfft=1024)

plt.figure()
plt.plot(f,Pxx)
plt.grid
plt.xlabel('Frequency')
plt.ylabel('Power')
plt.show()
```

This spectrum now resembles a real data spectrum in the earth sciences, and the spectral peaks are set against a substantial background noise level. The peak of the highest frequency even disappears into the noise and cannot be distinguished from maxima that are attributed to noise. Both spectra can be compared on the same plot (Fig. 5.6):

```
f,Pxx = signal.periodogram(x,fs=1,nfft=1024)
f,Pxxns = signal.periodogram(xns,fs=1,nfft=1024)

plt.figure()
plt.subplot(1,2,1)
plt.plot(f,Pxx)
plt.grid
plt.xlabel('Frequency')
plt.ylabel('Power')
plt.subplot(1,2,2)
plt.plot(f,Pxxns)
plt.grid
plt.xlabel('Frequency')
plt.ylabel('Power')
plt.show()
```

Next, we explore the influence of a linear trend on a spectrum. Long-term trends are common features in earth science data. We can see that this trend is misinterpreted as a very long period by the FFT, which produces a large peak with a frequency close to zero (Fig. 5.7):

```
f,Pxx = signal.periodogram(x,fs=1,nfft=1024)
f,Pxxt = signal.periodogram(xt,fs=1,nfft=1024)

plt.figure()
plt.subplot(1,2,1)
```

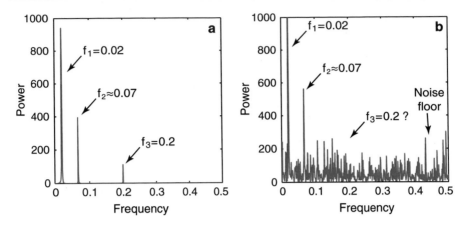

Fig. 5.6 Comparison of the auto-spectra for **a** noise-free and **b** noisy synthetic signals with periods of $\tau_1 = 50$ ($f_1 = 0.02$), $\tau_2 = 15$ ($f_2 \approx 0.07$), and $\tau_3 = 5$ ($f_3 = 0.2$). The highest-frequency peak disappears completely into the background noise and cannot be distinguished from peaks attributed to Gaussian noise.

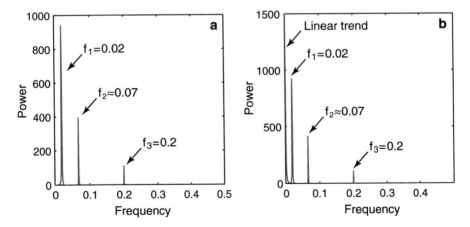

Fig. 5.7 Comparison of the auto-spectra for **a** the original noise-free signal with periods of $\tau_1 = 50$ ($f_1 = 0.02$), $\tau_2 = 15$ ($f_2 \approx 0.07$), and $\tau_3 = 5$ ($f_3 = 0.2$) and **b** the same signal superimposed by a linear trend. The linear trend is misinterpreted by the FFT as a very long period with a high amplitude.

```
plt.plot(f,Pxx)
plt.grid
plt.xlabel('Frequency')
plt.ylabel('Power')
plt.subplot(1,2,2)
plt.plot(f,Pxxt)
plt.grid
```

```
plt.xlabel('Frequency')
plt.ylabel('Power')
plt.show()
```

In order to eliminate the long-term trend, we use detrend(). This function removes linear trends, which are defined either as a single straight-line fit from the sequence x or as a continuous, piecewise linear trend from x with one or more breakpoints defined by the user:

```
xdt = signal.detrend(xt)
```

```
plt.figure()
plt.subplot(2,1,1)
plt.plot(t,x,'b-',t,xt,'r-')
plt.grid
plt.axis(np.array([0,200,-4,4]))
plt.subplot(2,1,2)
plt.plot(t,x,'b-',t,xdt,'r-')
plt.grid
plt.axis(np.array([0,200,-4,4]))
plt.show()
```

The resulting spectrum no longer shows the low-frequency peak:

```
f,Pxx = signal.periodogram(x,fs=1,nfft=1024)
f,Pxxdt = signal.periodogram(xdt,fs=1,nfft=1024)
```

```
plt.figure()
plt.subplot(1,2,1)
plt.plot(f,Pxx)
plt.grid
plt.xlabel('Frequency')
plt.ylabel('Power')
plt.subplot(1,2,2)
plt.plot(f,Pxxdt)
plt.grid
plt.xlabel('Frequency')
plt.ylabel('Power')
plt.show()
```

Some data sets contain a high-order trend that can be removed by fitting a higher-order polynomial to the data and subtracting the corresponding $x(t)$ values.

We now use two sine waves with identical periodicities $\tau = 5$ (equivalent to $f = 0.2$) and amplitudes equal to two in order to compute the cross-spectrum of two time series. The sine waves show a relative phase shift of 1. In the argument of the second sine wave, this corresponds to $2\pi/5$, which is one fifth of the full wavelength of $\tau = 5$:

```
t = np.linspace(1,1000,1000)
x = 2*np.sin(2*np.pi*t/5)
y = 2*np.sin(2*np.pi*t/5+2*np.pi/5)

plt.figure()
plt.plot(t,x,'b-',t,y,'r-')
plt.axis(np.array([0,50,-2,2]))
plt.grid
plt.show()
```

The cross-spectrum is computed with csd(), which uses Welch's method for computing power spectra (Fig. 5.8). Pxy is complex and contains both amplitude and phase information:

```
f,Pxy = signal.csd(x,y,fs=1,nfft=1024)

plt.figure()
plt.plot(f,np.abs(Pxy))
plt.grid
plt.xlabel('Frequency')
```

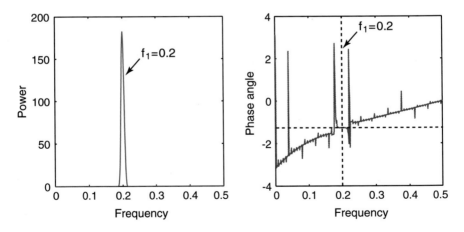

Fig. 5.8 Cross-spectrum of two sine waves with identical periodicities of $\tau = 5$ (equivalent to $f = 0.2$) and amplitudes of 2. The sine waves show a relative phase shift of $t = 1$. In the argument of the second sine wave, this corresponds to $2\pi/5$, which is one fifth of the full wavelength of $\tau = 5$. **a** The magnitude shows the expected peak at $f = 0.2$. **b** The corresponding phase difference in radians at this frequency is 1.2566, which equals $(-1.2566 \cdot 5)/(2 \cdot p) = -1.0000$, which is the relative phase shift of 1 between the two synthetic sine waves.

```
plt.ylabel('Power')
plt.show()
```

The function `csd()` specifies the number of FFT points `nfft` used to calculate the cross power spectral density, which is `1024` in our example. The sampling frequency `fc` is 1 in this example. The coherence of the two signals is one for all frequencies since we are working with noise-free data:

```
f,Cxy = signal.coherence(x,y,fs=1,nfft=1024)

plt.figure()
plt.plot(f,Cxy)
plt.grid
plt.xlabel('Frequency')
plt.ylabel('Coherence')
plt.show()
```

We use `coherence()`, which specifies the number of FFT points `nfft=1024`. The complex part of `Pxy` is required for computing the phase shift between the two signals using `angle()`:

```
phase = np.angle(Pxy)

plt.figure()
plt.plot(f,phase)
plt.grid
plt.xlabel('Frequency')
plt.ylabel('Phase angle')
plt.show()
```

The phase shift at a frequency of $f = 0.2$ (period $\tau = 5$) can be interpolated from the phase spectrum.

```
ph = np.interp(0.2,f,phase)
print(ph)
```

which produces the output

```
1.256632874776265
```

The phase spectrum is normalized to one full period $\tau = 2\pi$, and the phase shift of -1.2566 related to a period of 2π therefore becomes $((-1.2572 \cdot 5)/(2 \cdot \pi) = -1.0004 = -1.2566 \cdot 5)/(2 \cdot \pi) = -1.0000$ related to the true period of $\tau = 5$, which is the relative phase shift of 1 between the two synthetic sine waves.

We now use two sine waves with different periodicities to illustrate cross-spectral analysis. Both signals x and y have a periodicity of 5 but a phase shift of 1:

```
t = np.linspace(1,1000,1000)
x = np.sin(2*np.pi*t/15) + \
    0.5*np.sin(2*np.pi*t/5)
y = 2*np.sin(2*np.pi*t/50) + \
    0.5*np.sin(2*np.pi*t/5 + \
    2*np.pi/5)

plt.figure()
plt.plot(t,x,'b-',t,y,'r-')
plt.axis(np.array([0,100,-3,3]))
plt.grid
plt.show()
```

We can now compute the cross-spectrum Pxy, which clearly shows the common period of $\tau = 5$ (or the frequency of $f = 0.2$):

```
f,Pxy = signal.csd(x,y,fs=1,nfft=1024)

plt.figure()
plt.plot(f,np.abs(Pxy))
plt.grid
plt.xlabel('Frequency')
plt.ylabel('Power')
plt.show()
```

The coherence shows a high value that is close to one at $f = 0.2$:

```
f,Cxy = signal.coherence(x,y,fs=1,nfft=1024)

plt.figure()
plt.plot(f,Cxy)
plt.grid
plt.xlabel('Frequency')
plt.ylabel('Coherence')
plt.show()
```

The complex part of the cross-spectrum Pxy is required to calculate the phase shift between the two sine waves:

```
f,Pxy = signal.csd(x,y,fs=1,nfft=1024)
phase = np.angle(Pxy)
```

```
plt.figure()
plt.plot(f,phase)
plt.grid
plt.show()
```

The phase shift at a frequency of $f = 0.2$ (period $\tau = 5$) is

```
ph = np.interp(0.2,f,phase)
print(ph)
```

which produces the output

```
1.256632874776265
```

The phase spectrum is normalized to one full period $\tau = 2\pi$, and the phase shift of -1.2572 related to a period of 2π therefore equals $(-1.2572 \cdot 5)/(2 \cdot \pi) = -1.0004$ related to a period of 5, which is again the relative phase shift of 1 between the two synthetic sine waves.

5.5   Interpolating and Analyzing Unevenly Spaced Data

We can now use our experience in analyzing evenly spaced data to run a spectral analysis on unevenly spaced data. Such data are very common in the earth sciences, for example, in the field of paleoceanography, where deep-sea cores are typically sampled at constant depth intervals. The transformation of evenly spaced length-variable data into time-variable data in an environment with changing length–time ratios results in unevenly spaced time series. Numerous methods exist for interpolating unevenly spaced sequences of data or time series. The aim of these *interpolation techniques* for $x(t)$ data is to calculate the x-values for an equally spaced t-vector from the unevenly spaced $x(t)$ measurements. *Linear interpolation* predicts the x-values by drawing a straight line between two neighboring measurements and calculating the x-value at the appropriate point along the line. However, this method has its limitations because it assumes linear transitions in the data, which introduces a number of artifacts (including the loss of high-frequency components of the signal) and limits the data range to that of the original measurements.

Cubic spline interpolation is another method for interpolating data that are unevenly spaced. Cubic splines are piecewise continuously differentiable curves that require at least four data points for each step. An advantage of the method is that it preserves the high-frequency information contained in the data. However, steep gradients in the data sequence—which typically occur adjacent to extreme minima and maxima—could cause spurious amplitudes in the interpolated time series. Since all these (and other) interpolation techniques might introduce artifacts to the data, it is always advisable 1) to keep the total number of data points constant before and after interpolation, 2) to report the method employed for estimating the evenly spaced data sequence, and 3) to explore the effect of interpolation on the variance of the data.

Following this brief introduction to interpolation techniques, we can apply the most popular linear and cubic spline interpolation techniques to unevenly spaced data. Once we have interpolated the data, we can use the spectral tools that were previously applied to evenly spaced data (Sects. 5.3 and 5.4). We must first load the two time series:

```
%reset -f

import numpy as np
from scipy import signal
from scipy import interpolate
import matplotlib.pyplot as plt

series1 = np.loadtxt('series1.txt')
series2 = np.loadtxt('series2.txt')
```

Both synthetic data sets contain a two-column matrix with 339 rows. The first column contains ages in *kiloyears* (*kyrs*), which are unevenly spaced. The second column contains oxygen-isotope values measured on calcareous microfossils (foraminifera). The data sets contain 100 kyr, 40 kyr, and 20 kyr cyclicities, respectively, and they are overlain by Gaussian noise. In the 100 kyr frequency band, the second data series has shifted by 5 kyrs with respect to the first data series. To plot the data we type

```
plt.figure()
plt.plot(series1[0:,0],series1[0:,1])
plt.plot(series2[0:,0],series2[0:,1])
plt.show()
```

The spacing of the first data series can be visualized using

```
intv1 = np.diff(series1[0:,0])

plt.figure()
plt.plot(intv1)
plt.show()
```

The plot shows that the spacing varies around a mean interval of 3 kyrs, with a standard deviation of ca. 1 kyr. The minimum and maximum values for the time axis

```
print(np.amin(series1[0:,0]))
print(np.amax(series1[0:,0]))
```

of $t_{min} = 0$ and $t_{max} = 997$ kyrs provide some information about the temporal range of the data. The second data series

```
intv2 = np.diff(series2[0:,0])

plt.figure()
plt.plot(intv2)
plt.show()

print(np.amin(series2[0:,0]))
print(np.amax(series2[0:,0]))
```

has a similar range (from 0 to 997 kyrs). We see that both series have a mean spacing of 3 kyrs and range from 0 to ~ 1,000 kyrs. We now interpolate the data to an evenly spaced time axis. In so doing, we follow the rule that the number of data points should not be increased. The new time axis runs from 0 to 996 kyrs, with 3 kyr intervals:

```
t = np.arange(0.,996.+3.,3.)
```

We can now interpolate the two time series to this axis with *linear* and *spline* interpolation methods using `interp1d()`:

```
fun = interpolate.interp1d(series1[0:,0],
    series1[0:,1],kind='linear')
series1L = fun(t)
fun = interpolate.interp1d(series1[0:,0],
    series1[0:,1],kind='cubic')
series1S = fun(t)

fun = interpolate.interp1d(series2[0:,0],
    series2[0:,1],kind='linear')
series2L = fun(t)
fun = interpolate.interp1d(series2[0:,0],
    series2[0:,1],kind='cubic')
series2S = fun(t)
```

In the `linear` interpolation method, the interpolant is the straight line between neighboring data points, while in the spline interpolation method, the interpolant is a piecewise polynomial between these data points. The method `cubic` with `interp1d()` uses a piecewise cubic spline interpolation, that is, the interpolant is

a third-degree polynomial. The results are compared by plotting the first series
before and after interpolation:

```
plt.figure()
plt.plot(series1[0:,0],series1[0:,1],
    'ko',markersize=1)
plt.plot(t,series1L,'b-',linewidth=0.5)
plt.plot(t,series1S,'r-',linewidth=0.5)
plt.show()
```

We can already observe some major artifacts at ca. 370 kyrs. Whereas the linearly interpolated points are always within the range of the original data, the
spline interpolation method produces values that are unrealistically high or low
(Fig. 5.9). The results can be compared by plotting the second data series:

```
plt.figure()
plt.plot(series2[0:,0],series2[0:,1],
    'ko',markersize=1)
plt.plot(t,series2L,'b-',linewidth=0.5)
plt.plot(t,series2S,'r-',linewidth=0.5)
plt.show()
```

In this series, only a few artifacts can be observed. The *SciPy* package also contains `PchipInterpolator()`, which uses a *Piecewise Cubic Hermite Interpolating Polynomial* to perform a shape-preserving piecewise cubic interpolation. The
function avoids the typical artifacts of the splines because it preserves the original
shape of the data series.

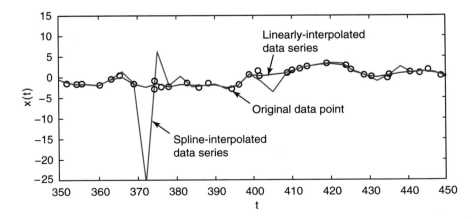

Fig. 5.9 Interpolation artifacts. Whereas the linearly interpolated points are always within the
range of the original data, the spline interpolation method results in unrealistic high and low values.

We can apply the function used above to calculate the power spectrum by computing the FFT for 256 data points with a sampling frequency of 1/3 kyr^{-1}:

```
f,Pxx = signal.periodogram(series1L,fs=1/3,nfft=256)

plt.figure()
plt.plot(f,Pxx)
plt.grid
plt.xlabel('Frequency')
plt.ylabel('Power')
plt.show()
```

Major peaks occur at frequencies of approximately 0.01, 0.025, and 0.05 kyr^{-1}, which correspond approximately to the 100, 40, and 20 kyr cycles., respectively. Analyzing the second time series

```
f,Pxx = signal.periodogram(series2L,fs=1/3,nfft=256)

plt.figure()
plt.plot(f,Pxx)
plt.grid
plt.xlabel('Frequency')
plt.ylabel('Power')
plt.show()
```

also yields major peaks at frequencies of 0.01, 0.025, and 0.05 kyr^{-1} (Fig. 5.10). We now compute the cross-spectrum for both data series:

```
f,Pxy = signal.csd(series1L,series2L,fs=1/3,
    nperseg=128,nfft=256)

plt.figure()
plt.plot(f,np.abs(Pxy))
plt.grid
plt.xlabel('Frequency')
plt.ylabel('Power')
plt.show()
```

The correlation (as indicated by the high value for the coherence) is quite convincing:

```
f,Cxy = signal.coherence(series1L,series2L,fs=1/3,
    nperseg=128,nfft=256)
```

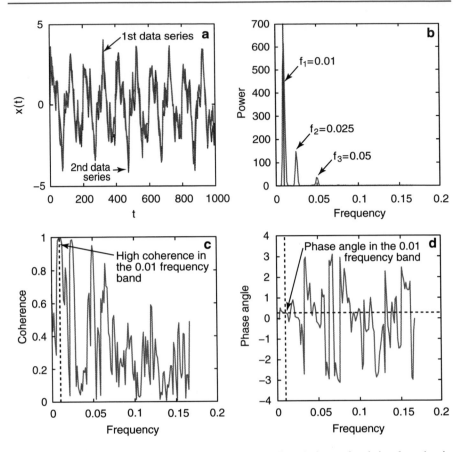

Fig. 5.10 Result from cross-spectral analysis of the two linearly interpolated signals: **a** the signals in the time domain, **b** the cross-spectrum of both signals, **c** the coherence of the signals in the frequency domain, and **d** the phase spectrum in radians.

```
plt.figure()
plt.plot(f,Cxy)
plt.grid
plt.xlabel('Frequency')
plt.ylabel('Coherence')
plt.show()
```

We can observe a fairly high coherence at frequencies of 0.01, 0.025, and 0.05 kyr^{-1}. The complex part of Pxy is required for calculating the phase difference for each frequency:

```
phase = np.angle(Pxy)
```

```
plt.figure()
plt.plot(f,phase)
plt.grid
plt.xlabel('Frequency')
plt.ylabel('Phase angle')
plt.show()
```

The phase shift at a frequency of $f = 0.01$ kyr^{-1} is calculated using

```
ph = np.interp(0.01,f,phase)
print(ph)
```

which produces the output

```
0.2777602833050877
```

The phase spectrum is normalized to a full period $\tau = 2\varpi$ and the phase shift of -0.2796 therefore equals $-0.2796 \cdot 100\,\text{kyrs})/(2 \cdot \pi) = -4.45\text{kyrs}$, which corresponds roughly to the phase shift of 5 kyrs that was introduced to the second data series with respect to the first series.

5.6   Evolutionary Power Spectrum

The amplitude of spectral peaks usually varies with time. This is particularly true for paleoclimate time series. Paleoclimate records usually show trends not only in the mean and variance, but also in the relative contributions of rhythmic components, such as the Milankovitch cycles in marine oxygen-isotope records. Evolutionary power spectra have the ability to map such changes in the frequency domain. The *evolutionary* or *windowed power spectrum* is a modification of the method introduced in Sect. 5.3 that computes the spectrum of overlapping segments of the time series. These overlapping segments are relatively short compared with the windowed segments used by the Welch method (Sect. 5.3), which is used to increase the signal-to-noise ratio of power spectra. The evolutionary power spectrum method therefore uses the *Short-Time Fourier Transform* (STFT) instead of the Fast Fourier Transform (FFT). The output of the evolutionary power spectrum is the short-term, time-localized frequency content of the signal. Various methods can be used to display the results. For instance, time and frequency can be plotted on the *x*- and *y*-axes, respectively, or vice versa, with the color of the plot being dependent on the height of the spectral peaks.

As an example, we use a data set that is similar to those used in Sect. 5.5. The data series contains three main periodicities of 100, 40, and 20 kyrs, respectively, as well as additive Gaussian noise. However, the amplitudes change through time, and this example can therefore be used to illustrate the advantage of the evolutionary power spectrum method. In our example, the 40 kyr cycle appears only after

ca. 450 kyrs, whereas the 100 and 20 kyr cycles are present throughout the time
series. We first load and display the data from the file *series3.txt* (Fig. 5.11):

```
%reset -f

import numpy as np
from scipy import signal
from scipy import interpolate
import matplotlib.pyplot as plt

series3 = np.loadtxt('series3.txt')

plt.figure()
plt.plot(series3[0:,0],series3[0:,1])
plt.xlabel('Time (kyr)')
plt.ylabel('d18O (per mille)')
plt.show()
```

Since both the standard and the evolutionary power spectrum methods require
evenly spaced data, we interpolate the data to an evenly spaced time vector t, as
demonstrated in Sect. 5.5:

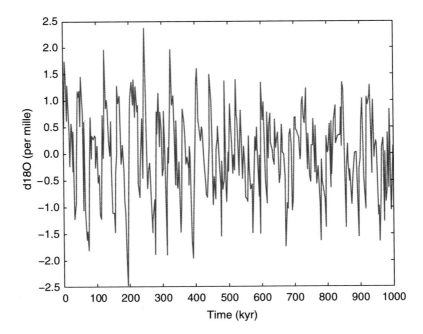

Fig. 5.11 Synthetic data set containing the three main periodicities of 100, 40, and 20 kyrs as
well as additive Gaussian noise. Whereas the 100 and 20 kyr cycles are present throughout the
time series, the 40 kyr cycle only appears at around 450 kyrs.

```
t = np.arange(0.,1000.+3.,3.)
fun = interpolate.interp1d(series3[0:,0],
    series3[0:,1],kind='linear',
    fill_value='extrapolate')
series3L = fun(t)
```

We then compute a non-evolutionary power spectrum for the full length of the
time series (Fig. 5.12). This exercise helps us to compare the differences between
the results of the standard and the evolutionary power spectrum methods:

```
f,Pxx = signal.periodogram(series3L,fs=1/3,nfft=256)

plt.figure()
plt.plot(f,Pxx)
plt.grid
plt.xlabel('Frequency')
plt.ylabel('Power')
plt.show()
```

The auto-spectrum shows major peaks at 100, 40 and 20 kyr cyclicities, as well as
some noise. However, the power spectrum does not provide any information about

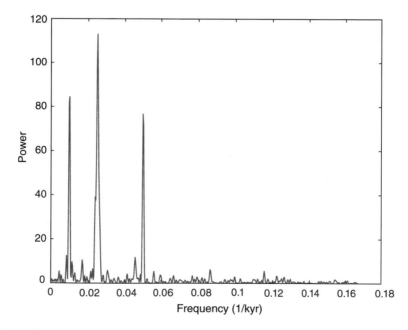

Fig. 5.12 Power spectrum of the complete time series, which shows major peaks at 100, 40 and
20 kyrs. However, the plot does not provide any information on the temporal behavior of the
cyclicities.

fluctuations in the amplitudes of these peaks. The non-evolutionary power spectrum simply represents the average of the spectral information contained in the data.

We now use `spectrogram()` to map the changes in the power spectrum over time. By default, the time series is divided into eight segments, each with a 50% overlap. Each segment is windowed with a Hamming window in order to suppress spectral leakage (Sect. 5.3). The function `spectrogram()` uses similar input arguments to those used in `periodogram()` in Sect. 5.3. We then compute the evolutionary power spectrum for a window of 64 data points with a 50 data point overlap. The STFT is computed for `nfft=256`. Since the spacing of the interpolated time vector is 3 kyrs, the sampling frequency is 1/3 kyr^{-1}:

```
f,t,s = signal.spectrogram(series3L,
    fs=1/3,nperseg=64,
    noverlap=50,nfft=256)

plt.pcolormesh(t,f,s,shading='gouraud')
plt.xlabel('Time (kyr)')
plt.ylabel('Frequency (1/kyr)')
plt.show()
```

We use `pcolormesh()` to display the output of `spectrogram()` in a color plot (Fig. 5.13) that displays yellow horizontal stripes that represent major maxima

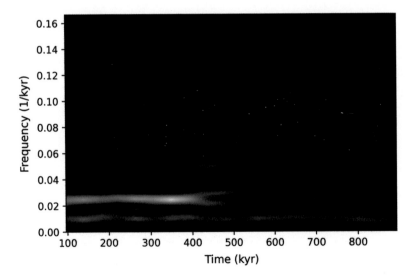

Fig. 5.13 Evolutionary power spectrum using `spectrogram()`, which computes the short-time Fourier transform (STFT) of overlapping segments of the time series. We use a Hamming window of 64 data points with a 50-data-point overlap. The STFT is computed for `nfft=256`. Since the spacing of the interpolated time vector is 3 kyrs, the sampling frequency is 1/3 kyr^{-1}. We use `pcolor()` to display the output of `spectrogram()` in a color plot that displays yellow vertical stripes, which represent major maxima at frequencies of 0.01 and 0.05 kyr^{-1} (i.e., every 100 and 20 kyrs). The plot shows the onset of the 40 kyr cycle at around 450 kyrs.

at frequencies of 0.01 and 0.05 kyr^{-1} (i.e., every 100 and 20 kyrs, respectively). There is also a 40 kyr cycle (corresponding to a frequency of 0.025 kyr^{-1}), but this only occurs after ca. 450 kyrs, as documented by the vertical red stripe in the lower half of the graph.

5.7   Lomb–Scargle Power Spectrum

The power spectrum methods introduced in the previous sections require evenly spaced data. In the earth sciences, however, time series are often unevenly spaced. Although interpolating the unevenly spaced data to a grid of evenly spaced times is one way of overcoming this problem (Sect. 5.5), interpolation introduces numerous artifacts to the data in both the time and frequency domains. As a result, an alternative method of time series analysis has become increasingly popular in the earth sciences: the *Lomb–Scargle algorithm* (e.g., Scargle 1981, 1982, 1989, 1990; Schulz and Stattegger 1998; Press et al. 2007).

The Lomb–Scargle algorithm only evaluates the data of the time series at the times t_i that are actually measured. Assuming a series $x(t)$ of n data points, the Lomb–Scargle normalized periodogram P_x as a function of angular frequency $\omega = 2\pi f > 0$ is given by

$$P_x(\omega) = \frac{1}{s^2} \left\{ \frac{\left(\sum_j (x_i - \bar{x}) \cos\left(\omega(t_j - \tau)\right) \right)^2}{\sum_j \cos^2\left(\omega(t_j - \tau)\right)} + \frac{\left(\sum_j (x_i - \bar{x}) \sin\left(\omega(t_j - \tau)\right) \right)^2}{\sum_j \sin^2\left(\omega(t_j - \tau)\right)} \right\}$$

where

$$\bar{x} = \frac{1}{n} \sum_{i=1}^{n} x_i$$

and

$$s^2 = \frac{1}{n-1} \sum_{i=1}^{n} (x_i - \bar{x})^2$$

are the arithmetic mean and the variance of the data, respectively (e.g., Scargle 1981, 1982, 1989, 1990, Press et al. 2007) (Sect. 3.2). The constant τ, which is defined by the relationship

$$\tan(2\omega\tau) = \frac{\sum_j \sin(2\omega t_j)}{\sum_j \cos(2\omega t_j)}$$

is an offset that makes $P_x(\omega)$ independent of shifting the t_i values by any constant amount. Scargle (1982) showed that this particular choice of the offset τ results in a solution for $P_x(\omega)$ that is identical to a least-squares fit of sine and cosine functions to the data series $x(t)$:

$$x(t) = A\cos(\omega t) + B\sin(\omega t)$$

The least-squares fit of harmonic functions to data series in conjunction with spectral analysis had previously been investigated by Lomb (1976), and the method is hence called the normalized Lomb–Scargle Fourier transform. The term *normalized* refers to the factor s^2 in the dominator of the equation for the periodogram.

Scargle (1982) showed that the Lomb–Scargle periodogram has an exponential probability distribution with a mean equal to one assuming that the noise is Gaussian distributed. The probability that $P_x(\omega)$ lies between some positive quantity z and $z+dz$ is $\exp(-z)dz$. If we scan m independent frequencies, the probability that none of them has a value larger than z is $(1-\exp(-z))^m$. We can therefore compute the false-alarm probability of the null hypothesis (i.e., the probability that a given peak in the periodogram is not significant) (Press et al. 2007):

$$P(> z) \equiv 1 - (1 - e^{-z})^m$$

Press et al. (2007) also suggested using the Nyquist criterion (Sect. 5.2) to determine the number of independent frequencies m assuming that the data are evenly spaced. In this case, the appropriate value for the number of independent frequencies is $m = 2n$, where n is the length of the time series.

The following Python code is based on the equations above for the Lomb–Scargle normalized periodogram. We first load the synthetic data that were generated to illustrate the use of the evolutionary (or windowed) power spectrum method in Sect. 5.6. The data contain periodicities of 100, 40, and 20 kyrs as well as additive Gaussian noise and are unevenly spaced about the time axis. We define two new vectors t and x that contain the original time vector and the synthetic oxygen-isotope data sampled at times t:

```
%reset -f

import math
import numpy as np
import scipy.signal as signal
import matplotlib.pyplot as plt

t = np.loadtxt('series3.txt',usecols=0)
x = np.loadtxt('series3.txt',usecols=1)
```

We then generate a frequency axis f. Since the Lomb–Scargle method is not able to deal with the frequency of zero (i.e., with an infinite period), we start at a frequency value that is equivalent to the spacing of the frequency vector. The parameter ofac is the oversampling factor that influences the resolution of the frequency axis about the N(frequencies)=N(datapoints) case. We also need the highest frequency fhi that can be analyzed by the Lomb–Scargle algorithm, which is commonly the Nyquist frequency fnyq that would be obtained if the data points were evenly spaced over the same time interval. The following code uses the input argument hifac, which is defined by Press et al. (2007) as hifac=fhi/fnyq,

```python
intv = np.mean(np.diff(t))
ofac = 4
hifac = 1
f = np.arange(((2*intv)**(-1))/(len(x)*ofac),hifac*
    (2*intv)**(-1)+((2*intv)**(-1))/
    (len(x)*ofac),((2*intv)**(-1))/
    (len(x)*ofac))
```

where int is the mean sampling interval. We can now compute the normalized Lomb–Scargle periodogram px as a function of the angular frequency wrun by translating the equations above for τ and P_x to Python code:

```python
px = np.zeros(np.shape(f))
for k in range(1,len(f)):
    wrun = 2*np.pi*f[k]
    tau = (1/2*wrun)*math.atan2(sum(np.sin(2*wrun*t)),
        sum(np.cos(2*wrun*t)))
    px[k] = 1/(2*np.var(x,ddof=1))* \
      ((sum(np.multiply((x-np.mean(x)),
        np.cos(wrun*(t-tau)))))**2 /
        sum((np.cos(wrun*(t-tau)))**2) +
        (sum(np.multiply((x-np.mean(x)),
        np.sin(wrun*(t-tau)))))**2 /
        sum((np.sin(wrun*(t-tau)))**2))
```

The significance level for any peak in the power spectrum px can now be computed. The variable prob indicates the false-alarm probability of the null hypothesis. A low prob therefore indicates a highly significant peak in the power spectrum:

```python
prob = 1-(1-np.exp(-px))**(2*len(x))
```

We now plot the power spectrum and the probabilities (Fig. 5.14):

```python
plt.figure()
plt.plot(f,px)
plt.xlabel('Frequency')
plt.ylabel('Power')
plt.show()
```

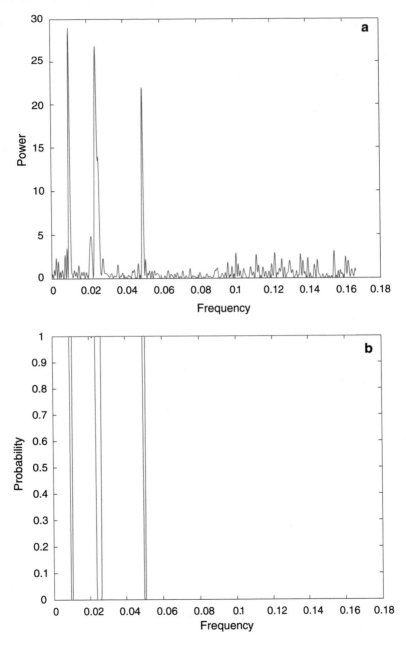

Fig. 5.14 a Lomb–Scargle power spectrum and **b** the false-alarm probability of the null hypothesis. The plot suggests that the 100, 40 and 20 kyr cycles are highly significant.

```
plt.figure()
plt.plot(f,prob)
plt.xlabel('Frequency')
plt.ylabel('Probability')
plt.show()
```

The two plots suggest that all three peaks are highly significant since the false-alarm probability is extremely low at cyclicities of 100, 40, and 20 kyrs.

An alternative way of displaying the significance levels was suggested by Press et al. (2007). In this method, the equation for the false-alarm probability of the null hypothesis is inverted in order to compute the corresponding confidence levels. As an example, we choose a confidence level of 95%. However, this number can also be replaced by a vector of several confidence levels, such as confid=[0.90 0.95 0.99]. We can now type

```
m = int(np.floor(0.5*ofac*hifac*len(x)))
effm = 2.*m/ofac
confid = np.array([0.9,0.95,0.99])
levels = np.log((1.-confid**(1./effm))**(-1))
```

where m is the true number of independent frequencies and effm is the effective number of frequencies using the oversampling factor ofac. The second plot displays the spectral peaks and the corresponding confidence levels:

```
plt.figure()
plt.plot(f,px)
for k in range(0,3):
    print(confid[k])
    print(levels[k])
    plt.plot(f,levels[k]*np.ones(len(f)),
        linestyle='--',
        color=[0.1,0.5,0.8])
plt.xlabel('Frequency')
plt.ylabel('Power')
plt.show()
```

All three spectral peaks at frequencies of 0.01, 0.025, and 0.05 kyr^{-1} exceed the 95% confidence level, which suggests that they represent important cyclicities. We have therefore obtained similar results to those obtained using the periodogram method. However, an advantage of the Lomb–Scargle method is that it does not require any interpolation of unevenly spaced data and that it permits quantitative significance testing.

The *SciPy* package includes lombscargle() for computing the Lomb–Scargle periodogram using frequencies w in radians. Typing

```
w = 2*np.pi*f
px = signal.lombscargle(t,x,w,normalize=True)

plt.figure()
plt.plot(f,px)
plt.xlabel('Frequency')
plt.ylabel('Power')
plt.show()
```

yields the exact same results. We use `normalized=True` as the spectrum type in order to obtain a Lomb–Scargle normalized periodogram, which is normalized by dividing px by `2*var(x)`, as in our Python code above.

5.8   Wavelet Power Spectrum

Section 5.6 demonstrated the use of a modification to the power spectrum method for mapping changes in cyclicity over time. A similar modification could in theory be applied to the Lomb–Scargle method, which would have the advantage of being able to be applied to unevenly spaced data. However, both methods assume that the data are composites of sine and cosine waves that are globally uniform in time and that have infinite time spans. The evolutionary power spectrum method divides the time series into overlapping segments and computes the Fourier transform of these segments. In order to avoid spectral leakage, the data are multiplied by windows that are smooth bell-shaped curves with positive values (Sect. 5.3). The higher the temporal resolution of the evolutionary power spectrum, the lower the accuracy of the result. Moreover, short time windows contain a large number of high-frequency cycles, whereas the low-frequency cycles are underrepresented.

In contrast to the Fourier transform, the *wavelet transform* uses base functions (*wavelets*) that have smooth ends per se (Lau and Weng 1995; Mackenzie et al. 2001). Wavelets are small packets of waves that are defined by a specific frequency and that decay toward either end. Since wavelets can be stretched and translated across both frequency and time with a flexible resolution, they can easily map changes in the time–frequency domain.

A wavelet transformation mathematically decomposes a signal $x(t)$ into elementary functions $\psi_{a,b}(t)$ derived from a *mother wavelet* $\psi(t)$ via dilation and translation,

$$\psi_{a,b} = \frac{1}{a^{1/2}} \psi\left(\frac{t-b}{a}\right)$$

where b denotes the position (translation) and a (>0) denotes the scale (dilation) of the wavelet (Lau and Weng 1995). The wavelet transform of the signal $x(t)$ about the mother wavelet $\psi(t)$ is defined as the convolution integral

$$W(b,a) = \frac{1}{a^{1/2}} \int \psi * \left(\frac{t-b}{a}\right) x(t)\, dt$$

where $\psi*$ is the complex conjugate of ψ. There are many mother wavelets available in the literature, such as the classic *Haar* wavelet, the *Morlet* wavelet, and the *Daubechies* wavelet. The most popular wavelet in the geosciences is the Morlet wavelet, which was introduced by French geophysicist Jean Morlet (1931–2007) and is defined by

$$\psi_0(\eta) = \pi^{-1/4} \exp\left(i\omega_0 \eta\right) \exp\left(-\eta^2/2\right)$$

where η is the time and ω_0 is the wave number (Torrence and Compo 1998). The wave number is the number of oscillations within the wavelet itself. We can easily compute a discrete version of the Morlet wavelet wave by translating the equation above to Python code, where eta is the non-dimensional time and w0 is the wave number. Changing w0 produces wavelets with different wave numbers. Note that it is important not to use i for an index in for loops since it is used here for the imaginary unit (Fig. 5.15):

```
%reset -f

import numpy as np
import scipy.signal as signal
from scipy import interpolate
import matplotlib.pyplot as plt

eta = np.arange(-10,10+0.1,0.1)
w0 = 6
wave = np.multiply(np.multiply(np.pi**(-1/4),
    np.exp(1j*w0*eta)),np.exp(-eta**2/2))

plt.figure()
plt.plot(eta,wave)
plt.xlabel('Position')
plt.ylabel('Scale')
plt.show()
```

In order to familiarize ourselves with wavelet power spectra, we calculate a continuous wavelet transform (CWT) of the synthetic data used in Sect. 5.6. We first load the synthetic data contained in file *series3.txt* while bearing in mind that the data contain periodicities of 100, 40, and 20 kyrs as well as additive Gaussian noise and that they are unevenly spaced about the time axis:

```
series3 = np.loadtxt('series3.txt')
```

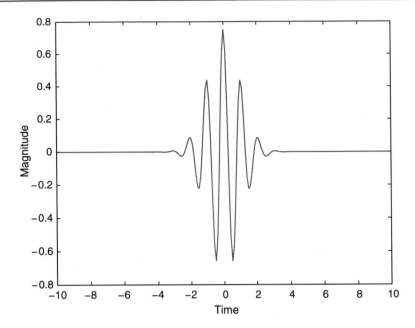

Fig. 5.15 Morlet mother wavelet with wave number 6.

As with the Fourier transform (Sects. 5.3 to 5.6) and in contrast to the Lomb–Scargle algorithm (Sect. 5.7), the wavelet transform requires evenly spaced data, and we therefore interpolate the data using `interp1d()`:

```
t = np.arange(0.,996.+3.,3.)
fun = interpolate.interp1d(series3[0:,0],
    series3[0:,1],kind='linear')
series3L = fun(t)
```

If our time series has a trend, we need to remove this trend before running the wavelet analysis (see Sect. 5.4):

```
series3L = signal.detrend(series3L)
```

We then use `cwt()` with the data series `series3L` and the inverse of our sampling interval as the sampling frequency. In our example, the sampling interval is 3, and the sampling frequency is therefore 1/3. The output of `cwt()` using a *Morlet* wavelet is the wavelet transform `cwtmatr`:

```
fs = 1/3
freqs = np.linspace(0.001,fs/2,1000)
w0 = 6.
widths = w0*fs/(2.*np.pi*freqs)
```

```
cwtmatr = signal.cwt(series3L,
    signal.morlet2,widths,w=w0)
```

We now display the wavelet power spectrum. We can change the limits of the axes, the colormap, and many other things. We can see cycles with frequencies of 0.01/ kyr (corresponding to a period of 100 kyrs), 0.025/kyr (40 kyrs), and 0.05/kyr (20 kyrs). The great thing about wavelet power spectra is that they display an evolutionary spectrum, that is, the appearance and disappearance of cycles through time:

```
plt.contourf(t,freqs,np.abs(cwtmatr),
    cmap='viridis',
    shading='gouraud')
plt.colorbar()
plt.xlabel('Time (kyr)')
plt.ylabel('Frequency (1/kyr)')
plt.xlim([0,1000])
plt.ylim([0.,0.07])
plt.show()
```

We now display the same plot with periods instead of frequencies on the y-axis. We can see cycles with a period of 100 kyrs, 40 kyrs, and 20 kyrs (Fig. 5.16):

```
periods = 1/freqs
plt.contourf(t,periods,np.abs(cwtmatr),
    cmap='viridis',
    shading='gouraud')
plt.colorbar()
plt.xlabel('Time (kyr)')
plt.ylabel('Frequency (1/kyr)')
plt.xlim([0,1000])
plt.ylim([0.,200])
plt.show()
```

We can also display the same plot using pcolormesh() instead of contourf() (Fig. 5.17):

```
plt.pcolormesh(t,periods,np.abs(cwtmatr),
    cmap='viridis',
    shading='gouraud')
plt.colorbar()
plt.xlabel('Time (kyr)')
plt.ylabel('Period (kyr)')
plt.xlim([0,1000])
plt.ylim([0.,200])
plt.show()
```

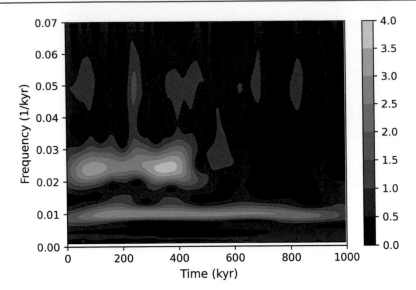

Fig. 5.16 Wavelet power spectrum for the synthetic data series contained in *series_3.txt*. The plot clearly shows major periodicities at frequencies of 0.01, 0.025, and 0.05 kyr^{-1}, which correspond to the 100, 40, and 20 kyr cycles, respectively. The 100 kyr cycle is present throughout the entire time series, whereas the 40 kyr cycle only appears at around 450 kyrs. The 20 kyr cycle is relatively weak but is probably present throughout the entire time series. The wavelet power spectrum was calculated using the continuous 1D wavelet transform `cwt`.

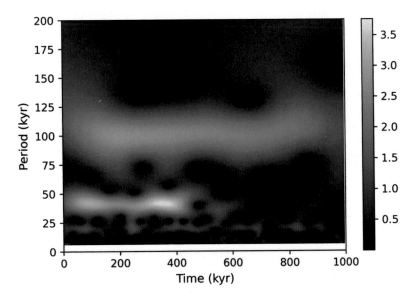

Fig. 5.17 Wavelet power spectrum for the synthetic data series contained in *series_3.txt*. The result is similar to Fig. 5.16 except that here, `pcolormesh()` was used to visualize the wavelet power spectrum.

Compared with the windowed power spectrum method, the wavelet power spectrum clearly shows a much higher resolution on both the time and frequency axes. Instead of dividing the time series into overlapping segments and computing the power spectrum for each segment, the wavelet transform uses short packets of waves that better map temporal changes in the cyclicities. However, the disadvantage of both the windowed power spectrum and the wavelet power spectrum is that they require evenly spaced data. The Lomb–Scargle method overcomes this problem but has limitations in its ability to map temporal changes in the frequency domain (as is also the case with the power spectrum method).

5.9 Detecting Abrupt Transitions in Time Series

This section from the MATLAB-based book MATLAB Recipes for Earth Sciences (Trauth et al. 2021) could not be translated to Python (Fig. 5.18 and 5.19).

Fig. 5.18 This figure from the MATLAB-based book MATLAB Recipes for Earth Sciences (Trauth 2021) could not be translated to Python.

Fig. 5.19 This figure from the MATLAB-based book MATLAB Recipes for Earth Sciences (Trauth 2021) could not be translated to Python.

5.10   Aligning Stratigraphic Sequences

This section from the MATLAB-based book MATLAB Recipes for Earth Sciences (Trauth et al. 2021) could not be translated to Python (Fig. 5.20 and 5.21).

Fig. 5.20 This figure from the MATLAB-based book MATLAB Recipes for Earth Sciences (Trauth 2021) could not be translated to Python.

Fig. 5.21 This figure from the MATLAB-based book MATLAB Recipes for Earth Sciences (Trauth 2021) could not be translated to Python.

5.11   Nonlinear Time Series Analysis (by N. Marwan)

The methods described in the previous sections were used to detect linear relationships in data. However, natural processes on the Earth are often complex and chaotic, and methods based on linear techniques may therefore yield unsatisfactory results. New techniques for nonlinear data analysis that are derived from chaos theory have become increasingly popular in recent decades. Such methods have been used to describe nonlinear behavior, for example, by defining the scaling laws and fractal dimensions of natural processes (Turcotte 1997; Kantz and Schreiber 1997). However, most methods of nonlinear data analysis require either long or stationary data series, and these requirements are rarely satisfied in the earth sciences. While most nonlinear techniques work well on synthetic data, these methods are unable to describe the nonlinear behavior of real data.

During recent decades, *recurrence plots* have become highly popular in science and engineering as a new method of nonlinear data analysis (Eckmann 1987; Marwan et al. 2007). Recurrence is a fundamental property of dissipative dynamical systems. Although small disturbances in such systems can cause exponential divergence in the systems' states, after some time, the systems return to a state that is close to a former state before again passing through a similar evolution. Recurrence plots allow such recurrent behavior of dynamical systems to be visually portrayed. The method is now widely accepted as a useful tool in the nonlinear analysis of short and nonstationary data sets.

Phase Space Portrait

The behavior of a complex dynamical system (e.g., the process of sedimentation in a lake) is determined by many variables, such as the properties of the water body and the sediment beneath the lake, the organisms living in the lake and its surroundings, precipitation (with high rainfall causing increased weathering and erosion in the catchment area), evaporation (with higher evaporation resulting, e.g., in higher potassium concentrations in the sediments), and wind speed (which can transport more solid particles into the lake). Analyzing the properties of such a dynamical system and the time-varying interaction between the system's components requires keeping a record of these state variables through time. Since there are usually no ancient records available of a system's state variables (e.g., precipitation, evaporation, wind speed), we must instead rely on indirect indicators (proxies) that have been preserved in natural archives. A proxy record is obtained by sampling a single variable of a multi-dimensional dynamical system, which is equivalent to projecting the dynamics of a complex system onto a single axis. Considering the example of sediment in a paleolake, the sampled variable could be a time series $x(t)$ of potassium concentrations along a lake sediment core that represents a natural archive of past states of the lake system.

One way to unfold the dynamics of a multi-dimensional system from a one-dimensional time series $x(t)$ is through time-delay *embedding*, which preserves the dynamic characteristics of the system (Packard et al. 1980). Such a one-dimensional time series—which is based on a single variable – contains information about the system and its state variables from which a complete reconstruction of the entire system can be made. In other words, since the potassium concentration is a result of a complex interplay between unknown environmental variables (e.g., precipitation, evaporation, wind velocity), analyzing the temporal variations in this environmental proxy helps us to understand the properties of the paleolake and the time-varying interactions between its components. The embedding of the time series $x(t)$ in a three-dimensional ($m = 3$) *phase space*, for example, means that three successive values $x(t)$, $x(t + \tau)$, and $x(t + 2\tau)$ with a temporal separation τ are represented as a point in a three-dimensional coordinate system (i.e., the phase space) (Packard et al. 1980). The *phase space portrait* displays the embedded time series of observations as a trajectory $\vec{y}(t) = [x(t), x(t + \tau), x(t + 2\tau)]^T$ within the phase space. The phase space trajectory represents the path over which the system's state evolves through time. The reconstructed phase space (which is reconstructed via embedding) is not exactly the same as the original phase space (the true variables that describe the lake), but its topological properties are preserved provided that the *embedding dimension* is sufficiently large (Packard et al. 1980; Takens 1981).

As an example, we now explore the phase space portrait of a periodic process. We first create the position vector x1 and the velocity vector x2:

```
%reset -f

import numpy as np
from scipy import interpolate
import matplotlib.pyplot as plt
import scipy.spatial.distance as distance

t = np.arange(0,3*np.pi+np.pi/10,np.pi/10)
x1 = np.sin(t)
x2 = np.cos(t)
```

The phase space portrait

```
plt.figure()
plt.plot(x1,x2)
plt.xlabel('x_1')
plt.ylabel('x_2')
plt.show()
```

is a circle, which suggests an exact recurrence of each state after one complete cycle (Fig. 5.22). Through time-delay embedding, we can reconstruct this phase space portrait using only a single observation (e.g., a velocity vector) and a time delay of five, which corresponds to one quarter of the period of a pendulum:

```
tau = 5
plt.figure()
plt.plot(x2[:-tau],x2[tau:])
plt.xlabel('x_1')
plt.ylabel('x_2')
```

As we can see, the reconstructed phase space is almost the same as the original phase space. We next compare this phase space portrait with that of a typical nonlinear system: the *Lorenz system* (Lorenz 1963). Weather patterns often do not change

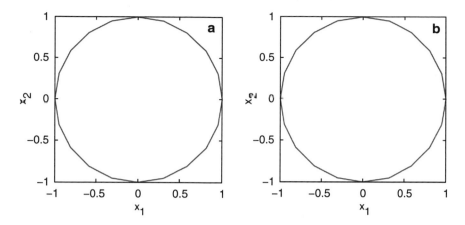

Fig. 5.22 a Original and **b** reconstructed phase space portrait of a periodic system. The reconstructed phase space is almost the same as the original phase space.

in a predictable manner. In 1963, Edward Lorenz introduced a simple three-dimensional model for describing the chaotic behavior exhibited by turbulence in the atmosphere. The variables that define the Lorenz system are the intensity of atmospheric convection, the temperature difference between ascending and descending currents, and the distortion of the vertical temperature profiles from linearity. Small variations in the initial conditions can cause dramatically divergent weather patterns—a behavior often referred to as the *butterfly effect*. The dynamics of the Lorenz system are described by three coupled nonlinear differential equations:

$$\frac{dx}{dt} = s(y(t) - x(t))$$

$$\frac{dy}{dt} = -x(t)z(t) + rx(t) - y(t)$$

$$\frac{dz}{dt} = x(t)y(t) - bz(t)$$

Integrating the differential equations yields a simple Python code for computing the *xyz* triplets of the Lorenz system. We use s=10, r=28, and b=8/3 as system parameters for controlling the chaotic behavior, and the time delay is dt=0.01. The initial values for the position vectors are x1=8, x2=9, and x3=25. However, these values can be changed to any other values, which of course then also changes the behavior of the system:

```
dt = .01
s = 10
r = 28
```

```
b = 8/3
x1, x2, x3 = 8, 9, 25
n = 5000
x = np.zeros((n,3))
for i in range(0, n):
    x1 = x1 + (-s*x1*dt) + (s*x2*dt)
    x2 = x2 + (r*x1*dt) - (x2*dt) - (x3*x1*dt)
    x3 = x3 + (-b*x3*dt) + (x1*x2*dt)
    x[i,:] = [x1, x2, x3]
```

Typical traces of a variable (e.g., the first variable) can be viewed by plotting x[:,0] over time (Fig. 5.23):

```
t = np.linspace(0.01,50,n)

plt.figure()
plt.plot(t,x[:,0])
plt.xlabel('Time')
plt.ylabel('Temperature')
plt.show()
```

We next plot the phase space portrait for the Lorenz system (Fig. 5.24):

```
plt.figure()
ax = plt.axes(projection='3d')
ax.plot(x[:,0],x[:,1],x[:,2])
ax.view_init(30,70-90)
ax.set_xlabel('x_1')
ax.set_ylabel('x_2')
```

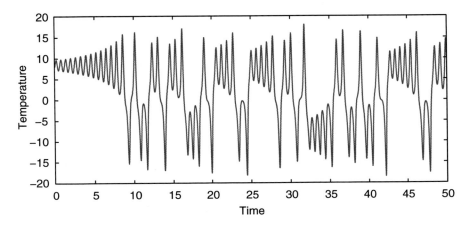

Fig. 5.23 Lorenz system. We used s=10, r=28, and b=8/3 as system parameters. The time delay is dt=0.01.

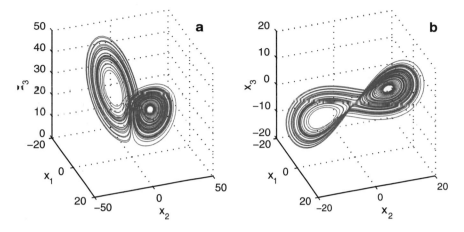

Fig. 5.24 a Phase space portrait of the Lorenz system. In contrast to the simple periodic system, the trajectories of the Lorenz system obviously do not precisely follow the previous course, but they are very close to it. **b** The reconstruction of the phase space portrait using only the first state and a time delay of 6 reveals a topologically similar phase portrait to a with the two typical ears.

```
ax.set_zlabel('x_3')
plt.show()
```

In contrast to the simple periodic system described above, the trajectories of the Lorenz system obviously do not precisely follow the previous course, but they are very close to it. Moreover, if we follow two very close segments of the trajectory, we see that they run into different regions of the phase space over time. The trajectory obviously circles around one fixed point in the phase space and then circles around another fixed point after a random time period. The curious orbit of the phase states around fixed points is known as the *Lorenz attractor*.

These observed properties are typical of chaotic systems. While small disturbances in such a system cause exponential divergences in its state, the system returns approximately to a previous state through a similar course. Reconstructing the phase space portrait using only the first state and a time delay of six

```
tau = 6
plt.figure()
ax = plt.axes(projection='3d')
ax.plot(x[0:-2*tau-1,0],
    x[0+tau:-tau-1,0],
    x[0+2*tau:-1,0])
ax.view_init(60,100-90)
ax.set_xlabel('x_1')
ax.set_ylabel('x_2')
ax.set_zlabel('x_3')
plt.show()
```

reveals a similar phase portrait with the two typical *ears* (Fig. 5.24). The characteristic properties of chaotic systems can also be observed in this reconstruction.

The time delay and the embedding dimension need to be chosen from a previous analysis of the data. The delay can be calculated with the help of the autocovariance (or autocorrelation) sequence. For our example of a periodic oscillation,

```
t = np.arange(0,3*np.pi+np.pi/10,np.pi/10)
x = np.sin(t)
```

we compute and plot the autocorrelation sequence

```
C = np.ones(len(x)-2)
for i in range(1,len(x)-2):
    r = np.corrcoef(x[0:-i],x[i:])
    C[i] = r[0,1]

plt.figure()
plt.plot(range(0,len(x)-2),C)
plt.xlabel('Delay')
plt.ylabel('Autocorrelation')
plt.grid()
```

We now choose a delay such that the autocorrelation sequence for the first time period equals zero. In our case, this is five, which is the value that we used in our example of phase space reconstruction. The appropriate embedding dimension can be calculated by using the false nearest neighbor method or (more simply) by using recurrence plots, which are introduced in the next subsection. The embedding dimension is gradually increased until the majority of the diagonal lines are parallel to the line of identity.

Either the phase space trajectory or its reconstruction serves as the basis for several measures that are defined in nonlinear data analysis, such as *Lyapunov exponents*, *Rényi entropies*, and *Rényi dimensions*. The book by Kantz and Schreiber (1997) on nonlinear data analysis is recommended for gaining more detailed information on these methods. Either phase space trajectories or their reconstructions are also necessary for constructing recurrence plots (Fig. 5.25).

Recurrence Plots

A common feature of dynamical systems is the property of *recurrence*. Patterns in the recurring states of a system reflect typical system characteristics whose descriptions contribute to our understanding of the system's dynamics. In our example, a recurrence of changes in the state variables of precipitation, evaporation, and wind speed could lead to similar (albeit not identical) conditions in the

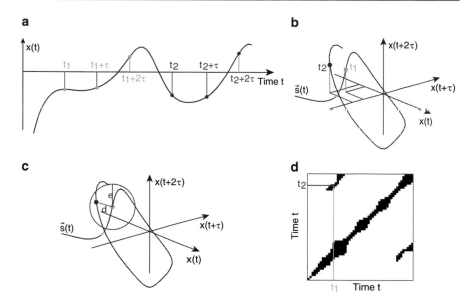

Fig. 5.25 Principle of recurrence plots (modified from Trauth et al. 2019). The dynamics of a multi-dimensional system are untangled from a one-dimensional time series $x(t)$ (black line in panel A) via time-delay embedding. The embedding of the time series $x(t)$ in a three-dimensional ($m = 3$) coordinate system (a phase space; shown in panel B), e.g., means that the three successive values of $x(t)$, $x(t+\tau)$, and $x(t+2\tau)$ with a temporal separation of τ are represented as a single point (t) within the phase space (Iwanski and Bradley 1998, Webber and Zbilut 2005, Marwan et al. 2007). Recurrence plots (RPs; panel D), which were first introduced by Eckmann et al. (1987), are graphic displays of such recurring states within a system and are calculated from the distance (e.g., the Euclidean distance d; shown in panel C) between all pairs of phase space vectors $s(t_1)$ and $s(t_2)$ below a threshold value e (also shown in panel C) (Marwan et al. 2007).

lake (e.g., the depth and size of the lake, the alkalinity and the salinity of the lake water, the species assemblage in the water body, or diagenesis in the sediment). A recurrence plot (RP) is a graphical display of such recurring states of the system that is calculated from the distance (e.g., Euclidean) between all pairs of phase space vectors $\vec{y}(t)$ within a threshold value ε (Marwan et al. 2007):

$$R_{i,j} = \{ \begin{matrix} 0, \ \| x_i - x_j \| > \varepsilon \\ 1, \ \| x_i - x_j \| \leq \varepsilon \end{matrix}$$

If the distance between two states at times i and j is smaller than a given threshold ε, the value of the recurrence matrix R is one; otherwise, it is zero. This analysis is therefore a pairwise test of all states. For n states, we compute n^2 tests. The recurrence plot is then the two-dimensional display of the n-by-n matrix, where black pixels represent $R_{i,j} = 1$ and white pixels indicate $R_{i,j} = 0$, and with a coordinate system that represents two time axes. Such recurrence plots can help to create a preliminary characterization of the dynamics of a system or to find transitions and interrelationships within a system (cf. Fig. 5.26).

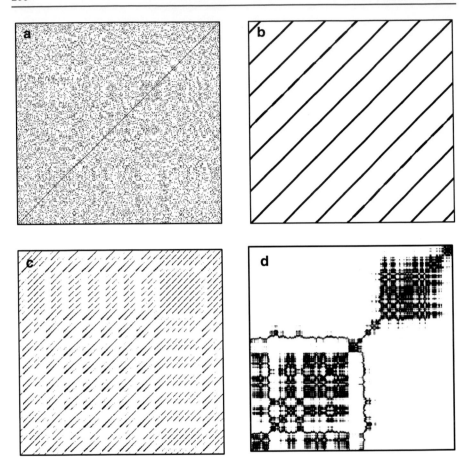

Fig. 5.26 Recurrence plots representing typical dynamical behaviors: **a** stationary uncorrelated data (white noise), **b** periodic oscillation, **c** chaotic data (Roessler system), and **d** non-stationary data with abrupt changes.

As a first example, we load the synthetic time series that contains the 100 kyr, 40 kyr, and 20 kyr cycles used in the previous sections. Since the data are unevenly spaced, we need to linearly interpolate them to an evenly spaced time axis:

```
series1 = np.loadtxt('series1.txt')
t = np.arange(0.,996.+3.,3.)
fun = interpolate.interp1d(series1[0:,0],
    series1[0:,1],kind='linear')
series1L = fun(t)
```

We start with the assumption that the phase space is only one-dimensional. We use `pdist()` to calculate a list D of pairwise distances between all points of the phase space trajectory. The function `squareform()` converts the list D to a symmetric distance matrix S:

```
n = len(series1L)
D = distance.pdist(np.tile(series1L,(2,1)).T)
S = distance.squareform(D);
```

We can now plot the distance matrix for the data set,

```
plt.figure()
plt.imshow(S,
    origin='lower')
plt.xlabel('Time')
plt.ylabel('Time')
plt.show()
```

where a colorbar provides a quantitative measure of the distances between states (Fig. 5.27). We now apply a threshold ε to the distance matrix in order to generate the black/white recurrence plot (Fig. 5.28):

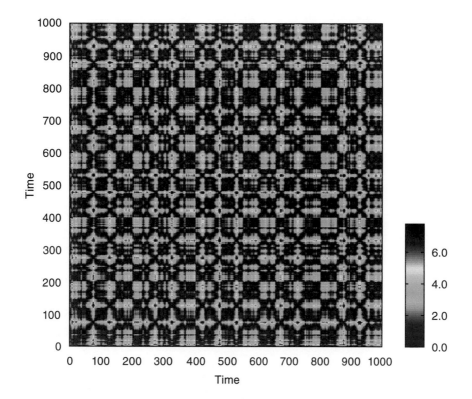

Fig. 5.27 Display of the distance matrix from the synthetic data, which provides a quantitative measure of the distances between states at particular times. Blue colors indicate small distances, and red colors represent large distances.

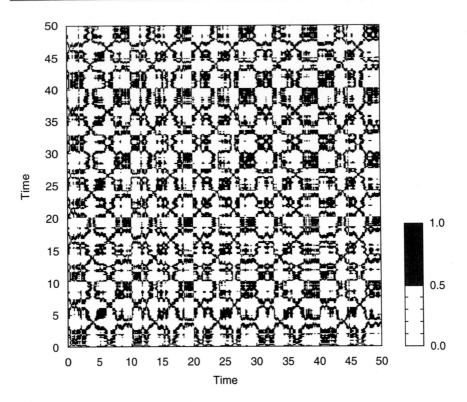

Fig. 5.28 Recurrence plot for synthetic data derived from the distance matrix (as shown in Fig. 5.27) after applying a threshold of $\varepsilon = 1$.

```python
plt.figure()
plt.imshow(S<1,
    cmap='gray_r',
    origin='lower')
plt.xlabel('Time')
plt.ylabel('Time')
plt.show()
```

Both plots reveal periodically recurring patterns. The distances between these patterns represent the cycles contained in the time series. The most important periodic patterns have periods of 200 and 100 kyrs, respectively. The 200 kyr period is the most obvious due to the superposition of the 100 and 40 kyr cycles, which are common divisors of 200 kyrs. Moreover, there are smaller substructures within the recurrence plot that have periods of 40 and 20 kyrs, respectively.

As a second example, we now apply the method of recurrence plots to the Lorenz system. We again generate *xyz* triplets from the coupled differential equations:

```python
dt = .01
s = 10
```

```
r = 28
b = 8/3
x1, x2, x3 = 21, 9, 25
n = 4000
x = np.zeros((n,3))

for i in range(0,n):
    x1 = x1 + (-s*x1*dt) + (s*x2*dt)
    x2 = x2 + (r*x1*dt) - (x2*dt) - (x3*x1*dt)
    x3 = x3 + (-b*x3*dt) + (x1*x2*dt)
    x[i,:] = [x1, x2, x3]
```

We then choose the resampled first component of this system and reconstruct a phase space trajectory using an embedding of $m = 3$ and $\tau = 2$:

```
t = np.linspace(0.01,40,n)
xs = x[0:-1:5,0]

m = 3
tau = 2

n = len(xs)
n2 = n - tau*(m-1)
```

The original data series had a length of 4,000 data points, which was reduced to 800 data points (equivalent to 40 s), but due to the time-delay method, the reconstructed phase space trajectory has an actual length of 796 data points. We can create the phase space trajectory using

```
y = np.zeros((n2,m))
for mi in range(0,m):
    y[:,mi] = xs[np.arange(0,n2) + tau*mi]
```

and the distance matrix using pdist() and squareform():

```
D = distance.pdist(y)
S = distance.squareform(D)

plt.figure()
plt.imshow(S<10,
    cmap='gray_r',
    origin='lower')
plt.xlabel('Time')
plt.ylabel('Time')
plt.show()
```

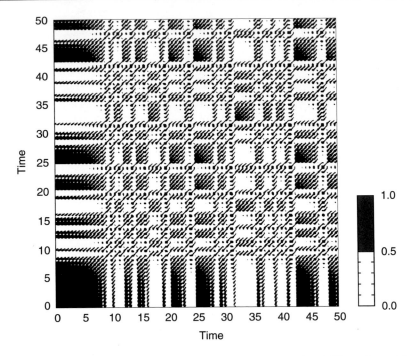

Fig. 5.29 Recurrence plot of the Lorenz system using a threshold of $\varepsilon = 2$. The regions with regular diagonal lines reveal unstable periodic orbits, which are typical of chaotic systems.

This recurrence plot reveals many short diagonal lines (Fig. 5.29), which represent periods of time during which the phase space trajectory runs parallel to earlier or later sequences in this trajectory, that is, periods of time during which the states and dynamics are similar. The distances between these diagonal lines represent the periods of the cycles, which—in contrast to the periods of a harmonic oscillation—vary and are not constant (Fig. 5.24).

Recurrence Quantification

The structure of recurrence plots can also be described by a suite of quantitative measures in a *recurrence quantification analysis* (RQA). Several measures are based on the distribution of the lengths of diagonal or vertical lines as well as on the *local proximity configuration*. These measures can be used to trace hidden transitions within a dynamical process. As an example, we next consider two measures: the recurrence rate and the transitivity coefficient. The *recurrence rate* is the density of points in the recurrence plot and corresponds to the recurrence probability of the system. The *transitivity coefficient* has its roots in graph theory and characterizes the regularity (or complexity) of the system.

For our example, we consider a time series of normally distributed (Gaussian) noise with a transition in the mean at time 1,000 and a change in autocorrelation prior to this transition (starting at time 800). Importantly, neither the mean nor the variances change in the precursor region, and this change thus can also not be found using conventional methods:

```
np.random.seed(20)
n = 1000
x = np.random.uniform(0,1,n)

x = np.concatenate((np.zeros(n),np.ones(n)))
x = x + np.random.normal(0,.3,len(x))
for i in range(n-round(n/5),n):
    x[i] = .9*x[i-1]+.5*x[i]+.02
x = x/np.std(x,ddof=1)

t = np.arange(0,len(x))

plt.figure()
plt.plot(t, x)
plt.xlabel('Time')
```

We first embed the time series and then calculate the recurrence plot:

```
n = len(x)
tau = 2
m=2
n2 = n - tau*(m-1)

s = np.zeros((n2,m))
for mi in range(0,m):
    s[:,mi] = x[np.arange(0,n2) + tau*mi]

D = distance.pdist(s)
S = distance.squareform(D)
R = S<.5

plt.imshow(R,
    extent=[0,n2,0,n2],
    cmap='gray_r',
    origin='lower')
plt.xlabel('Time')
plt.ylabel('Time')
plt.show()
```

The RP shows two blocks of randomly distributed points, each of which marks times when the system randomly returns to a similar state. In random noise, similar states frequently recur, but without any regularity, except for the main diagonal line (i.e., the line of identity). The two blocks of the RP mark the transition at time 1,000. The states before this change point differ strongly from the states after the change point, as is represented by the almost empty lower-right and upper-left quadrants of the RP. However, the preceding increase in autocorrelation between times 800 and 1,000 is not clearly visible in the RP even though it appears slightly denser. In this situation, quantifying the structures in the RP could prove beneficial.

To calculate the recurrence rate, we can simply compute the mean of the matrix R,

```
RR = np.mean(R)
print(RR)
```

which yields

```
0.10654648642636631
```

The probability that the system will return to a randomly selected previous state is therefore about 10%.

The transitivity coefficient is a graph-theoretical measure of the probability that three connected network nodes (triples) are completely interconnected (i.e., that they form a closed triangle):

$$\tau = \frac{\sum_{i,j,k}^{N} A_{ij}A_{ij}A_{ik}}{\sum_{i,j,k=1}^{N} A_{ij}A_{ik}(1 - \delta_{ik})}$$

This measure can be intuitively understood with respect to recurrences in the phase space. We identify a recurrence of states by a network link: Close points on the phase space trajectory are connected by a link. Three connected points form a triple, but only if all three points recur close to one another and thus form a triangle. Such a triangular configuration remains along the phase space trajectory if the dynamic is highly regular (i.e., recurring states remain recurring over a long period of time). However, if the dynamic is chaotic, then parts of the phase space trajectory that were initially close subsequently diverge, and the triangular configuration breaks down even though the corresponding triple nodes might remain interconnected for some time. The probability of finding triangles is therefore higher for regular dynamics and lower for chaotic dynamics. This explanation is, of course, rather simplified, but a theoretically substantiated explanation can be found in Donner et al. (2011).

In order to calculate the probability that triples also form triangles, we need to compute the number of connected triples and the number of triangles, which can be achieved directly from the recurrence plot by excluding the main diagonal:

```
A = R - np.eye(len(R))
```

The number of triangles and triples is then

```
numTripl = np.sum(np.matmul(A,A))
numTria = np.trace(np.matmul(A,np.matmul(A,A)))
```

and finally, the transitivity coefficient is the ratio

```
Trans = numTria/numTripl
print(Trans)
```

which yields

```
0.6459486011860506
```

This number means that the system does not have regular dynamics (which would yield a transitivity coefficient close to one).

In order to detect changes in the dynamics, the RQA can be computed for moving windows. The RQA of moving windows can be calculated in two ways: either (1) by calculating the RQA measures for windows that move along a single, global RP or (2) by calculating individual RPs for windows that move along the original time series. If nonstationarities or trends are not part of the main focus of the analysis, approach (2) helps to find transitions that ignore these nonstationarities. However, if detecting overall (global, long-scale) changes is of interest (e.g., to test for nonstationarity), the recurrence conditions should be kept constant over time (thereby allowing the global RP of the whole time series to be considered), and approach (1) should be selected (Marwan 2011). The selected window size needs to be small enough to ensure a good temporal resolution but large enough to cover typical variations (e.g., several cycles) in order to find recurrences.

In our example, we choose a moving window length of 150 data points and an overlap of 20% (Fig. 5.30):

```
w = 150
```

We then calculate the recurrence rate RR and the transitivity coefficient Trans within these moving windows (Fig. 5.31 and 5.32):

```
Trans = np.zeros(len(R)-w)
RR = np.zeros(len(R)-w)
for i in range(0,len(R)-w,int(w/5)):
    subR = R[i:i+w,i:i+w]
    RR[i] = np.mean(subR)
    subA = A[i:i+w,i:i+w]
    numTripl = np.sum(np.matmul(subA,subA))
    numClosTria = np.trace(np.matmul(subA,
        np.matmul(subA,subA)))
    Trans[i] = numClosTria/numTripl
```

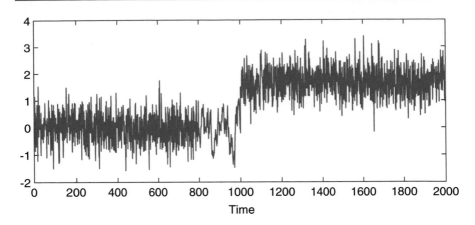

Fig. 5.30 Time series of the synthetic data used in the example of quantitative measures of recurrence plots.

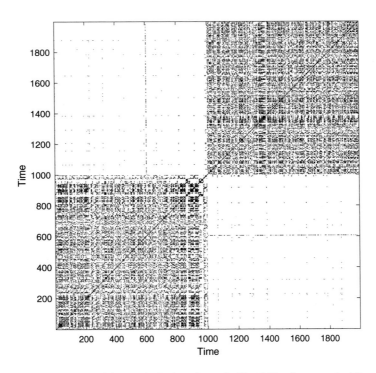

Fig. 5.31 Recurrence plot of the synthetic data shown in Fig. 5.30 using an embedding of $m = 5$ and $\tau = 3$ and applying a threshold of $\varepsilon = 1.2$.

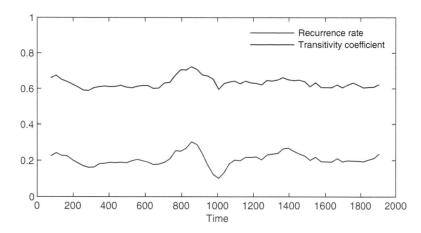

Fig. 5.32 Recurrence rate and transitivity coefficient for the synthetic data in Fig. 5.30 using a moving window of 150 data points and an overlap of 20%.

```python
plt.figure()
plt.plot(t[int(w/2):int(w/2)+len(RR):int(w/5)],
    RR[0:-1:int(w/5)],
    label='Recurrence rate')
plt.plot(t[int(w/2):int(w/2)+len(RR):int(w/5)],
    Trans[0:-1:int(w/5)],
    label='Transitivity coefficient')
plt.xlabel('Time')
plt.legend()
```

The RQA measures show a clear transition within the higher autocorrelation section. At this point, the measures have higher values, which reflects the higher density of black dots, the changing neighborhood configuration due to the autocorrelation, and the change in the dynamics of the system. The presence of these higher values can be used to detect such a change in autocorrelation (e.g., as a precursor to a *tipping point*) and to ultimately predict the tipping point itself. The two empty quadrants interfere with the moving window directly at the change point that occurs at time 1,000, thereby leading to a sudden decrease in the *RR* measure.

In order to reliably interpret the variations in the recurrence measures, a statistical test should be applied (Marwan 2011). Other, more complex measures that quantify additional aspects of the dynamics (e.g., predictability or laminar phases) are included in the Cross Recurrence Plot Toolbox for MATLAB, which is available at

```
http://tocsy.pik-potsdam.de/CRPtoolbox/
```

Bivariate and multivariate extensions of recurrence plots allow nonlinear correlation tests and synchronization analyses to be carried out. A detailed introduction to methods based on recurrence plots can be found on the following website:

```
http://www.recurrence-plot.tk
```

Recurrence plots have already been used to analyze many problems in the earth sciences. Comparing the dynamics of modern precipitation data with paleo-rainfall data inferred from annual layered lake sediments in the northwestern Argentine Andes serves as a good example of such analyses (Marwan et al. 2003). In this example, the recurrence plot method was applied to red-color intensity transects across varved lake sediments that were approximately 30 kyrs old (Sect. 8.7). Comparing the recurrence plots from the sediments with those from modern precipitation data revealed that the reddish layers document the more intense rainy seasons that occur during La Niña years. Linear techniques were unable to link the increased flux of reddish clays with either the El Niño or the La Niña phase of the El Niño/Southern Oscillation. Moreover, recurrence plots helped to support the hypothesis that longer rainy seasons, increased precipitation, and the stronger influence of the El Niño/Southern Oscillation caused an increase in the number of landslides 30 kyrs ago (Marwan et al. 2003; Trauth et al. 2003). A second earth science example—which includes a comprehensive description of recurrence plots and recurrence quantification analysis—used the technique to classify different types of variability and transitions in records from lake sediment cores in order to achieve an improved understanding of the response of the biosphere to climate change and especially the response of humans in the southern Ethiopian Rift (Trauth et al. 2019).

Recommended Reading

Blackman, RB, Tukey, JW (1958) The Measurement of Power Spectra. Dover NY

Brückner E (1890) Klimaschwankungen seit 1700 nebst Bemerkungen über die Klimaschwankungen der Diluvialzeit. Geographische Abhandlungen 4:153–484

Cooley JW, Tukey JW (1965) An algorithm for the machine calculation of complex fourier series. Math Comput 19(90):297–301

Donner RV, Heitzig J, Donges JF, Zou Y, Marwan N, Kurths J (2011) The geometry of chaotic dynamics—A complex network perspective. European Physical Journal B 84:653–672

Eckmann JP, Kamphorst SO, Ruelle D (1987) Recurrence plots of dynamical systems. Europhys Lett 5:973–977

Hann J (1901) Lehrbuch der Meteorologie. Tauchnitz, Leipzig

Iwanski J, Bradley E (1998) Recurrence plot analysis: To embed or not to embed? Chaos 8:861–871

Kantz H, Schreiber T (1997) Nonlinear time series analysis. Cambridge University Press, Cambridge

Köppen W (1931) Grundriss der Klimakunde –, 2nd edn. de Gruyter, Berlin

Lau KM, Weng H (1995) Climate signal detection using wavelet transform: How to make a time series sing. Bull Am Meteor Soc 76:2391–2402

Lorenz EN (1963) Deterministic nonperiodic flow. Journal of Atmospheric Sciences 20:130–141

Mackenzie D, Daubechies I, Kleppner D, Mallat S, Meyer Y, Ruskai MB, Weiss G (2001) Wavelets: Seeing the Forest and the Trees. Beyond Discovery, National Academy of Sciences, December 2001, available online at http://www.beyonddiscovery.org

Marwan N (2011) How to avoid potential pitfalls in recurrence plot based data analysis. International Journal of Bifurcation and Chaos 21:1003–1017

Marwan N, Trauth MH, Vuille M, Kurths J (2003) Nonlinear time-series analysis on present-day and pleistocene precipitation data from the NW Argentine Andes, Clim Dyn 21:317–332

Marwan N, Romano MC, Thiel M, Kurths J (2007) Recurrence plots for the analysis of complex systems. Phys Rep 438:237–329

Mudelsee M (2014) Climate time series analysis: Classical statistical and bootstrap methods, 2nd edn. Springer Verlag, Berlin Heidelberg New York

Packard NH, Crutchfield JP, Farmer JD, Shaw RS (1980) Geometry from a time series. Phys Rev Lett 45:712–716

Press WH, Teukolsky SA, Vetterling WT, Flannery BP (2007) Numerical recipes: The art of scientific computing –, 3rd edn. Cambridge University Press, Cambridge

Scargle JD (1981) Studies in astronomical time series analysis. I. modeling random processes in the time domain. Astrophys J Suppl Ser 45:1–71

Scargle JD (1982) Studies in astronomical time series analysis. II. statistical aspects of spectral snalysis of unevenly spaced data. Astrophys J 263:835–853

Scargle JD (1989) Studies in astronomical time series analysis. III. Fourier transforms, autocorrelation functions, and cross-correlation functions of Unevenly spaced data. Astrophys J 343:874–887

Scargle JD (1990) Studies in astronomical time series analysis. IV. modeling chaotic and random processes with linear filters. Astrophys J 359:469–482

Schulz M, Stattegger K (1998) SPECTRUM: Spectral analysis of unevenly spaced paleoclimatic time series. Comput Geosci 23:929–945

Solomon S, Qin D, Manning M, Chen Z, Marquis M, Averyt KBM. (2007) Tignor. In: Miller, H.L. (Ed.), Contribution of Working Group I to the Fourth Assessment Report of the Intergovernmental Panel on Climate Change. Cambridge University Press, Cambridge, United Kingdom and New York, NY, USA

Takens F (1981) Detecting strange attractors in turbulence. Lect Notes Math 898:366–381

Torrence C, Compo GP (1998) A practical guide to wavelet analysis. Bull Am Meteor Soc 79:61–78

Trauth MH, Bookhagen B, Marwan N, Strecker MR (2003) Multiple landslide clusters record quaternary climate changes in the NW Argentine Andes. Palaeogeogr Palaeoclimatol Palaeoecol 194:109–121

Trauth MH, Asrat A, Duesing W, Foerster V, Kraemer KH, Marwan N, Maslin MA, Schaebitz F (2019) Classifying past climate change in the Chew Bahir basin, southern Ethiopia, using recurrence quantification analysis. Climate Dynamics, 53:2557–2572, code ClimDyn_RP_RQA.zip or https://tinyurl.com/yyqjyoq4

Trauth MH, Asrat A, Cohen A, Duesing W, Foerster V, Kaboth-Bahr S, Kraemer KH, Lamb H, Marwan N, Maslin MA, Schaebitz F (2021) Recurring types of variability and transitions in the ~620 kyr record of climate change from the Chew Bahir basin, southern Ethiopia. Quaternary Science Reviews, code QSR_RP_RQA or https://tinyurl.com/s6abkknz

Turcotte DL (1997) Fractals and chaos in geology and geophysics. Cambridge University Press, Cambridge

Webber CL, Zbilut JP (2005) Recurrence quantification analysis of nonlinear dynamical systems. In: Riley MA, Van Orden GC, Tutorials in contemporary nonlinear methods for the behavioral sciences.

Welch PD (1967) The use of fast fourier transform for the estimation of power spectra: A method based on time averaging over short, Modified Periodograms. IEEE Trans. Audio Electroacoustics AU-15:70–73

Further Reading

Ansari AR, Bradley RA (1960) Rank-Sum Tests for Dispersion. Annals of Mathematical Statistics, 31:1174–1189. [Open access]

Campisano CJ et al (2017) The Hominin Sites and Paleolakes Drilling Project: High-Resolution Paleoclimate Records from the East African Rift System and Their Implications for Understanding the Environmental Context of Hominin Evolution. PaleoAnthropology 2017:1–43

Grenander U (1958) Bandwidth and variance in estimation of the spectrum. J Roy Stat Soc B 20:152–157

Holschneider M (1995) Wavelets, an analysis tool. Oxford University Press, Oxford

Killick R, Fearnhead P, Eckley IA (2012) Optimal detection of changepoints with a linear computational cost. J Am Stat Assoc 107:1590–1598

Lavielle M (2005) Using penalized contrasts for the change-point problem. Signal Process 85:1501–1510

Lepage Y (1971) A combination of Wilcoxon's and Ansari-Bradley's statistics. Biometrika 58:213–217

Lomb NR (1977) Least-Squared frequency analysis of unequally spaced data. Astrophysics and Space Sciences 39:447–462

Mann HB, Whitney DR (1947) On a test of whether one of two random variables is stochastically larger than the other. Ann Math Stat 18:50–60

Marwan N, Thiel M, Nowaczyk NR (2002) Cross recurrence plot based synchronization of time series. Nonlinear Process Geophys 9(3/4):325–331

Mudelsee M (2000) Ramp function regression: A tool for quantifying climate transitions. Comput Geosci 26:293–307

Mudelsee M, Stattegger M (1997) Exploring the structure of the mid-Pleistocene revolution with advanced methods of time-series analysis. Int J Earth Sci 86:499–511

Muller RA, MacDonald GJ (2000) Ice ages and astronomical causes—data. Springer Verlag, Berlin Heidelberg New York, Spectral Analysis and Mechanisms

Paliwal KK, Agarwal A, Sinha SS (1982) A modification over Sakoe and Chiba's dynamic time warping algorithm for isolated word recognition. Signal Process 4:329–333

Romano M, Thiel M, Kurths J, von Bloh W (2004) Multivariate recurrence plots. Phys Lett A 330(3–4):214–223

Sakoe H, Chiba S (1978) Dynamic programming algorithm optimization for spoken word recognition. IEEE Transactions on Acoustics, Speech, and Signal Processing ASSP-26:43–49

Schuster A (1898) On the investigation of hidden periodicities with application to a supposed 26 day period of meteorological phenomena. Terr Magn Atmos Electr 3:13–41

Trauth MH (2015) MATLAB Recipes for Earth Sciences –, 4th edn. Springer Verlag, Berlin Heidelberg New York

Trauth MH, Larrasoaña JC, Mudelsee M (2009) Trends, rhythms and events in Plio-Pleistocene African climate. Quatern Sci Rev 28:399–411

Trauth MH, Foerster V, Junginger A, Asrat A, Lamb H, Schaebitz F (2018) Abrupt or gradual? Change point analysis of the late pleistocene-Holocene climate record from chew Bahir, Southern Ethiopia. Quatern Res 90:321–330

Trulla LL, Giuliani A, Zbilut JP, Webber CL Jr (1996) Recurrence quantification analysis of the logistic equation with transients. Phys Lett A 223(4):255–260

Weedon G (2003) Time-series analysis and cyclostratigraphy—examining stratigraphic records of environmental change. Cambridge University Press, Cambridge

Signal Processing

6

6.1 Introduction

Signal processing involves techniques for manipulating a signal in order to minimize the effects of noise, to correct all kinds of unwanted distortions, and to separate out various components of interest. Most signal processing algorithms include the design and realization of filters. A *filter* can be described as a system that transforms signals. *System theory* provides the mathematical background for filter design and realization. A filter has an input and an output, with the *output signal y(t)* being modified with respect to the *input signal x(t)* (Fig. 6.1). The *signal transformation* can be carried out through a mathematical process known as *convolution* or, if filters are involved, as *filtering*.

This chapter deals with the design and realization of *digital filters* with the help of a computer. However, many natural processes resemble *analog filters* that record natural events in geological archives. A single rainfall event is not documented in lake sediments because the short-duration and low-amplitude signal of the event in the sediments (e.g., a short-term change in the chemical composition of the sediments) is strongly attenuated in the well-mixed water body before it is archived in the sediments. Bioturbation also introduces serious signal distortions (e.g., changes in the amplitude and phase of the signal), such as in deep-sea sediments, where it leads to apparent phase shifts in the oxygen-isotope signal of individual foraminifera species (e.g., Trauth 2013). In addition to such natural filters, the field collection and sampling of geological data alters and smooths the data with respect to its original form. For example, a finite-sized sediment sample is integrated over a certain period of time, which smooths the natural signal. Similarly, the measurement of magnetic susceptibility in a sediment core with the help of a loop sensor

Supplementary Information The online version contains supplementary material available at https://doi.org/10.1007/978-3-031-07719-7_6.

Input signal   Signal transformation   Output signal

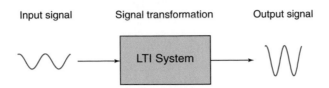

Fig. 6.1 Schematic of a linear time-invariant (LTI) system. The input signal is transformed into an output signal.

introduces substantial smoothing since the loop response is integrated over a section of the core.

The characteristics of these natural filters are often difficult to determine, whereas numerical filters are designed with well-defined characteristics. In addition, artificial filters are time invariant in most cases, whereas natural filters (e.g., mixing within the water body of a lake or bioturbation at the water–sediment interface) may vary with time. An easy way of describing or predicting the effect of a filter is by exploring the filter output from a simple input signal, such as a sine wave, a square wave, a sawtooth function, a ramp function, or a step function. Although there is an endless variety of such input signals, most systems or filters are described by their impulse response, that is, the output resulting from the input of a unit impulse.

This chapter proceeds with Sect. 6.2 on generating periodic signals, trends, and noise, which takes its cue from Sect. 5.2 of the previous chapter. Section 6.3 then considers linear time-invariant systems and provides a mathematical background for filters. The subsequent Sects. 6.4 to 6.9 deal with the design, realization, and application of linear time-invariant filters. Section 6.10 then considers the use of adaptive filters, which were originally developed for use in the telecommunication industry. These adaptive filters automatically extract noise-free signals from duplicate measurements of the same object. Such filters can be used in a large number of applications, such as in removing noise from duplicate paleoceanographic time series and in improving the signal-to-noise ratio of parallel color intensity transects across varved lake sediments (see Chap. 5, Fig. 5.1). Adaptive filters are also widely used in geophysics for noise canceling. For signal processing, we use the *NumPy* (https://numpy.org), *Matplotlib* (https://matplotlib.org), and *SciPy* (https://scipy.org) packages, which contain all the necessary routines.

6.2   Generating Signals

Python provides numerous tools for generating basic signals that can be used to illustrate the effects of filters. In Chap. 5, we generated a signal by adding together three sine waves with different amplitudes and periods. In the following example, the time array is transposed in order to generate column arrays:

```
%reset -f

import numpy as np
import matplotlib.pyplot as plt

t - np.linspace(1,100,100)
x = 2 *np.sin(2*np.pi*t/50) + \
        np.sin(2*np.pi*t/10) + \
    0.5*np.sin(2*np.pi*t/5)

plt.figure()
plt.plot(t,x)
plt.axis(np.array([0,100,-4,4]))
plt.show()
```

Frequency-selective filters are used to remove specific frequency bands from the data. As an example, we can design a filter that suppresses the portion of the signal with a periodicity of $\tau = 10$ while leaving the other two cycles unaffected. The effects of such filters on simple periodic signals can also be used to predict signal distortions of natural filters.

A *step function* is another basic input signal that can be used to explore filter characteristics. Step functions describe the transition from a value of one toward a value of zero at a specific time. The function stairs() draws a stairstep graph of the elements of x. The function needs edges e, which we calculate from t:

```
t = np.linspace(1,100,100)
x = np.ones(100)
x[51:100] = np.zeros(49)

e = np.zeros(101)
e[0] = np.diff(t[0:2])/2
e[1:101] = t+np.diff(t[0:2])/2

plt.figure()
plt.stairs(x,e)
plt.axis(np.array([0,100,-4,4]))
plt.show()
```

This signal can be used to study the effects of a filter on a sudden transition. An abrupt climate change could be regarded as an example. Most natural filters tend to smooth such a transition and to distribute it over a longer time period.

A *unit impulse* is a third important signal type that we use in the following examples. This signal equals zero at all times except at a single data point, where it equals one. The function stem() plots the data sequence x as stems from the *t*-axis using circles for the data values:

```
t = np.linspace(1,100,100)
x = np.zeros(100)
x[48] = 1
x[49:100] = np.zeros(50)

plt.figure()
plt.stem(t,x)
plt.axis(np.array([0,100,-4,4]))
plt.show()
```

The unit impulse is the most popular synthetic signal used to study the performance of a filter. The output of the filter (i.e., the impulse response) describes the characteristics of the filter very well. Moreover, the output of a linear time-invariant filter can be described by the superposition of impulse responses that have been scaled by multiplying the output of the filter by the amplitude of the input signal.

6.3   Linear Time-Invariant Systems

Filters can be described as systems with an input $x(t)$ and an output $y(t)$. We therefore first describe the characteristics of systems in general before subsequently considering filters. Important characteristics of a system include

- *Continuity*—A system with continuous inputs $x(t)$ and outputs $y(t)$ is a continuous system. Most natural systems are continuous. However, after sampling natural signals, we obtain discrete data series and model these natural systems as discrete systems with discrete inputs and outputs.
- *Linearity*—For linear systems, the output $y(t)$ of the linear combination of several input signals $x_i(t)$ where

$$x(t) = k_1 x_1(t) + k_2 x_2(t)$$

is the same as the linear combination of the outputs $y_i(t)$:

$$y(t) = k_1 y_1(t) + k_2 y_2(t)$$

Important properties of linearity are scaling and additivity (*superposition*), which allow the input and output to be multiplied by a constant k_i either before or after transformation. Superposition allows additive components of the input to be extracted and transformed separately. Fortunately, many natural systems follow a linear pattern of behavior. Complex linear signals (e.g., additive harmonic components) can be separated out and transformed independently. Milankovitch cycles provide an example of linear superposition in paleoclimate records, although there is an ongoing debate about the validity of this theory. Numerous nonlinear systems that do not possess the properties of scaling and additivity also exist in nature. An example of a linear system is

```
%reset -f

import numpy as np
import matplotlib.pyplot as plt

x = np.linspace(1,100,100)
y = 2*x

plt.figure()
plt.plot(x,y)
plt.show()
```

where x is the input signal and y is the output signal. An example of a nonlinear system is

```
x = np.linspace(1,100,100)
y = x**2

plt.figure()
plt.plot(x,y)
plt.show()
```

- *Time invariance*—The system output $y(t)$ does not change as a result of a delay in the input $x(t+i)$; instead, the system characteristics remain constant over time. Unfortunately, natural systems often change their characteristics over time. For instance, benthic mixing (or bioturbation) depends on various environmental variables (e.g., nutrient supply), and the system's properties consequently vary over time. In such a case, it is difficult to determine the actual input of the system using the output (e.g., when extracting the actual climate signal from a bioturbated sedimentary record).
- *Invertibility*—An invertible system is a system in which the original input signal $x(t)$ can be reproduced from the system's output $y(t)$. This is an important property if unwanted signal distortions are to be corrected, in which case the known system is inverted and the output then used to reconstruct the undisturbed input. For example, a core logger that measures magnetic susceptibility with a loop sensor integrates the signal over a specific core interval; the sensitivity is highest at the position of the loop and decreases both down-core and up-core. This system is invertible, that is, we can compute the input signal $x(t)$ from the output signal $y(t)$ by inverting the system. The inverse of the linear system above is

```
x = np.linspace(1,100,100)
y = 0.5*x
```

```
plt.figure()
plt.plot(x,y)
plt.show()
```

where x is the input signal and y is the output signal. A nonlinear system

```
x = np.linspace(-100,100,201)
y = x**2
```

```
plt.figure()
plt.plot(x,y)
plt.show()
```

is not invertible. Since this system yields equal responses for different inputs, such as y=4 for inputs x=−2 and x=+2, the input x cannot be reconstructed from the output y. A similar situation can also occur in linear systems, such as

```
x = np.linspace(1,100,100)
y = np.zeros(x.shape)
```

```
plt.figure()
plt.plot(x,y)
plt.show()
```

The output y is zero for all inputs x, and the output therefore does not contain any information about the input.

- *Causality*—The system response only depends on present and past inputs $x(0)$, $x(-1)$, …, whereas future inputs $x(+1)$, $x(+2)$, … have no effect on the output $y(0)$. All real-time systems (e.g., telecommunication systems) must be causal since they cannot have future inputs available to them. All systems and filters in Python are indexed as causal. In the earth sciences, however, numerous non-causal filters are used. Filtering images and signals that have been extracted from sediment cores is an example for which future inputs are available at the time of filtering. Output signals must be delayed after filtering in order to compensate for the differences between causal and non-causal indexing.
- *Stability*—A system is stable if the output $y(t)$ of a finite input $x(t)$ is also finite. Stability is critical in filter design, where filters often have the disadvantage of provoking divergent outputs. In such cases, the filter design must be revised and improved.

Linear time-invariant (LTI) systems are very popular as a special type of filter. Such systems have all the advantages described above and are also easy to design and use in many applications. Sections 6.4 to 6.9 below describe the design, realization, and application of LTI-type filters for extracting specific frequency components from signals. These filters are mainly used to reduce the noise level in signals. Unfortunately, however, many natural systems do not behave as LTI systems because the signal-to-noise ratio often varies with time. Section 6.10 describes the application of adaptive filters that automatically adjust their characteristics in a time-variable environment.

6.4   Convolution, Deconvolution, and Filtering

Convolution is a mathematical description of a system transformation. Filtering is an application of the convolution process. A running mean of length five provides an example of such a simple filter. The output of an arbitrary input signal is

$$y(t) = \frac{1}{5} \sum_{k=-2}^{2} x(t-k)$$

The output $y(t)$ is simply the average of the five input values $x(t-2)$, $x(t-1)$, $x(t)$, $x(t+1)$, and $x(t+2)$. In other words, all five consecutive input values are multiplied by a factor of 1/5 and summed to form $y(t)$. In this example, all input values are multiplied by the same factor, that is, they are equally weighted. The five factors used in the operation above are therefore called filter weights b_k. The filter can be represented by the array

```
%reset -f

import numpy as np
import scipy.signal as signal
import matplotlib.pyplot as plt
from numpy.random import default_rng

b = np.array([0.2,0.2,0.2,0.2,0.2])
print(b)
```

which consists of the five identical filter weights 0.2. Since this filter is symmetric, it does not shift the signal on the time axis; rather, the only function of this filter is to smooth the signal. Running means of a given length are often used to smooth signals, mainly for cosmetic reasons. Modern spreadsheet software usually contains running means as a function for smoothing data series. The effectiveness of a smoothing filter increases with increasing filter length.

The weights that a filter of an arbitrary length uses can be varied. As an example, let us consider an asymmetric filter of five filter weights:

```
b = np.array([0.05,0.08,0.14,0.26,0.47])
print(b)
```

The sum of all of the filter weights is one, and the filter therefore does not introduce any additional variance to the signal. However, since the filter is highly asymmetric, it shifts the signal along the time axis, that is, it introduces a phase shift.

The general mathematical representation of the filtering process is the convolution

$$y(t) = \sum_{k=-N_1}^{N_2} b_k \cdot x(t-k)$$

where b_k is the array of *filter weights* and N_1+N_2 is the *filter order*, which is the length of the filter reduced by one. Filters with five weights (as in our example) have an order of four. In contrast to this format, Python uses the engineering standard for indexing filters, that is, filters are always defined as causal. The convolution used by Python is therefore

$$y(t) = \sum_{k=0}^{N} b_k \cdot x(t-k)$$

where N is the order of the filter. Several of the frequency-domain tools provided by Python cannot simply be applied to non-causal filters that have been designed for applications in the earth sciences. Phase corrections are therefore commonly carried out in order to simulate non-causality. For example, frequency-selective filters—which are introduced in Sect. 6.9—can be applied using `filtfilt()`, which provides zero-phase forward and reverse filtering.

The functions `convolve()` and `lfilter()`, which provide digital filtering in Python, are best illustrated in terms of a simple running mean. The n elements of the array $x(t_1)$, $x(t_2)$, $x(t_3)$, ..., $x(t_n)$ are replaced with the arithmetic means of subsets of the input array. For instance, a running mean over three elements computes the mean of inputs $x(t_{n-1})$, $x(t_n)$, $x(t_{n+1})$ in order to obtain the output $y(t_n)$. We can illustrate this simply by generating a random signal,

```
t = np.linspace(1,100,100)
rng = default_rng(0)
x1 = rng.standard_normal((100))
```

designing a filter that averages three data points of the input signal,

```
b1 = np.array([1,1,1])/3
print(b1)
```

and convolving the input array with the filter

```
y1 = np.convolve(x1,b1)
```

The elements of b1 are the weights of the filter. In our example, all filter weights are the same and are equal to 1/3. Note that `convolve()` yields an array that has a length of $n+m-1$, where m is the length of the filter:

```
m1 = len(b1)
```

We can explore the contents of our workspace in order to check the length of the input and output of `convolve()`. Typing

```
np.who()
```

yields

```
Name            Shape           Bytes           Type
==============================================================

b               5               40              float64
t               100             800             float64
x1              100             800             float64
b1              3               24              float64
y1              102             816             float64

Upper bound on total bytes  =       2480
```

Here, we see that the actual input series `x1` has a length of 100 data points, whereas the output series `y1` has two additional elements. Convolution generally introduces $(m-1)/2$ data points at each end of the data series. In order to compare input and output signals, we therefore clip the output signal at either end:

```
y1 = y1[1:101]
```

A more general way of correcting the phase shifts of `convolve()` is by typing

```
y1 = y1[np.arange(int((m1-1)/2),len(y1)-int((m1-1)/2))]
print(y1)
```

which of course only works for an odd number of filter weights. An alternative way of correcting the phase shifts is to use `same` for the input argument `shape` in `convolve()` in order to return the most central 100 data points of the convolution. As a result, `y1` has the same size as the input signal `x1`:

```
y1 = np.convolve(x1,b1,'same')
```

We can then plot both input and output signals for comparison using `legend()` to display a legend for the plot:

```
plt.figure()
line1, = plt.plot(t,x1,color='b',label='x1(t)')
line2, = plt.plot(t,y1,color='r',label='y1(t)')
plt.legend(handles=[line1,line2])
plt.show()
```

This plot illustrates the effect that the running mean has on the original input series. The output y1 is substantially smoother than the input signal x1. If we increase the length of the filter, we obtain an even smoother signal output y2:

```
b2 = np.array([1,1,1,1,1])/5
m2 = len(b2)

y2 = np.convolve(x1,b2,'same')

plt.figure()
line1, = plt.plot(t,x1,color='b',label='x1(t)')
line2, = plt.plot(t,y1,color='r',label='y1(t)')
line3, = plt.plot(t,y2,color='g',label='y2(t)')
plt.legend(handles=[line1,line2,line3])
plt.show()
```

The reverse of convolution is deconvolution, which is often used in signal processing to reverse the effects of a filter. We use the first example of a random signal x1, design a filter that averages three data points of the input signal, and convolve the input array with the filter, which yields the output y1:

```
t = np.linspace(1,100,100)
rng = default_rng(0)
x1 = rng.standard_normal(((100))

b1 = np.array([1,1,1])/3
print(b1)

y1 = np.convolve(x1,b1)
```

We use deconvolve() to reverse the convolution and compare the deconvolution result x1d with the original signal x1:

```
x1d,rem = signal.deconvolve(y1,b1)

plt.figure()
line1, = plt.plot(t,x1,color='b',label='x1(t)')
line2, = plt.plot(t,x1d,color='r',linestyle=':',label='x1d(t)')
plt.legend(handles=[line1,line2])
plt.show()
```

As we can see, there is no difference between x1d and x1. However, there is a difference if we add noise to the signal and deconvolve the result by typing

```
y1n = y1 + 0.05*rng.standard_normal(len(y1))
x1nd,rem = signal.deconvolve(y1n,b1)

plt.figure()
line1, = plt.plot(t,x1,
    color='b',
    label='x1(t)')
line2, = plt.plot(t,x1nd,
    color='r',
    linestyle=':',
    label='x1nd(t)')
plt.legend(handles=[line1,line2])
plt.show()
```

The next section provides a broader definition of filters.

6.5   Comparing Functions for Filtering Data Series

The filters described in the previous section are very simple examples of *nonrecursive filters*, in which the filter output $y(t)$ depends only on the filter input $x(t)$ and the filter weights b_k. Prior to introducing a broader description of linear time-invariant filters, we replace convolve() with lfilter(), which can also be used for *recursive filters*. In this case, the output $y(t_n)$ depends not only on the filter input $x(t_n)$, but also on previous elements of the output $y(t_{n-1})$, $y(t_{n-2})$, $y(t_{n-3})$, and so on (Sect. 6.6). We first use convolve() for nonrecursive filters in order to compare the results of convolve() and lfilter():

```
%reset -f

import numpy as np
import scipy.signal as signal
import matplotlib.pyplot as plt
from numpy.random import default_rng

t = np.linspace(1,100,100)
rng = default_rng(0)
x3 = rng.standard_normal((100))
```

We then design a filter that averages five data points of the input signal:

```
b3 = np.array([1,1,1,1,1])/5
m3 = len(b3)
```

The input signal can be convolved using convolve():

```
y3 = np.convolve(x3,b3,'same')
```

We next follow a similar procedure with lfilter() and compare the result with that obtained using convolve(). In contrast to convolve() without using same, lfilter() yields an output array with the same length as the input array. Unfortunately, lfilter() assumes that the filter is causal. This is of great importance in electrical engineering, where filters are often applied in real time. In the earth sciences, however, the entire signal is available in most applications at the time of processing the data. The data series is filtered by

```
y4 = signal.lfilter(b3,1,x3)
```

and the phase correction is then carried out using

```
y4[0:len(y4)-m3+1] = y4[np.arange(int((m3-1)/2),
    len(y4)-int((m3-1)/2))]
y4[len(y4)-m3+1:len(y4)] = np.zeros((1,m3-1))
```

which works only for an odd number of filter weights. This command simply shifts the output by $(m-1)/3$ toward the lower end of the t-axis and then fills the data to the end with zeros. Comparing the ends of both outputs illustrates the effect of this correction, where

```
print(y3[0:5])
print(y4[0:5])
```

yields

```
[0.1268096  0.14778963 0.04065575
 0.08782872 0.3750497 ]
[0.1268096  0.14778963 0.04065575
 0.08782872 0.3750497 ]
```

which is the lower end of the output. We can see that both arrays y3 and y4 contain the same elements. We now explore the upper end of the data array, where

```
print(y3[94:])
print(y4[94:])
```

yields the output

```
[ 0.05479578  0.02424718 -0.0969001
 -0.42716122 -0.63345184 -0.66565375]
[ 0.05479578  0.02424718  0.
  0.          0.          0.        ]
```

The arrays are identical up to element y(end−m3+1), but then the second array
(y4) contains zeros instead of true data values. Plotting the results with

```
plt.figure()
fig,(ax1,ax2) = plt.subplots(2)
ax1.plot(t,x3,color='h')
ax1.plot(t,y3,color='r')
ax2.plot(t,x3,color='b')
ax2.plot(t,y4,color='r')
plt.show()
```

or in a single plot

```
plt.figure()
plt.plot(t,x3,color='b')
plt.plot(t,y3,color='g')
plt.plot(t,y4,color='r',
    linestyle=':')
plt.show()
```

reveals that the results from using convolve() are identical with those from using
lfilter() except at the upper end of the data array. These observations are important for our next steps in signal processing, particularly if we are interested in
leads and lags between various components of signals.

6.6   Recursive and Nonrecursive Filters

We now expand the nonrecursive filters by adding a recursive component such that
the output $y(t_n)$ depends not only on the filter input $x(t_n)$, but also on previous output values $y(t_{n-2})$, $y(t_{n-2})$, $y(t_{n-3})$, and so on. This filter requires not only the nonrecursive filter weights b_i, but also the recursive filters weights a_i (Fig. 6.2), and it
can be described by the *difference equation*:

$$y(t) = \sum_{k=-N_1}^{N_2} b_k \cdot x(t-k) - \sum_{k=1}^{M} a_k \cdot y(t-k)$$

This is a non-causal version of the difference equation, but Python again uses the
causal indexing

$$y(t) = \sum_{k=0}^{N} b_k \cdot x(t-k) - \sum_{k=1}^{M} a_k \cdot y(t-k)$$

with the known problems in the design of zero-phase filters. The larger of the two
quantities M on one hand and $N_1 + N_2$ or N on the other hand is the order of the filter.

Input signal x(t)

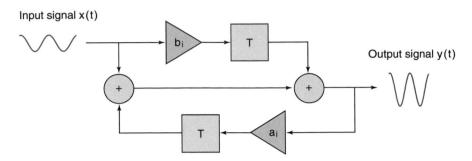

Output signal y(t)

Fig. 6.2 Schematic of a linear time-invariant filter with an input $x(t)$ and an output $y(t)$. The filter is characterized by its weights a_i and b_i and by the delay elements T. Nonrecursive filters require only nonrecursive weights b_i, whereas recursive filters also require recursive filter weights a_i.

We use the same synthetic input signal x5 as in the previous example to illustrate the performance of a recursive filter:

```
%reset -f

import numpy as np
import scipy.signal as signal
import matplotlib.pyplot as plt
from numpy.random import default_rng

t = np.linspace(1,100,100)
rng = default_rng(0)
x5 = rng.standard_normal((100))
```

This input is then filtered using a recursive filter with a set of weights a5 and b5,

```
b5 = np.array([0.0048,0.0193,0.0289,0.0193,0.0048])
a5 = np.array([1.0000,-2.3695,2.3140,-1.0547,0.1874])

m5 = len(b5)

y5 = signal.lfilter(b5,a5,x5)
```

and the output y5 is corrected for the phase

```
y5[0:len(y5)-m5+1] = y5[np.arange(int((m5-1)/2),
    len(y5)-int((m5-1)/2))]
y5[len(y5)-m5+1:len(y5)] = np.zeros((1,m5-1))
```

We can now plot the results:

```
plt.figure()
plt.plot(t,x5,color='b')
plt.plot(t,y5,color='r')
plt.show()
```

This filter clearly changes the signal. The output contains only low-frequency components, and all higher frequencies have been eliminated. A comparison of the periodograms for the input and the output reveals that all frequencies above $f = 0.1$ (corresponding to a period of $\tau = 10$) have been suppressed:

```
f,Pxx = signal.periodogram(x5,fs=1,nfft=128)
f,Pyy = signal.periodogram(y5,fs=1,nfft=128)

plt.figure()
plt.plot(f,Pxx,color='b')
plt.plot(f,Pyy,color='r')
plt.grid
plt.xlabel('Frequency')
plt.ylabel('Power')
plt.show()
```

We have now designed a frequency-selective filter, that is, a filter that eliminates certain frequencies while leaving other frequencies relatively unaffected. The next section introduces tools that are used both to characterize a filter in the time and frequency domains and to predict the effect of a frequency-selective filter on arbitrary signals.

6.7   Impulse Response

The impulse response is a very convenient way of describing the characteristics of a filter (Fig. 6.3). The impulse response h is useful in LTI systems, in which the convolution of the input signal $x(t)$ with h is used to obtain the output signal $y(t)$:

$$y(t) = \sum_{k=-N_1}^{N_2} h_k \cdot x(t - k)$$

The values of the impulse response h can be shown to be identical to the filter weights in nonrecursive filters, but not in recursive filters. The convolution equation above is often written in a short form:

$$y(t) = h(t) * x(t)$$

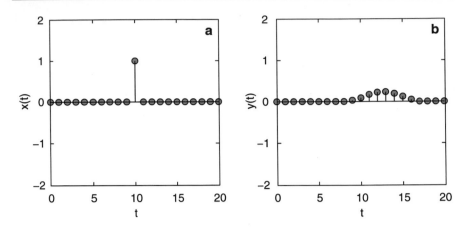

Fig. 6.3 Transformation of **a** a unit impulse for computing **b** the impulse response of a system. The impulse response is often used to describe and predict the performance of a filter.

In many examples, convolution in the time domain is replaced with a simple multiplication of the Fourier transforms $H(f)$ and $X(f)$ in the frequency domain:

$$Y(f) = H(f) \cdot X(f)$$

The output signal $y(t)$ in the time domain is then obtained via a reverse Fourier transform of $Y(f)$. Signals are often convolved in the frequency domain rather than in the time domain due to the relative simplicity of the multiplication. However, the Fourier transformation itself introduces a number of artifacts and distortions, and convolution in the frequency domain is therefore not without problems. In the following examples, we apply the convolution only in the time domain.

First, we generate a unit impulse:

```
%reset -f

import numpy as np
import scipy.signal as signal
import matplotlib.pyplot as plt

t = np.linspace(0,20,21)
x6 = np.zeros(21)
x6[10] = 1

plt.figure()
plt.stem(t,x6,basefmt='k-')
plt.axis(np.array([0,20,-2,2]))
plt.show()
```

The function stem() plots the data sequence x6 as stems from the *x*-axis that terminate with circles for the data value. This can be a better way of plotting digital data than using the continuous lines generated by plot(). We now feed x6 into the filter and explore the output y6. The impulse response is identical to the weights of nonrecursive filters:

```
b6 = np.array([1,1,1,1,1])/5
m6 = len(b6)


y6 = signal.lfilter(b6,1,x6)
```

We again correct y6 for the phase shift of lfilter(), although this might not be important in the present example:

```
y6[0:len(y6)-m6+1] = y6[np.arange(int((m6-1)/2),
    len(y6)-int((m6-1)/2))]
y6[len(y6)-m6+1:len(y6)] = np.zeros((1,m6-1))
```

We obtain an output array y6 of the same length and phase as the input array x6. We now plot the results for comparison:

```
plt.figure()
plt.stem(t,x6)
plt.stem(t,y6,
    linefmt='r-',
    markerfmt='ro',
    basefmt='k-')
plt.axis(np.array([0,20,-2,2]))
plt.show()
```

In contrast to plot(), stem() accepts only one data series, and the second series y6 is therefore overlaid on the same plot by using stem() twice. The effect of the filter can be clearly seen in the plot: It averages the unit impulse over a length of five elements. Furthermore, the values of the output y6 equal the filter weights of a6. In our example, these values are 0.2 for all elements of a6 and y6.

For a recursive filter, however, the output y6 does not match the filter weights. Once again, we first generate an impulse:

```
t = np.linspace(0,20,21)
x7 = np.zeros(21)
x7[10] = 1
```

An arbitrary recursive filter with weights a7 and b7 is then designed:

```
b7 = np.array([0.0048,0.0193,0.0289,0.0193,0.0048])
a7 = np.array([1.0000,-2.3695,2.3140,-1.0547,0.1874])

m7 = len(b7)

y7 = signal.lfilter(b7,a7,x7)

y7[0:len(y7)-m7+1] = y7[np.arange(int((m7-1)/2),
    len(y7)-int((m7-1)/2))]
y7[len(y7)-m7+1:len(y7)] = np.zeros((1,m7-1))
```

The stem plot of the input x2 and the output y2 shows an interesting impulse response:

```
plt.figure()
plt.stem(t,x7)
plt.stem(t,y7,linefmt='r-',markerfmt='ro',basefmt='k-')
plt.axis(np.array([0,20,-2,2]))
plt.show()
```

The signal is smeared over a broader area and is also shifted toward the right. This filter therefore not only affects the amplitude of the signal, but also shifts the signal toward lower or higher values. Such phase shifts are usually unwanted characteristics of filters, although shifts along the time axis might be of particular interest in some applications.

6.8   Frequency Response

We next investigate the frequency response of a filter, that is, the effect of a filter on the amplitude and phase of a signal (Fig. 6.4). The frequency response $H(f)$ of a filter is the Fourier transform of the impulse response $h(t)$. The absolute value of the complex frequency response $H(f)$ is the magnitude response of the filter $A(f)$:

$$A(f) = |H(f)|$$

The argument of the complex frequency response $H(f)$ is the phase response of the filter:

$$\Phi(f) = \arg(H(f))$$

Since Python filters are all causal, it is difficult to explore the phase of signals using the corresponding functions included in the Signal Processing Toolbox. The user's guide for this toolbox simply recommends that the filter output be delayed in the time domain by a fixed number of samples, as we did in the previous examples. As another example, a sine wave with a period of 20 and an amplitude of 2 is used as an input signal:

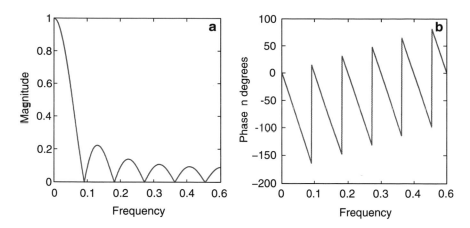

Fig. 6.4 **a** Magnitude and **b** phase response of a running mean over eleven elements.

```
%reset -f

import numpy as np
import scipy.signal as signal
from scipy import interpolate
import matplotlib.pyplot as plt

t = np.linspace(1,100,100)
x8 = 2 * np.sin(2*np.pi*t/20)
```

A running mean over eleven elements is designed, and this filter is applied to the input signal x8:

```
b8 = np.ones(11)/11;
m8 = len(b8)

y8 = signal.lfilter(b8,1,x8)
```

The phase of the output y8 is corrected for causal indexing:

```
y8[0:len(y8)-m8+1] = y8[np.arange(int((m8-1)/2),
    len(y8)-int((m8-1)/2))]
y8[len(y8)-m8+1:len(y8)] = np.zeros((1,m8-1))
```

Both the input and output of the filter are plotted:

```
plt.figure()
plt.plot(t,x8,color='b')
```

```
plt.plot(t,y8,color='r')
plt.show()
```

The filter clearly reduces the amplitude of the sine wave. Whereas the input signal x8 has an amplitude of 2, the output y8 has an amplitude of

```
print(np.max(y8))
```

which yields

```
1.1479548208500083
```

The filter reduces the amplitude of a sine with a period of 20 by

```
print(1-np.max(y8[40:60])/2)
```

which yields

```
0.42602258957499606
```

that is, it reduces the amplitude by approximately 43%. Elements 40 to 60 are used for computing the maximum value of y8 in order to avoid edge effects. Nevertheless, the filter does not affect the phase of the sine wave, that is, both the input and output are in phase.

However, the same filter has a different impact on a different input signal. Let us design another sine wave x9 with a similar amplitude but with a different period of 15:

```
t = np.linspace(1,100,100)
x9 = 2 * np.sin(2*np.pi*t/15)
```

Applying a similar filter and correcting the output y9 for the phase shift of lfil-ter() yields

```
t = np.linspace(1,100,100)
x9 = 2 * np.sin(2*np.pi*t/15)

b9 = np.ones(11)/11;
m9 = len(b9)

y9 = signal.lfilter(b9,1,x9)

y9[0:len(y9)-m9+1] = y9[np.arange(int((m9-1)/2),
    len(y9)-int((m9-1)/2))]
y9[len(y9)-m9+1:len(y9)] = np.zeros((1,m9-1))
```

The output y9 is again in phase with the input x9, but the amplitude is dramatically reduced compared with that of the input,

```
plt.figure()
plt.plot(t,x9,color='b')
plt.plot(t,y9,color='r')
plt.show()

print(1-np.max(y9[40:60])/2)
```

which yields

```
0.6768410326431586
```

The running mean over eleven elements reduces the amplitude of this signal by ~68%. More generally, the filter response clearly depends on the frequency of the input. The frequency components of a more complex signal that contains multiple periodicities are affected in a different way. The frequency response of a filter

```
b10 = np.ones(11)/11;
```

can be computed using freqz(). This function uses filter weights a and b and the number of frequency points worN at which the frequency response was computed:

```
f,h = signal.freqz(b10,a=1,worN=512,fs=1)
```

The function freqz() returns the complex frequency response h of the digital filter b10 together with the frequencies f. Next, we calculate and display the magnitude of the frequency response:

```
magnitude = np.abs(h)

plt.figure()
plt.plot(f,magnitude)
plt.xlabel('Frequency')
plt.ylabel('Magnitude')
plt.show()
```

This plot can be used to predict the effect of the filter for any frequency of an input signal. We can interpolate the magnitude of the frequency response in order to calculate the increase or decrease in a signal's amplitude for a specific frequency. As an example, the interpolation of magnitude for a frequency of 1/20,

```
fun = interpolate.interp1d(f,magnitude)
print(1-fun(1/20))
```

which yields

```
0.4260303441186485
```

results in the expected ~ 43% reduction in the amplitude of the sine wave with a period of 20. The sine wave with a period of 15 experiences an amplitude reduction of

```
fun = interpolate.interp1d(f,magnitude)
print(1-fun(1/15))
```

which yields

```
0.675055631124985
```

that is, an amplitude reduction of approximately 68%, which is similar to the value observed previously. It is very important that such a running mean eliminate certain frequencies, such as those for which magnitude=0. As an example, applying the filter to a signal with a period of approximately 1/0.09082 completely eliminates the signal. Furthermore, since the magnitude of the frequency response is the absolute value of the complex frequency response h, the magnitude response is actually negative between ~0.09082 and ~0.1816, between ~0.2725 and ~0.3633, and between ~0.4546 and the Nyquist frequency. All signal components that have frequencies within these intervals are mirrored on the t-axis. As an example, we try a sine wave with a period of 7 (e.g., a frequency of approximately 0.1429), which is within the first interval with a negative magnitude response:

```
t = np.linspace(1,100,100)
x10 = 2 * np.sin(2*np.pi*t/7)

b10 = np.ones(11)/11;
m10 = len(b10)

y10 = signal.lfilter(b10,1,x10)

y10[0:len(y10)-m10+1] = y10[np.arange(int((m10-1)/2),
    len(y10)-int((m10-1)/2))]
y10[len(y10)-m10+1:len(y10)] = np.zeros((1,m10-1))

plt.figure()
plt.plot(t,x10,color='b')
plt.plot(t,y10,color='r')
plt.show()
```

The sine wave with a period of 7 experiences an amplitude reduction of

```
fun = interpolate.interp1d(f,magnitude)
print(1-fun(1/7))
```

which yields

```
0.7957488527349983
```

that is, a reduction of approximately 80%, but it also changes its sign, as we can see from the plot. Eliminating certain frequencies and flipping the signal have important consequences when interpreting causality in the earth sciences. These filters should therefore be avoided completely even though they are offered as standards in spreadsheet programs. As an alternative, filters with a specific frequency response should be used, such as a Butterworth lowpass filter (Sect. 6.9).

The frequency response can be calculated for all kinds of filters. Indeed, it is a valuable tool for predicting the effects of a filter on signals in general. The phase response can also be calculated from the complex frequency response h of the filter (Fig. 6.4):

```
phase = 180*np.angle(h)/np.pi

plt.figure()
plt.plot(f,phase)
plt.xlabel('Frequency')
plt.ylabel('Phase in degrees')
plt.show()
```

The phase angle phase is plotted in degrees. We observe frequent jumps in this plot, which are an artifact of arctan2() within angle(). We can *unwrap* the phase response in order to eliminate the jumps that are greater than or equal to 180° with the help of unwrap():

```
plt.figure()
plt.plot(f,np.unwrap(phase))
plt.xlabel('Frequency')
plt.ylabel('Phase in degrees')
plt.show()
```

In our example, this has no effect since no jumps occur that are greater than or equal to 180°. Since the filter has a linear phase response phase, no shifts occur in the frequency components of the signals relative to one another. We would therefore not expect any distortions of the signal in the frequency domain. The phase shift of the filter on a specific period can be computed using

```
fun = interpolate.interp1d(f,180*np.angle(h)/np.pi)
print(fun(1/20)*20/360)
```

which yields

```
-5.0
```

and using

```
fun = interpolate.interp1d(f,180*np.angle(h)/np.pi)
print(fun(1/15)*15/360)
```

which yields

```
-5.0
```

for sine waves with periods of 20 and 15, respectively. Since Python uses causal indexing for filters, the phase needs to be corrected in a similar way to the delayed output of the filter. In our example, we used a filter with a length of eleven. We must therefore correct the phase by $(11-1)/2 = 5$, which suggests a zero-phase shift for the filter for both frequencies.

This also works for recursive filters. Consider a simple sine wave with a period of 8 and the previously employed recursive filter:

```
t = np.linspace(1,100,100)
x11 = 2*np.sin(2*np.pi*t/8)

b11 = np.array([0.0048,0.0193,0.0289,0.0193,0.0048])
a11 = np.array([1.0000,-2.3695,2.3140,-1.0547,0.1874])

m11 = len(b11)

y11 = signal.lfilter(b11,a11,x11)
```

We correct the output for the phase shift introduced by causal indexing and plot both input and output signals:

```
y11[0:len(y11)-m11+1] = y11[np.arange(int((m11-1)/2),
    len(y11)-int((m11-1)/2))]
y11[len(y11)-m11+1:len(y11)] = np.zeros((1,m11-1))

plt.figure()
plt.plot(t,x11,color='b')
plt.plot(t,y11,color='r')
plt.show()
```

The magnitude is reduced by

```
print(1-np.max(y11[40:60])/2)
```

which yields

```
0.6464745957914514
```

which is also supported by the magnitude response:

```
f,h = signal.freqz(b11,a11,worN=512,fs=1)

magnitude = np.abs(h)

plt.figure()
plt.plot(f,magnitude)
plt.xlabel('Frequency')
plt.ylabel('Magnitude')
plt.show()

fun = interpolate.interp1d(f,magnitude)
print(1-fun(1/8))

0.6461584154466922
```

The phase response

```
phase = 180*angle(h)/pi;

plt.figure()
plt.plot(f,np.unwrap(phase))
plt.xlabel('Frequency')
plt.ylabel('Phase in degrees')
plt.show()

fun = interpolate.interp1d(f,180*np.angle(h)/np.pi)
print(fun(1/8)*8/360)

2.9442611936390373
```

must again be corrected for causal indexing. Since the sampling interval was one and the filter length is five, we must add $(5-1)/2 = 2$ to the phase shift of 2.9443. This suggests a corrected phase shift of 4.9442, which is exactly the delay seen in the plot:

```
plot(t,x11,t,y11), axis([30 40 -2 2])
```

The next section provides an introduction to the design of filters with a desired frequency response. These filters can be used to amplify or suppress different components of arbitrary signals.

6.9   Filter Design

We now aim to design filters with a specific frequency response. We first generate a synthetic signal x12 with two periods, 50 and 5. The power spectrum of the signal shows the expected peaks at frequencies of 0.02 and 0.20:

```
%reset -f

import numpy as np
import scipy.signal as signal
import matplotlib.pyplot as plt

t = np.linspace(1,1000,1000)
x12 = 2*np.sin(2*np.pi*t/50) + \
    0.5*np.sin(2*np.pi*t/5)

plt.figure()
plt.plot(t,x12)
plt.axis(np.array([0,100,-4,4]))
plt.show()

f,Pxx = signal.periodogram(x12,fs=1,nfft=128)

plt.figure()
plt.plot(f,Pxx)
plt.grid
plt.xlabel('Frequency')
plt.ylabel('Power')
plt.show()
```

The Butterworth filter design technique is widely used to create filters of any order with a lowpass, highpass, bandpass, and bandstop configuration (Fig. 6.5). In our example, we would like to design a five-order lowpass filter with a cutoff frequency of 0.10. The inputs of butter() are the order of the filter, the cutoff frequency, the filter type btype, and the sampling frequency fs. The outputs from butter() are the filter weights a12 and b12:

```
b12,a12 = signal.butter(5,0.1,btype='lowpass',fs=1)
```

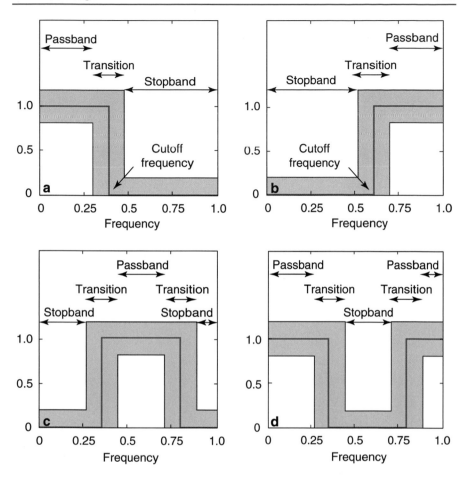

Fig. 6.5 Frequency responses for the fundamental types of frequency-selective filters. **a** A low-pass filter for suppressing the high-frequency component of a signal. In the earth sciences, such filters are often used to suppress high-frequency noise in a low-frequency signal. **b** Highpass filters for removing all low frequencies and trends in natural data. **c–d** Bandpass and bandstop filters for extracting or suppressing a certain frequency band. The solid line in all graphics depicts the ideal frequency response of a frequency-selective filter, while the blue band shows the tolerance of a low-order design of such a filter. In practice, the frequency response lies within the blue band. Modified from von Grünigen (2014).

The frequency characteristics of the filter show a relatively smooth transition from the passband to the stopband, but the advantage of the filter is its low order:

```
f,h = signal.freqz(b12,a12,worN=1024,fs=1)

plt.figure()
plt.plot(f,np.abs(h))
```

```
plt.xlabel('Frequency')
plt.ylabel('Magnitude')
plt.show()
```

We can again apply the filter to the signal by using `lfilter()`. Frequency-selective filters such as lowpass, highpass, bandpass, and bandstop filters are designed to suppress certain frequency bands, but phase shifts should be avoided. The function `filtfilt()` provides zero-phase-shift forward and reverse digital filtering. After filtering in the forward direction, the filtered sequence is reversed and runs back through the filter. The magnitude of the signal is not affected by this operation since it is either 0% or 100% of the initial amplitude depending on the frequency. Any phase shifts introduced by the filter are canceled out by the forward and reverse application of the same filter. This function also helps in overcoming the problems with the causal indexing of filters in Python by eliminating the phase differences between the causal and non-causal versions of the same filter. Filtering and then plotting the results clearly illustrates the effects of the filter:

```
xf12 = signal.filtfilt(b12,a12,x12)

plt.figure()
plt.plot(t,x12,color='b')
plt.plot(t,xf12,color='r')
plt.axis(np.array([0,100,-4,4]))
plt.show()
```

We might now wish to design a new filter with a more rapid transition from passband to stopband. Such a filter requires a higher order, that is, it needs a larger number of filter weights a12 and b12. We now create a 15-order Butterworth filter as an alternative to the filter above:

```
b13,a13 = signal.butter(15,0.1,btype='lowpass',fs=1)

f,h = signal.freqz(b13,a13,worN=1024,fs=1)

plt.figure()
plt.plot(f,np.abs(h))
plt.xlabel('Frequency')
plt.ylabel('Magnitude')
plt.show()
```

The frequency response is clearly improved. The entire passband is relatively flat at a value of 1.0, whereas the stopband is approximately zero everywhere. We next modify our input signal by introducing a third period of 5. This signal is then used to illustrate the operation of a Butterworth bandstop filter:

```
t = np.linspace(1,1000,1000)
x14 = 2*np.sin(2*np.pi*t/50) + \
        np.sin(2*np.pi*t/10) + \
     0.5*np.sin(2*np.pi*t/5)


f,Pxx = signal.periodogram(x14,fs=1,nfft=128)

plt.figure()
plt.plot(f,Pxx)
plt.show()
```

The new Butterworth filter is a bandstop filter. The stopband of the filter lies between the frequencies 0.05 and 0.15. It can therefore be used to suppress the period of 10, which corresponds to a frequency of 0.1:

```
b14,a14 = signal.butter(5,(0.05,0.15),
     btype='bandstop',fs=1)
xf14 = signal.filtfilt(b14,a14,x12)

f,Pxx = signal.periodogram(x14,fs=1,nfft=128)

plt.figure()
plt.plot(f,Pxx)
plt.show()

plt.figure()
plt.plot(t,x14,color='b')
plt.plot(t,xf14,color='r')
plt.axis(np.array([0,100,-4,4]))
plt.show()
```

The plots show the effect of this filter. The frequency band between 0.05 and 0.15—and therefore also the frequency of 0.1—has been successfully removed from the signal.

6.10   Adaptive Filtering

The fixed filters used in the previous sections lead to the basic assumption that the signal degradation is known and does not change with time. However, *a priori* knowledge of the signal and noise statistical characteristics is not usually available in most applications. In addition, both the noise level and the variance of the genuine signal can be highly nonstationary with respect to time, such as with stable isotope records during a glacial–interglacial transition. Fixed filters therefore cannot be used in a nonstationary environment in which there is no knowledge of the signal-to-noise ratio.

Adaptive filters, which are widely used in the telecommunication industry, could help in overcoming these problems. An adaptive filter is an inverse modeling process that iteratively adjusts its own coefficients automatically without requiring any *a priori* knowledge of the signal or the noise. Operating an adaptive filter involves 1) a filtering process whose purpose is to produce an output in response to a sequence of data and 2) an adaptive process that provides a mechanism for adaptively controlling the filter weights (Haykin 2003).

In most practical applications, the adaptive process is oriented toward minimizing an estimation error. The estimation error e at an instant i is defined as the difference between the desired response d_i and the actual filter output y_i, which is the filtered version of a signal x_i, as shown by

$$e_i = d_i - y_i$$

where $i = 1, 2, \ldots, n$ and n is the length of the input data array. In the case of a nonrecursive filter that is characterized by an array of filter weights W with f elements, the filter output y_i is given by the inner product of the transposed array W and the input array X_i:

$$y_i = W^T \cdot X_i$$

The choice of desired response d that is used in the adaptive process depends on the application. Traditionally, d is a combination signal that is comprised of a signal s and random noise n_0. The signal x contains noise n_1 that is uncorrelated with the signal s but correlated in some unknown way with the noise n_0. In noise-canceling systems, the practical objective is to produce a system output y that is a best fit in the least-squares sense to the desired response d.

Different approaches have been developed for solving this multivariate minimum error optimization problem (e.g., Widrow and Hoff 1960; Widrow et al. 1975; Haykin 1991). The selection of one algorithm over another is influenced by various factors, including the rate of convergence (i.e., the number of adaptive steps required for the algorithm to converge close enough to an optimum solution), the misadjustment (i.e., the measure of the amount by which the final value of the mean squared error deviates from the minimum squared error of an optimal filter; e.g., Wiener 1949; Kalman and Bucy 1961), and the tracking (i.e., the capability of the filter to work in a nonstationary environment and to track changing statistical characteristics of the input signal) (Haykin 1991).

The simplicity of the least-mean-squares (LMS) algorithm—which was originally developed by Widrow and Hoff (1960)—has made it the benchmark against which other adaptive filtering algorithms are tested. For applications in the earth sciences, we use a variant of this filter to extract the noise from two signals S and X, both of which contain the same signal s, but with uncorrelated noises n_1 and n_2 (Hattingh 1988). As an example, consider a simple duplicate set of measurements of the same material, such as two parallel stable isotope records from the same foraminifera species. We would expect two time series, each with n elements and that contain the same desired signal overlain by different, uncorrelated noise. The first record is used as the primary input S,

$$S = (s_1, s_2, \ldots, s_n)$$

and the second record is used as the reference input X:

$$X = (x_1, x_2, \ldots, x_n)$$

As demonstrated by Hattingh (1988), the desired noise-free signal can be extracted by filtering the reference input X using the primary input S as the desired response d. The minimum error optimization problem is solved by the least-mean-square norm. The mean squared error e_i^2 is a second-order function of the weights in the nonrecursive filter. The dependence of e_i^2 on the unknown weights W may be seen as a multidimensional paraboloid with a uniquely defined minimum point. The weights that correspond to the minimum point on this error surface define the optimal Wiener solution (Wiener 1949). The value that is computed for the weight array W using the LMS algorithm represents an estimator whose expected value approaches the Wiener solution as the number of iterations approaches infinity (Haykin 1991). Gradient methods are used to reach the minimum point on the error surface. To simplify the optimization problem, Widrow and Hoff (1960) developed an approximation for the required gradient function that can be computed directly from the data. This leads to a simple relationship for updating the filter weight array W:

$$W_{i+1} = W_i + 2 \cdot u \cdot e_i \cdot X_i$$

The new set of filter weights W_{i+1} is based on the previous set W_i plus a term that is the product of a bounded step size (or convergence factor) u, a function of the input state X_i, and a function of the error e_i. In other words, the error e_i calculated from the previous step is fed back into the system in order to update filter coefficients for the next step (Fig. 6.6). The fixed convergence factor u regulates the speed and stability of adaption. A low value of u ensures a higher level of accuracy, but more data are needed to enable the filter to reach the optimum solution. In the modified version of the LMS algorithm by Hattingh (1988), this problem is overcome by processing the data in a loop such that the filter can have another chance to improve its own coefficients and adapt to changes in the data.

In the following example, which introduces canc(), each of these loop runs is called an iteration, and many loop runs are required if optimal results are to be achieved. This algorithm extracts the noise-free signal from two arrays x and s, which contain the correlated signals and uncorrelated noise. As an example, we generate two signals yn1 and yn2, which contain the same sine wave but different Gaussian noise:

```
%reset -f

import numpy as np
from numpy import linalg as lanalg
from canc import canc
from matplotlib import cm
import scipy.signal as signal
```

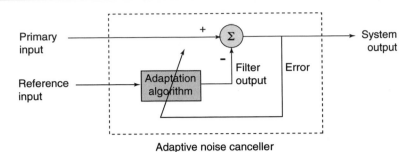

Adaptive noise canceller

Fig. 6.6 Schematic of an adaptive filter. Each iteration involves a new estimate of the filter weights W_{i+1} based on the previous set of filter weights W_i plus a term that is the product of a bounded step size u, a function of the filter input X_i, and a function of the error e_i. In other words, the error e_i calculated in the previous step is fed back into the system in order to update filter coefficients for the next step (modified from Trauth 1998).

```
import matplotlib.pyplot as plt
from numpy.random import default_rng

x = np.linspace(0,100,1001)
y = np.sin(x)
rng = default_rng(0)
yn1 = y + 0.5*rng.standard_normal(y.shape)
yn2 = y + 0.5*rng.standard_normal(y.shape)

plt.figure()
plt.plot(x,yn1,x,yn2)
plt.show()
```

The algorithm `canc()` formats both signals, feeds them into the filter loop, corrects the signals for phase shifts, and formats the signals for the output. The required inputs are the signals x and s, the step size u, the filter length 1, and the number of iterations `iter`. In our example, the two noisy signals are yn1 and yn2. We make an arbitrary choice of a filter with 1=5 filter weights. A value of u in the range of $0 < u < 1/\lambda_{max}$ (where λ_{max} is the largest eigenvalue of the autocorrelation matrix for the reference input) leads to reasonable results (Haykin 1991) (Fig. 6.7). The value of u is computed using

```
k = np.kron(yn1,np.transpose(yn1[None]))
w,v = lanalg.eig(k)
u = 1/np.max(w)
print(u)
```

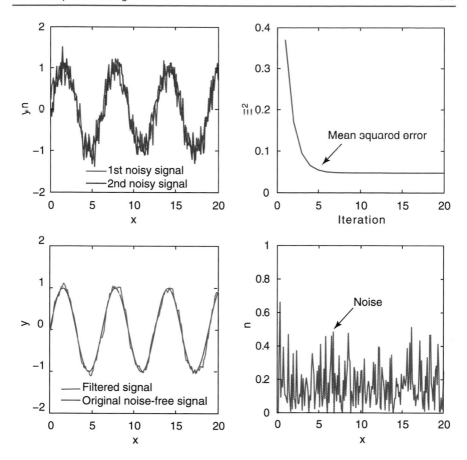

Fig. 6.7 Output of the adaptive filter. **a** The duplicate records corrupted by uncorrelated noise are fed into the adaptive filter with 5 weights with a convergence factor of 0.0019. After 20 iterations, the filter yields **b** the learning curve, **c** the noise-free record, and **d** the noise extracted from the duplicate records.

where kron() returns the Kronecker tensor product of yn1 and yn1' (which is a matrix formed by taking all possible products between the elements of yn1 and yn1') and eig() returns the eigenvector of k. This yields

```
0.0013492697467229055
```

We now run the adaptive filter canc() for 20 iterations and use the value of u from above:

```
z,e,mer,w = canc(yn1,yn2,0.0013,5,20)
```

The output variables from `canc()` are the filtered primary signal z, the extracted noise e, the mean squared error mer for the number of iterations it performed with step size u, and the filter weights w for each data point in yn1 and yn2. The plot of the mean squared error mer

```
plt.figure()
plt.plot(mer)
plt.show()
```

illustrates the performance of the adaptive filter, although the chosen step size (u=0.0013) clearly results in a relatively rapid convergence. In most examples, a smaller step size decreases the rate of convergence but improves the quality of the final result. We therefore reduce u and run the filter again with further iterations:

```
z,e,mer,w = canc(yn1,yn2,0.0001,5,20)
```

The plot of the mean squared error mer against the iterations

```
plt.figure()
plt.plot(mer)
plt.show()
```

now converges after about six iterations. In practice, the user should vary the input arguments u and 1 in order to obtain the optimum result. We can now compare the filter output with the original noise-free signal:

```
plt.figure()
plt.plot(x,y,'b',x,z,'r')
plt.show()
```

This plot demonstrates that the noise level of the signal has been reduced dramatically by the filter. Next, the plot

```
plt.figure()
plt.plot(x,e,'r')
plt.show()
```

shows the noise extracted from the signal. Using the last output from `canc()`, we can calculate and display the mean filter weights of the final iteration from w:

```
wmean = np.mean(w,axis=0)

plt.figure()
plt.plot(wmean)
plt.show()
```

The frequency characteristic of the filter provides a more illustrative representation of the effect of the filter,

```
f,h = signal.freqz(wmean,1,worN=1024,fs=1)

plt.figure()
plt.plot(f,np.abs(h))
plt.show()
```

which shows that the filter is a lowpass filter with a relatively smooth transition band. This means that it does not have the quality of a recursive filter that was designed, for example, using the Butterworth approach. However, the filter weights are calculated in an optimization process rather than being chosen arbitrarily.

The strength of an adaptive filter lies in its ability to filter a time series with a variable signal-to-noise ratio along the time axis. Since the filter weight array W is updated for each individual data point, these filters are even used in real-time applications, such as in telecommunication systems. We examine this behavior through an example in which the signal-to-noise ratio in the middle of the time series (x=500) is reduced from about 10% to zero:

```
x = np.linspace(0,100,1001)
y = np.sin(x)
rng = default_rng(0)
yn1 = y + 0.5*rng.standard_normal(y.shape)
yn2 = y + 0.5*rng.standard_normal(y.shape)
yn1[501:] = y[501:]
yn2[501:] = y[501:]

plt.figure()
plt.plot(x,yn1,x,yn2)
plt.show()
```

The value of u is again computed by

```
k = np.kron(yn1,np.transpose(yn1[None]))
w,v = lanalg.eig(k)
u = 1 / np.max(w)
print(u)
```

which yields

```
0.0016086744408774288
```

We now run the adaptive filter `canc()` for 20 iterations and use the value of u from above:

```
z,e,mer,w = canc(yn1,yn2,0.0013,5,20)
```

The plot of the mean squared error `mer` versus the number of performed iterations it with step size u

```
plt.figure()
plt.plot(mer)
plt.show()
```

illustrates the performance of the adaptive filter, although the chosen step size u=0.0016 clearly results in a relatively rapid convergence. Again, we can now compare the filter output with the original noise-free signal:

```
plot(x,y,'b',x,z,'r')
```

This plot reveals that the filter output y is almost the same as the noise-free signal x. The plot

```
plot(x,e,'r')
```

shows the noise extracted from the signal. Here, we can observe some signal components that have been removed by the filter in error from within the noise-free segment of the time series beyond the change point in the signal-to-noise ratio. Using the last output of `canc()`, we can calculate and display the filter weights w of the final iteration,

```
X = np.linspace(1,5,5)
Y = np.linspace(2,999,997)
X, Y = np.meshgrid(X, Y)
W = w[2:999,0:]

plt.figure()
fig,ax = plt.subplots(subplot_kw={"projection":"3d"})
surf = ax.plot_surface(X,Y,W,cmap=cm.coolwarm)
ax.view_init(60,100-90)
plt.show()
```

which nicely shows the adaptation of the filter weights before and after the change in the signal-to-noise ratio. We plot only the middle part of w because the edges 1:(1-1)/2 and end-(1-1)/2:end are all zero due to the filter length of 1. We can also use this example to demonstrate the effect that the values of u and 1 have on the performance of the adaptive filter. In theory, a smaller u leads to more accurate results, but the rapid adaptation to a changing signal-to-noise ratio does not work well. Larger values of 1 also yield better results, but the number of data points that are lost through the filtering process increases by $(1-1)/2$.

The application of this algorithm has been demonstrated with duplicate oxygen-isotope records from ocean sediments (Trauth 1998). This work by Trauth (1998) illustrates the use not only of the modified LMS algorithm, but also of another type of adaptive filter—the recursive least-squares (RLS) algorithm—in various different environments (Haykin 1991; Trauth 1998).

Recommended Reading

von Grünigen DH (2014) Digitale Signalverarbeitung mit einer Einführung in die kontinuierlichen Signale und Systeme, Fünfte, neu, bearbeitete. Fachbuchverlag Leipzig im Carl Hanser Verlag, Leipzig

Hattingh M (1988) A new data adaptive filtering program to remove noise from geophysical time- or space series data. Comput Geosci 14(4):467–480

Kalman R, Bucy R (1961) New results in linear filtering and prediction theory. ASME Tans Ser D Jour Basic Eng 83:95–107

Trauth MH (1998) Noise removal from duplicate paleoceanographic time-series: the use of adaptive filtering techniques. Math Geol 30(5):557–574

Trauth MH (2013) TURBO2: a MATLAB simulation to study the effects of bioturbation on paleoceanographic time series. Comput Geosci 61:1–10

Widrow B, Hoff M Jr (1960) Adaptive switching circuits. IRE WESCON Conv Rev 4:96–104

Widrow B, Glover JR, McCool JM, Kaunitz J, Williams CS, Hearn RH, Zeidler JR, Dong E, Goodlin RC (1975) Adaptive noise cancelling: principles and applications. Proc IEEE 63(12):1692–1716

Wiener N (1949) Extrapolation, interpolation and smoothing of stationary time series, with engineering applications. MIT Press, Cambridge, Mass (reprint of an article originally issued as a classified National Defense Research Report, February, 1942)

Further Reading

Alexander ST (1986) Adaptive signal processing: theory and applications. Springer, Berlin Heidelberg New York

Buttkus B (2000) Spectral analysis and filter theory in applied geophysics. Springer, Berlin Heidelberg New York

Cowan CFN, Grant PM (1985) Adaptive filters. Prentice Hall, Englewood Cliffs, New Jersey

Haykin S (2003) Adaptive filter theory. Prentice Hall, Englewood Cliffs, New Jersey

Sibul LH (1987) Adaptive signal processing. IEEE Press

Weeks M (2007) Digital signal processing using MATLAB and wavelets. Infinity Science Press, Jones and Bartlett Publishers, Boston Toronto London Singapore

Spatial Data

7

7.1 Introduction

Most data in the earth sciences are spatially distributed either as *vector data* (points, lines, polygons; e.g., river networks, coastlines) or as *raster data* (square grids, triangular grids; e.g., gridded topography). Vector data are generated by digitizing map objects, such as drainage networks or outlines of lithologic units. Raster data can be obtained directly from a satellite sensor output, but gridded data can also be interpolated from irregularly distributed field sampling (*gridding*).

Section 7.2 introduces vector data using coastline data as an example. The acquisition and handling of raster data are then illustrated using digital topographic data in Sects. 7.3 to 7.5. The availability and use of digital elevation data has increased considerably since the early 1990s. With a resolution of 5 arc min (about 9 km), ETOPO5 was one of the first data sets available for topography and bathymetry. In October 2001, it was replaced by ETOPO2, which has a resolution of 2 arc minutes (about 4 km), and in March 2009, ETOPO1 – which has a resolution of 1 arc min (about 2 km) – became available. There is also a data set for topography called GTOPO30, which was completed in 1996 and has a horizontal grid spacing of 30 arc seconds (about 1 km). More recently, the 30 and 90 m resolution data from the Shuttle Radar Topography Mission (SRTM) have replaced the older data sets in most scientific studies.

Supplementary Information The online version contains supplementary material available at https://doi.org/10.1007/978-3-031-07719-7_7.

Section 7.6 can be found in the MATLAB version of this book but could not be translated to Python. However, they are listed here with corresponding section numbers in order to create identical chapter numbering (and thereby also identical figure numbering and computer scripts) between the two versions of the book. The reader can thus use both books side by side to trace the translation of the scripts from MATLAB to Python (and back).

© The Author(s), under exclusive license to Springer Nature Switzerland AG 2022
M. H. Trauth, *Python Recipes for Earth Sciences*, Springer
Textbooks in Earth Sciences, Geography and Environment,
https://doi.org/10.1007/978-3-031-07719-7_7

The second part of this chapter deals with the computation of continuous surfaces from unevenly spaced data and with the statistics of spatial data in Sects. 7.7 and 7.8. In the earth sciences, most data are collected in an irregular pattern. Access to rock samples is often restricted to natural outcrops (e.g., shoreline cliffs and the walls of a gorge) or to anthropogenic outcrops (e.g., road cuttings, quarries, or drill cores). The sections on interpolating such unevenly spaced data illustrate the use of the most important gridding routines and outline potential pitfalls when using these methods. Sections 7.9 to 7.11 introduce various methods for statistically analyzing spatial data, including applying statistical tests to point distributions in Sect. 7.9, analyzing digital elevation models in Sect. 7.10, and providing an overview of geostatistics and kriging in Sect. 7.11. This chapter requires the *NumPy* (https://numpy.org), *Matplotlib* (https://matplotlib.org), *SciPy* (https://scipy.org), *Rasterio* (https://rasterio.readthedocs.io), *GDAL* (https://gdal.org/api/python.html), *scikit-image* (https://scikit-image.org), *PyGMT* (https://www.pygmt.org) (which requires *xarray* for work with labelled multi-dimensional arrays; https://docs.xarray.dev), and *Cartopy* (https://scitools.org.uk/cartopy) packages.

7.2 The Global Geography Database GSHHG

The *Global Self-consistent, Hierarchical, High-resolution Geography* (GSHHG) database is an amalgamation of two public domain databases that was created by Paul Wessel (SOEST, University of Hawaii, Honolulu, HI) and Walter Smith (NOAA Laboratory for Satellite Altimetry, Silver Spring, MD) (Wessel and Smith 1996). The GSHHG database consists of the older GSHHS shoreline database (Soluri and Woodson 1990; Wessel and Smith 1996), with the poor-quality Antarctica data having been replaced by more accurate data from Bohlander and Scambos (2007) and with rivers and borders having been taken from the CIA World Data Bank II (WDBII) (Gorny 1977). GSHHG data can be downloaded from the SOEST server

```
http://www.soest.hawaii.edu/pwessel/gshhg/index.html
```

in three different formats (as Generic Mapping Tools (GMT) files, as ESRI shapefiles, or as native binary files) and in a wide range of spatial resolutions. As an example, we can import the longitude/latitude coordinates from an ASCII file extracted from the data set using

```
%reset -f

import numpy as np
import cartopy.crs as ccrs
import matplotlib.pyplot as plt
import matplotlib.ticker as mticker
from cartopy.mpl.ticker import (LongitudeFormatter,
                LatitudeFormatter,
                LongitudeLocator,
```

```
        LatitudeLocator)

 data = np.loadtxt('coastline_data.txt')
```

There are different ways of displaying the data. As an example, we can use the plot() command from *Matplotlib* to create a line plot after having first defined the figure() and axes() properties, as follows:

```
 plt.figure()
 plt.axes(xlabel='Longitude',
     ylabel='Latitude',
     aspect='equal')
 plt.plot(data[0:,0],data[0:,1],
     color='k',
     linewidth=0.3)
 plt.show()
```

More advanced plotting functions are contained in the *CartoPy* package, which allows alternative versions of this plot to be generated (Fig. 7.1):

```
 ax = plt.axes(projection=ccrs.PlateCarree())
 plt.plot(data[0:,0],data[0:,1],
     color='k',
     linewidth=0.3)
```

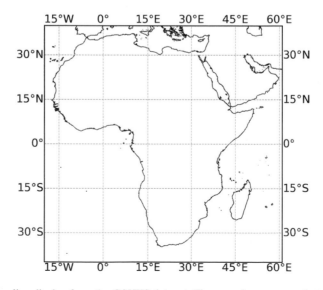

Fig. 7.1 Shoreline display from the GSHHS data set. The map shows an area between latitudes 0° and 15° north and between longitudes 40° and 50° east. This simple map was made using plot() with equal axis aspect radios (data from Wessel and Smith 1996).

```
ax.set_extent([-20,61,-40,40],
    ccrs.PlateCarree())
gl = ax.gridlines(crs=ccrs.PlateCarree(),
    draw_labels=True,
    linewidth=0.5,
    linestyle='--')
gl.xlocator = mticker.FixedLocator([-15,0,15,30,45,60])
gl.ylocator = mticker.FixedLocator([-30,-15,0,15,30])
plt.show()
```

Here, we use *Plate Carrée* projection, which is an equidistant cylindrical projection that maps meridians to evenly spaced vertical straight lines and that maps latitudes to evenly spaced horizontal straight lines. We use the function Fix-edLocator() to define the location of the ticks. The GSHHS data set is actually included in *CartoPy* using coastlines():

```
ax = plt.axes(projection=ccrs.PlateCarree())
ax.coastlines(resolution='10m',
    linewidth=0.5)
ax.set_extent([-20,61,-40,40],
    ccrs.PlateCarree())
gl = ax.gridlines(crs=ccrs.PlateCarree(),
    draw_labels=True,
    linewidth=0.5)
gl.xlocator = mticker.FixedLocator([-15,0,15,30,45,60])
gl.ylocator = mticker.FixedLocator([-30,-15,0,15,30])
plt.show()
```

We see that the data set in *coastline_data.txt* contains only the outline of the African continent because the data in the original GSHHS data set are sorted by continent. The data set provided in *CartoPy* also shows the coastlines of the area surrounding Africa.

7.3   The 1 Arc-Minute Gridded Global Relief Data ETOPO1

The *1 arc-minute global relief model of Earth's surface* (ETOPO1) is a global database of topographic and bathymetric data on a regular 1 arc-minute grid (about 2 km) (Amante and Eakins 2009). The ETOPO1 database supersedes the older ETOPO2 and ETOPO5 global relief grids, which are nevertheless still available. ETOPO1 is a compilation of data from a variety of sources. It can be downloaded from the NOAA National Centers for Environmental Information (NCEI) website:

```
https://www.ngdc.noaa.gov/mgg/global/global.html
```

We can download either the whole-world grids or custom grids for ice surfaces or bedrock. We create and extract a custom grid online from ETOPO1 using NCEI's WCS client:

```
http://maps.ngdc.noaa.gov/viewers/wcs-client/
```

We first choose the *ETOPO1* (bedrock) layer. We then select *Select with Coordinates* by clicking the *xy* icon in the upper-left corner of the map, and we specify an area of interest, such as a latitude of *North: 20* (20°N) and *South: −20* (20°S) and a longitude between *East: 30* (30°E) and *East: 60* (60°E), which covers the eastern African coast. Negative latitude and longitude values are assigned to the Southern Hemisphere and to the Western Hemisphere, respectively. Clicking *OK* marks the area of interest as a transparent yellow rectangle on the map. Next, we choose *Arc-GIS ASCII Grid* as the *Output Format* and download the file *etopo1_bedrock.asc*, which contains the following data:

```
ncols     1801
nrows     2401
xllcorner      29.991666666667
yllcorner      -20.008333333333
cellsize       0.016666666667
294 299 293 288 285 282 ...
237 241 245 266 264 274 ...
259 263 267 262 263 266 ...
310 306 300 294 296 291 ...
348 346 352 356 353 353 ...
381 383 381 381 382 381 ...
(cont'd)
```

The headers document the size of the data array (e.g., 1,801 columns and 2401 rows in our example), the coordinates of the lower-left corner (e.g., approximately $x = 20$ and $y = -20$), and the cell size (e.g., $\sim 0.0167 = 1/60$ of a degree latitude and longitude, or 1 arc minute). Older versions of the ETOPO1 data set as well as other similar data sets may also contain the −32,768 flag for data voids. We comment the header out by typing # at the beginning of the first five lines:

```
#ncols     1801
#nrows     2401
#xllcorner      29.991666666667
#yllcorner      -20.008333333333
#cellsize       0.016666666667
294 299 293 288 285 282 ...
237 241 245 266 264 274 ...
259 263 267 262 263 266 ...
310 306 300 294 296 291 ...
348 346 352 356 353 353 ...
381 383 381 381 382 381 ...
(cont'd)
```

We save the file as *etopo1_data.txt*, and we then load the data into the workspace:

```
%reset -f

import numpy as np
import pygmt as pygmt
import xarray as xr
from matplotlib import cm
import matplotlib.pyplot as plt
from matplotlib.colors import LightSource

ETOPO1 = np.loadtxt('etopo1_data_python.txt')
```

We next flip the array up and down using `flipud()` in order to change the indexing of the data array according to Python conventions:

```
ETOPO1 = np.flipud(ETOPO1)
```

Finally, we check that the data are now correctly stored in the workspace by printing the minimum and maximum elevations for the area:

```
print(np.max(ETOPO1))
print(np.min(ETOPO1))
```

In this example, the maximum elevation for the area is 5677 m, and the minimum elevation is −5859 m. The reference level is sea level (i.e., 0 m). We now define a coordinate system with `meshgrid()` using the information that the lower-left corner is at latitude 20°S and longitude 30°E. The resolution is 1 arc min, which corresponds to 1/60 of a degree:

```
lon = np.linspace(30,60,1801)
lat = np.linspace(-20,20,2401)
LON,LAT = np.meshgrid(lon,lat)
```

We now generate a colored surface from the elevation data using `pcolormesh()`:

```
plt.figure()
ax1 = plt.axes(aspect='equal')
plt.pcolormesh(LON,LAT,ETOPO1,
    cmap='jet',
    shading='gouraud')
plt.colorbar()
plt.show()
```

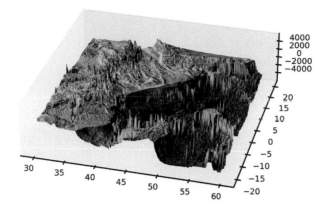

Fig. 7.2 Surface plot of the ETOPO1 elevation data. The plot uses shading, which is very popular in 3D computer graphics and creates a combined diffuse and specular reflection on surfaces (data from Amante and Eakins 2009).

This script opens a new figure window and generates a colored surface using the colormap jet and generates Gouraud shading using gouraud. The surface is highlighted by a set of color shades and is displayed in an overhead view. A spectacular way of plotting digital topography is as a surface plot with angled lighting using LightSource() from the *colors* module of the *Matplotlib* package (Fig. 7.2):

```python
plt.figure()
fig,ax2 = plt.subplots(subplot_kw={"projection":"3d"})
ve = 500
ax2.set_box_aspect((1,1,ve/1850))
ls = LightSource(270,45)
rgb = ls.shade(ETOPO1,
    cmap=cm.jet,
    vert_exag=ve/1850,
    blend_mode='hsv')
ax2.plot_surface(LON,LAT,ETOPO1,
    rstride=1,
    cstride=1,
    linewidth=0,
    facecolors=rgb,
    antialiased=False,
    shade=True)
ax2.tick_params(axis='both',
    labelsize=6)
ax2.view_init(30,15-90)
ax2.grid(False)
fig.savefig('etopo1_python.png',dpi=300)
plt.show()
```

The function LightSource() is used to create a light source from a specific azimuth and elevation. We use vert_exag for vertical exaggeration in order to emphasize vertical features of the surface plot, and we use blend_mode = 'hsv' to combine the colormapped data values with the illumination intensity. The actual surface plot is then created with surface() using the rgb object created by shade() as facecolors. It is important to note that rstride and cstride define the number of rows and columns of ETOPO1 that will be used to render the graph. If we choose 1 (i.e., full resolution), rendering takes very long, and we should therefore use a larger value (e.g., 5, as above) to resample the data set vertically and horizontally by this factor in order to reduce rendering time. We export the plot in a PNG format with a resolution of 300 dpi to the file *etopo1_python.png*.

We can also use makecpt(), Figure(), and grdview() from the *PyGMT* package, which makes use of the *xarray* package:

```
ds = xr.DataArray(data=ETOPO1,
    dims=['lat','lon'],
    coords={'lat':lat,'lon':lon})
frame = ['xa1f0.25','ya1f0.25','wSEnZ']
pygmt.makecpt(
    cmap='geo',
    series=f'-6000/4000/10',
    continuous=True)

fig = pygmt.Figure()
fig.grdview(grid=ds,
    perspective=[150,30],
    frame=frame,
    projection='M15c',
    zsize='4c',
    surftype='i',
    plane='-6000+gazure',
    shading=0)
fig.basemap(
    perspective=True,
    rose='jTL+w3c+l+o-2c/-1c')
fig.colorbar(perspective=True,
    position='JMR+o2c/0c+w10c',
    frame=['a2000'])
fig.show()
```

These script first uses makecpt() to choose the geo colormap from *PyGMT* that is appropriate for elevation data since this function relates land and sea colors to hypsometry and bathymetry, respectively. The function then uses grdview() to display the actual surface plot and to define the perspective, the frame, and the projection. We add a rose that points to geographic north with `basemap()` and a colorbar using `colorbar()`.

7.4   The 30 Arc-Second Elevation Model GTOPO30

The *Global 30 Arc-Second Elevation Data* (GTOPO30) is a 30 arc-second (approximately 1 km) global digital elevation data set that contains only elevation data and no bathymetry. The data set was developed by the Earth Resources Observation System Data Center and is available from the USGS EarthExplorer website at

```
https://earthexplorer.usgs.gov
```

The EarthExplorer website can be used to search for, preview, and download satellite images, aerial photographs, and cartographic products through the U.S. Geological Survey (USGS). The GTOPO30 data set is located in the *Digital Elevation* category. On this website, we first select the desired map section in the *Search Criteria* either by entering the coordinates of the four corners of the map or by zooming into the area of interest and selecting *Use Map*. As an example, we enter the coordinates of the Suguta Valley in the northern Kenya Rift (2°8'37.58"N 36°33'47.06"E). We then choose *GTOPO30* from the *Digital Elevation* collection as the *Data Set* and click *Results*, which produces a list of records together with a toolbar for previewing and downloading data. In our example, we find a single data set *GT30E020N40*, which was acquired on 1 December 1996 and that is centered at 15°N 40°E. We need to register with the USGS website, log in, and then download the data set as a 55.0 MB GeoTIFF and a 24.9 MB DEM, the latter of which is provided as a 26.1 MB compressed zip file. After decompressing the file *gt30e020n40_dem.zip*, we obtain eight files that contain the raw data and header files in various formats. The file also provides a GIF image of a shaded relief display of the data.

Importing the GTOPO30 data into the workspace is simple. We rename the files by removing *gt30* from the file name in the unzipped collection and use

```
%reset -f
```

```
import numpy as np
from matplotlib import cm
from matplotlib.colors import LightSource
import matplotlib.pyplot as plt
from matplotlib.colors import ListedColormap
```

```
with open('e020n40.dem') as fid:
    rectype = np.dtype('>i2')
    GTOPO30 = np.fromfile(fid, dtype=rectype)
GTOPO30 = np.reshape(GTOPO30,(6000,4800))
GTOPO30 = np.flipud(GTOPO30)
GTOPO30[GTOPO30<0] = 0
```

to import the data. This script reads the data from the tile *e020n40* (without a file extension) at full resolution (scale factor = 1) in the array GTOPO30, which has dimensions of 1,200-by-1,200 cells. The coordinate system is defined using the *lon/lat* limits listed above. The resolution is 30 arc seconds, which corresponds to 1/120 of a degree:

```
lon = np.linspace(20,60,4800)
lat = np.linspace(-10,40,6000)
LON,LAT = np.meshgrid(lon,lat)
```

We need to reduce the upper limits by 1/120 in order to obtain an array of similar dimensions to the GTOPO30 array. A grayscale image can be generated from the elevation data using pcolormesh(). The fourth power of the *colormap* gray_r is used to darken the map, and the colormap is then flipped vertically in order to obtain dark colors for high elevations and light colors for low elevations (instead of the other way around):

```
gray_r = cm.get_cmap('gray_r',256)
newcolors = gray_r(np.linspace(0,1,256))
newcolors = newcolors**4
newcmp = ListedColormap(newcolors)

plt.figure()
plt.axes(aspect='equal',
    xlim=(30,40),
    ylim=(-5,5))
plt.pcolormesh(LON,LAT,GTOPO30,
    cmap=newcmp,
    shading='auto')
plt.colorbar()
plt.show()
```

This script opens a new figure window and generates the gray surface, which is displayed in an overhead view. We again plot the digital topography as a surface plot with angled lighting using a code similar to the one used in Sect. 7.4:

```
plt.figure()
fig,ax2 = plt.subplots(subplot_kw={"projection":"3d"})
```

```
ax2.set_box_aspect((1,1,0.2))
ls = LightSource(270,45)
rgb = ls.shade(GTOPO30,
    cmap=cm.hsv,
    vert_exag=1,
    blend_mode='hsv')
ax2.plot_surface(LON[600:1800,1200:2400],
    LAT[600:1800,1200:2400],
    GTOPO30[600:1800,1200:2400],
    rstride=1,
    cstride=1,
    linewidth=0,
    facecolors=rgb,
    antialiased=False,
    shade=True)
ax2.tick_params(axis='both',labelsize=6)
ax2.grid(False)
plt.savefig('gtopo30_python.png',dpi=300)
plt.show()
```

As in Sect. 7.3, we can create a much more appealing graphic using the functions from the *PyGMT* package (Fig. 7.3):

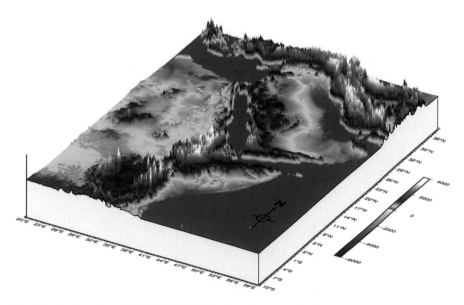

Fig. 7.3 Surface plot of the GTOPO30 elevation data created with functions from the *PyGMT* package using the colormap **geo**. The plot includes a colorbar and a rose that points to geographic north (data from the U.S. Geological Survey).

```
GTOPO30 = np.float64(GTOPO30)

ds = xr.DataArray(data=GTOPO30,
    dims=['lat','lon'],
    coords={'lat':lat,'lon':lon})
frame = ['xa1f0.25','ya1f0.25','wSEnZ']
pygmt.makecpt(
    cmap='geo',
    series=f'-6000/4000/10',
    continuous=True)

fig = pygmt.Figure()
fig.grdview(grid=ds,
    perspective=[150,30],
    frame=frame,
    projection='M15c',
    zsize='4c',
    surftype='i',
    plane='-6000+gazure',
    shading=0)
fig.basemap(
    perspective=True,
    rose='jTL+w3c+l+o10c/12c')
fig.colorbar(perspective=True,
    position='JMR+o2c/0c+w10c',
    frame=['a2000'])
fig.savefig('gtopo30_python.png',dpi=300)
fig.show()
```

This script first uses `makecpt()` to choose the geo colormap from *PyGMT* that is appropriate for elevation data. The script then uses `grdview()` to display the actual surface plot as well as to define the perspective, the frame, and the projection of the graphic. We add a compass rose that points to geographic north using `basemap()` and a colorbar using `colorbar()`. We again export the graph as a PNG file with a resolution of 300 dpi to the file *gtopo30_python.png*.

7.5   The Shuttle Radar Topography Mission SRTM

The *Shuttle Radar Topography Mission* (SRTM) was an 11-day mission of the Space Shuttle *Endeavour* in February 2000 that used a radar system to map the Earth's surface (Farr and Kobrick 2000, Farr et al. 2007). SRTM was an international project spearheaded by the National Geospatial-Intelligence Agency (NGA) and the National Aeronautics and Space Administration (NASA). Detailed information on the SRTM project (including a gallery of images and a user's forum) can be accessed on NASA's website:

```
http://www2.jpl.nasa.gov/srtm/
```

The data were processed at the Jet Propulsion Laboratory. They are distributed through the United States Geological Survey's (USGS) National Map Viewer and Download Platform:

```
https://www.usgs.gov/core-science-systems/ngp/tnm-delivery/
```

Alternatively, the raw data files can be downloaded from

```
http://dds.cr.usgs.gov/srtm/
```

This directory contains zipped files of SRTM DEMs from various areas of the world, which are processed by the SRTM global processor and sampled at resolutions of 1 arc second (SRTM-1, 30 m grid) and 3 arc seconds (SRTM-3, 90 m grid). As an example, we download the 1.7 MB file *s01e036.hgt.zip* from

```
http://dds.cr.usgs.gov/srtm/version2_1/SRTM3/Africa/
```

which contains SRTM-3 data of the Kenya Rift Valley in eastern Africa. All elevations are in meters and are referenced to the WGS84 EGM96 geoid, which is documented at

```
http://earth-info.nga.mil/GandG/wgs84/index.html
```

The name of this file refers to the longitude and latitude of the lower-left (southwest) pixel of the tile, that is, to a latitude one degree south and a longitude 36° east. SRTM-3 data contain 1201 lines and 1201 rows, with similar numbers of overlapping rows and columns. After downloading *S01E036.hgt.zip*, we first decompress this file and save *S01E036.hgt* in our working directory. The digital elevation model is provided as 16 bit signed integer data in a simple binary raster. The bit order is *big-endian* (Motorola's standard), which means that the most significant bit comes first. The data are imported into the workspace using

```
%reset -f
```

```
import numpy as np
import numpy.ma as ma
from scipy import signal
from matplotlib import cm
from matt Liplotlib.colors imporghtSource
```

```
import matplotlib.pyplot as plt
from matplotlib.colors import ListedColormap

with open('S01E036.hgt') as fid:
    rectype = np.dtype('>i2')
    SRTM = np.fromfile(fid,dtype=rectype)

SRTM = np.reshape(SRTM,(1201,1201))
SRTM = np.flipud(SRTM)
```

This script opens the file *s01e036.hgt* using open() and defines the file identifier
fid, which is then used to read the binaries from the file using the data type > i2
for big-endian 2 byte (or 16 bit) integers and to write them in the array SRTM using
fromfile(). The array first needs to be reshaped to (1201, 1201) using reshape()
and then flipped vertically using flipud(). We next change the data type to 64 bit
floating-point numbers using astype(), and the *–32,768* flag for data voids can be
replaced by the global mean elevation using mean():

```
SRTM = SRTM.astype(float)
SRTM[SRTM==-32768] = np.mean(SRTM)
```

Finally, we check whether the data are now correctly stored in the workspace by
printing the minimum and maximum elevations of the area:

```
np.max(SRTM)
np.min(SRTM)
```

In our example, the maximum elevation of the area is 3992 m above sea level, and
the minimum is 1504 m. A coordinate system can be defined using the information
that the lower-left corner is *s01e036*. The resolution is 3 arc seconds, which corresponds to 1/1200 of a degree:

```
lon = np.linspace(36,37,1201)
lat = np.linspace(-1,0,1201)
LON,LAT = np.meshgrid(lon,lat)
```

A shaded grayscale map can be generated from the elevation data using
pcolormesh(), which displays a shaded surface with simulated lighting:

```
gray_r = cm.get_cmap('gray_r',256)
newcolors = gray_r(np.linspace(0,1,256))
newcolors = newcolors**1
newcmp = ListedColormap(newcolors)
```

```
plt.figure()
plt.axes(aspect='equal')
plt.pcolormesh(LON,LAT,SRTM,
    cmap=newcmp,
    shading='auto')
plt.colorbar()
plt.show()
```

This script opens a new figure window and generates both a shaded relief map (using interpolated shading) and a gray colormap (which is displayed in an overhead view). Since SRTM data contain a large amount of noise, we first smooth the data via an arbitrary 9-by-9 pixel moving average filter using `convolve2d()`. Contrary to the discussion of running means in Chapter 6.8, the side effects of the running means are not important at this point. The new array is then stored in `SRTM_FILTERED`:

```
B = 1/81 * np.ones((9,9))
SRTM_FILTERED = signal.convolve2d(SRTM,B,mode='same')
```

The corresponding shaded relief map is generated using

```
plt.figure()
plt.axes(aspect='equal')
plt.pcolormesh(LON,LAT,SRTM_FILTERED,
    cmap=newcmp,
    shading='auto')
plt.colorbar()
plt.show()
```

Neighboring SRTM tiles overlap one line or column, which we must delete when merging SRTM data sets. We use the Baringo Basin in Kenya as an example, which is covered by two SRTM tiles. We first load the two SRTM data sets from the files *N00E036.hgt* and *N00E035.hgt*, as described previously:

```
with open('N00E035.hgt') as fid:
    rectype = np.dtype('>i2')
    SRTM1 = np.fromfile(fid,dtype=rectype)

with open('N00E036.hgt') as fid:
    rectype = np.dtype('>i2')
    SRTM2 = np.fromfile(fid,dtype=rectype)
```

We next reshape and flip the data sets up and down:

```
SRTM1 = np.reshape(SRTM1,(1201,1201))
SRTM1 = np.flipud(SRTM1)
SRTM2 = np.reshape(SRTM2,(1201,1201))
SRTM2 = np.flipud(SRTM2)
```

We eliminate the overlapping (identical) column in the first tile SRTM1 and concatenate it horizontally with the second tile SRTM2 using concatenate(). Then we replace the gaps with the global mean elevation:

```
SRTM = ma.concatenate([SRTM1[:,:1200],SRTM2],axis=1)

SRTM = SRTM.astype(float)
SRTM[SRTM==-32768] = np.mean(SRTM)
```

We next smooth the data using a running mean:

```
B = 1/25 * np.ones((5,5))
SRTM_FILTERED = signal.convolve2d(SRTM,B,mode='same')
```

We then define a coordinate grid and contour levels (see Sect. 7.7):

```
lon = np.linspace(35,37,2401)
lat = np.linspace(-1,0,1201)
LON,LAT = np.meshgrid(lon,lat)
v = np.array([700,800,900,1000,1100,1200,
    1300,1500,2500,3000])
```

The script below yields a great display of the data using a code similar to the one used in Sects. 7.3 and 7.4. Please note the use of set_box_aspect() to reduce the length of the z-axis:

```
plt.figure()
fig,ax2 = plt.subplots(subplot_kw={"projection":"3d"})
ax2.set_box_aspect((2,1,0.1))
ax2.set_xlim((35,37))
ax2.set_ylim((-1,0))
ax2.set_zlim((1500,3500))
ls = LightSource(270,45)
rgb = ls.shade(SRTM,
    cmap=cm.hsv,
    vert_exag=1,
    blend_mode='soft',
    vmin=0,
    vmax=np.max(SRTM_FILTERED))
ax2.plot_surface(LON,LAT,SRTM_FILTERED,
```

```
        rstride=1,
        cstride=1,
        linewidth=0,
        facecolors=rgb,
        antialiased=False,
        shade=True,
        vmin=1500,
        vmax-4000)
    ax2.tick_params(axis='both',
        labelsize=6)
    ax2.grid(False)
    plt.show()
```

We can also create and assign the colormap gist_earth by making it appropriate
for elevation data since it relates land and sea colors to hypsometry and bathymetry, respectively (Fig. 7.4):

```
plt.figure()
fig,ax2 = plt.subplots(subplot_kw={"projection":"3d"})
ax2.set_box_aspect((2,1,0.1))
ax2.set_xlim((35,37))
ax2.set_ylim((-1,0))
ax2.set_zlim((1500,3500))
ls = LightSource(270,45)
rgb = ls.shade(SRTM,
    cmap=cm.gist_earth,
    vert_exag=1,
```

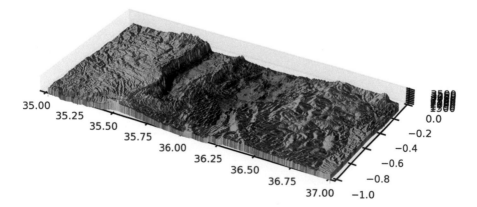

Fig. 7.4 Display from the filtered SRTM elevation data set. The plot uses shading, which is very
popular in 3D computer graphics and creates a combined diffuse and specular reflection on surfaces (data from Farr and Kobrick 2000, Farr et al. 2007).

```
          blend_mode='soft',
          vmin=0,
          vmax=np.max(SRTM_FILTERED))
  ax2.contour(LON,LAT,SRTM_FILTERED+1,
          levels=v,
          linewidths=0.3,
          colors='w',
          linestyles='solid')
  ax2.plot_surface(LON,LAT,SRTM_FILTERED,
          rstride=5,
          cstride=5,
          linewidth=0,
          facecolors=rgb,
          antialiased=False,
          shade=True,
          vmin=1500,
          vmax=4000)
  ax2.tick_params(axis='both',
          labelsize=6)
  ax2.grid(False)
  plt.savefig('srtm_python.png',dpi=300)
  plt.show()
```

This code also exports the graphics into a PNG format with a 300 dpi resolution:

```
  print -dpng -r300 srtmimage.png
```

The new file *srtmimage.png* has a size of 3.1 MB and can now be imported into another software package, such as Adobe Photoshop.

7.6   Exporting 3D Graphics to Create Interactive Documents

This section from the MATLAB-based book MATLAB Recipes for Earth Sciences (Trauth 2021) could not be translated to Python.

7.7   Gridding and Contouring

The previous data sets were all stored in evenly spaced two-dimensional arrays. However, most data in the earth sciences are obtained from irregular sampling patterns. The data are therefore unevenly spaced and need to be interpolated in order to allow a smooth and continuous surface to be computed from our measurements in the field. *Surface estimation* is typically carried out in two major steps

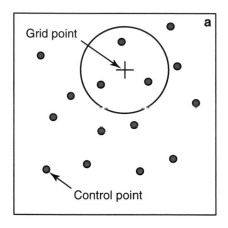 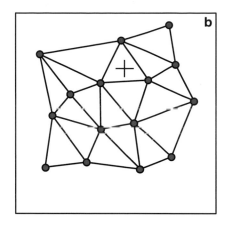

Fig. 7.5 Methods for selecting the control points to use for estimating the values at grid points. **a** Construction of a circle around the grid point (plus sign) with a radius defined by spatial autocorrelation of the z-values at the control points (small circles). **b** Use of a triangle mesh to select the control points for calculating the grid point, with the option of also including the vertices of the adjoining triangles. Modified from Swan and Sandilands (1995).

(Fig. 7.5): First, the number of *control points* needs to be selected, and second, the value of the variable of interest needs to be estimated for the *grid points*.

Control points are unevenly spaced field measurements, such as the thickness of sandstone units at different outcrops or the concentration of a chemical tracer in water wells. The data are generally represented as *xyz* triplets, where *x* and *y* are spatial coordinates and z is the variable of interest. In such cases, most gridding methods require continuous and unique data. However, spatial variables in the earth sciences are often discontinuous and not spatially unique: For example, the sandstone unit may be faulted or folded. Furthermore, gridding requires spatial autocorrelation, that is, the neighboring data points should be correlated with one another through a specific relationship. There is no point in making a surface estimation if the z variables are random and have no autocorrelation. After selecting the control points, a number of different methods are available for calculating the z-values at the evenly spaced grid points.

Various techniques exist for selecting the control points. Most methods make arbitrary assumptions about the autocorrelation of the z-variable. The *nearest neighbor criterion* includes all control points within a circular neighborhood of the grid point, and the radius of the circle is specified by the user (Fig. 7.5a). Since the degree of spatial autocorrelation likely decreases with increasing distance from the grid point, considering too many distant control points at once likely leads to erroneous results when computing values for these grid points. On the other hand, using radii that are too small may limit the total control points used in calculating the grid point values to a very small number, thereby resulting in a highly error-prone estimate of the modeled surface.

In another technique, all control points are connected in a triangular network (Fig. 7.5b). Every grid point is located within the triangular area formed by three control points. The z-value of the grid point is computed from the z-values of the control points at the vertices of the triangle. A modification of this form of gridding also uses the three points at the vertices of the three adjoining triangles. *Kriging* (which is introduced in Sect. 7.11) is an alternative approach for selecting control points and is often regarded as the ultimate method of gridding. Some users even use the term *geostatistics* as a synonym. Kriging is a method for quantifying spatial autocorrelation and hence for quantifying the dimension of the circle used to select the control points. More sophisticated versions of kriging use an elliptical area instead of a circle.

As mentioned above, the second step in surface estimation is the actual computation of the z-values for the grid points. The *arithmetic mean* of the measured z-values at the control points

$$\bar{z} = \frac{1}{n} \sum_{i=1}^{n} z_i$$

provides the easiest way of computing the values at the grid points. This is a particularly useful method if there are only a limited number of control points. If the study area is well covered by control points and the distance between these points is highly variable, the z-values of the grid points should be computed using a *weighted mean*. This involves weighting the z-values at the control points by the inverse of the distance d_i from the grid points:

$$\bar{z} = \frac{\sum\limits_{i=1}^{n} (z_i/d_i)}{\sum\limits_{i=1}^{n} (1/d_i)}$$

Depending on the spatial scaling relationship of the variable z, the inverse of the square root of the distance (rather than simply the inverse of the distance) may be used to weight the z-values. Fitting 3D *splines* to the control points offers another method for computing the grid point values and is commonly used in the earth sciences. Most routines used in surface estimation involve *bicubic polynomial splines*, that is, a third-degree 3D polynomial is fitted to at least six adjacent control points. The final surface is a composite that comprises different portions of these splines.

One of the advantages of square meshes is also one of their disadvantages: Due to the equidistance between these meshes, the memory requirement per area is the same everywhere (and thus also in areas with a low density of control points and a low variation of the variables of interest). One way of overcoming this problem is to use triangular meshes instead of square meshes. These triangular meshes are calculated by *triangulation* from the irregularly distributed control points. There are several different methods for calculating such *triangulated irregular networks* (TINs), one of which is *Delaunay triangulation*, which is used in Sect. 7.6. In this method, the points are linked to triangles such that no other points are contained within a circle on which the three vertices of the triangle are present.

The *SciPy* package contains several interpolation methods. We first use the bicubic spline, which interpolates the data in two dimensions using a spline, that is, a third-degree 3D polynomial is fitted to at least six adjacent control points, thereby resulting in a surface (and its first derivative) that is continuous. This gridding method is particularly well suited for producing smooth surfaces from noisy data sets with unevenly distributed control points. As an example, we use synthetic *xyz* data that represent the vertical distance between the surface of an imaginary stratigraphic horizon that has been displaced by a normal fault on the one hand and by a reference surface on the other hand. The foot wall of the fault shows roughly horizontal strata, whereas the hanging wall is characterized by two large sedimentary basins. The *xyz* data are irregularly distributed and thus need to be interpolated to a regular grid. These *xyz* data are stored as a three-column table in a file *normalfault.txt*:

```
4.3229698e+02     7.4641694e+01     9.7283620e-01
4.4610209e+02     7.2198697e+01     6.0655065e-01
4.5190255e+02     7.8713355e+01     1.4741054e+00
4.6617169e+02     8.7182410e+01     2.2842172e+00
4.6524362e+02     9.7361564e+01     1.1295175e-01
4.5526682e+02     1.1454397e+02     1.9007110e+00
4.2930233e+02     7.3175896e+01     3.3647807e+00
(cont'd)
```

The first and second column contain the coordinates x (between 420 and 470 of an arbitrary spatial coordinate system) and y (between 70 and 120), while the third column contains the vertical z-values. The data are loaded using

```
%reset -f

import numpy as np
from scipy import interpolate
from math import log10, floor
from matplotlib import cm
import matplotlib.pyplot as plt
from matplotlib.colors import LightSource

data = np.loadtxt('normalfault.txt')
```

Initially, we wish to create an overview plot of the spatial distribution of the control points. In order to label the points in the plot, numerical z-values of the third column are converted to character string representations with a maximum of two digits:

```
labels = np.zeros(len(data))
for i in range(0,78):
    labels[i] = round(data[i,2],
        -int(floor(log10(abs(0.1*data[i,2])))))
```

The 2D plot of our data is generated in two steps: First, the data are displayed as empty circles using `plot()`. Second, the data are labeled using `text()`, which adds text contained in `labels` to the *xy* locations. The value of 1 is added to all *x*-coordinates in order to produce a small offset between the circles and the text:

```
plt.figure()
for i in range(0,78):
    plt.plot(data[i,0],data[i,1],
        marker='o',
        markeredgecolor=(0.3,0.5,0.8),
        markerfacecolor='none',)
    plt.text(data[i,0]+1,data[i,1],labels[i],
        fontsize=8)
plt.show()
```

This plot helps us to define the axis limits for gridding and contouring. The function `meshgrid()` transforms the domain specified by vectors x and y into arrays XI and YI. The rows of the output array XI are copies of the vector x, and the columns of the output array YI are copies of the vector y. We choose 1.0 as the grid interval and therefore obtain 51 data points in either direction x and y:

```
x = np.linspace(420,470,51)
y = np.linspace(70,120,51)
XI,YI = np.meshgrid(x,y)
```

The *bicubic* spline interpolation is used to interpolate the unevenly spaced data at the grid points specified by XI and YI:

```
ZI = interpolate.griddata(data[:,0:2],data[:,2],(XI,YI),
    method='bicubic')
```

The simplest way of displaying the gridding results is as a contour plot using `contour()`. By default, the number and the values of the contour levels are chosen automatically. The choice of the contour levels depends on the minimum and maximum values of *z*:

```
plt.figure()
plt.contour(XI,YI,ZI)
plt.show()
```

Alternatively, the number of contours can be chosen manually (e.g., ten contour levels):

```
plt.figure()
plt.contour(XI,YI,ZI,10)
plt.show()
```

Contouring can also be performed at values specified in a vector v. Since the maximum and minimum values of z are

```
print(np.min(data[:,2]))
print(np.max(data[:,2]))
```

which yields

```
-27.435687
21.301825
```

we choose

```
v = np.linspace(-40,20,7)
```

The command

```
plt.figure()
fig,ax = plt.subplots()
c = ax.contour(XI,YI,ZI,v)
ax.clabel(c,inline=1,fontsize=8)
plt.show()
```

yields the QuadContourSet object c, which can be used as input in clabel(), which labels contours automatically. Filled contours are an alternative to the empty contours used above. This function is used together with colorbar(), which displays a legend for the plot. In addition, we can plot the locations (small circles) and z-values (contour labels) of the true data points (Fig. 7.6):

```
plt.figure()
plt.contourf(XI,YI,ZI)
for i in range(0,78):
    plt.plot(data[i,0],data[i,1],
        marker='o',
        markeredgecolor=(0,0,0),
        markerfacecolor='none',)
    plt.text(data[i,0]+1,data[i,1],labels[i],
```

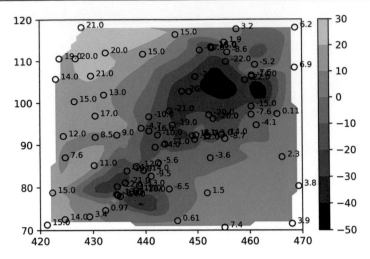

Fig. 7.6 Contour plot with the locations (small circles) and z-values (labels) of the control points.

```
        fontsize=8)
plt.colorbar()
plt.show()
```

A pseudocolor plot is generated using pcolor(). Black contours are also added at the same levels as in the example above:

```
plt.figure()
plt.pcolor(XI,YI,ZI)
for i in range(0,78):
    plt.plot(data[i,0],data[i,1],
        marker='o',
        markeredgecolor=(0,0,0),
        markerfacecolor='none',)
    plt.text(data[i,0]+1,data[i,1],labels[i],
        fontsize=8)
plt.contour(XI,YI,ZI,v,colors='k')
plt.colorbar()
plt.show()
```

The third dimension is added to the plot using plot_surface(). We can also use this example to introduce view_init(el,az) in order to specify the direction of viewing, where el is the elevation and az is the azimuth (or horizontal rotation) (both in degrees). The values el=30 and az=-127.5 define the oblique view of the plot,

Fig. 7.7 Three-dimensional
colored surface.

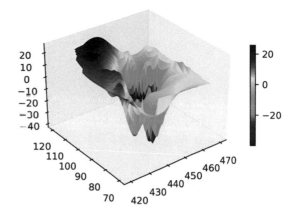

```
plt.figure()
fig,ax2 = plt.subplots(subplot_kw={"projection":"3d"})
surf = ax2.plot_surface(XI,YI,ZI,
    rstride=1,
    cstride=1,
    linewidth=0,
    cmap=cm.jet,
    antialiased=False,
    shade=True)
ax2.view_init(30,-127.5)
ax2.grid(False)
fig.colorbar(surf,shrink=0.5)
plt.show()
```

whereas `az=90` and `el=-90` are directly overhead and are the default 2D view (Fig. 7.7).

The bicubic spline interpolation described in this section provides a solution to most gridding problems. However, different applications in the earth sciences require different methods of interpolation, and each application has its own problems. The next section compares bicubic spline interpolation with other gridding methods and summarizes their strengths and weaknesses.

7.8   Comparison of Methods and Potential Artifacts

The first example in this section illustrates the use of the *bilinear interpolation* technique for gridding unevenly spaced data (Fig. 7.8–7.9). Bilinear interpolation is an extension of the one-dimensional technique of linear interpolation introduced in Sect. 5.5. In the two-dimensional case, linear interpolation is first performed in one direction, and then in the other direction. Although the bilinear method appears to be one of the simplest interpolation techniques and might thus not be expected to yield serious artifacts or distortions in the data, the opposite is true because the method has a number of disadvantages. As a result, other methods are preferred in many applications.

The sample data used in the previous section can again be loaded to study the effects of bilinear interpolation:

```
%reset -f

import numpy as np
from scipy import interpolate
from math import log10, floor
import matplotlib.pyplot as plt

data = np.loadtxt('normalfault.txt')
labels = np.zeros(len(data))
for i in range(0,len(data)):
    labels[i] = round(data[i,2],
        -int(floor(log10(abs(0.1*data[i,2])))))
```

We now choose the `linear` method while using `griddata()` to interpolate the data:

```
x = np.linspace(420,470,201)
y = np.linspace(70,120,201)
XI,YI = np.meshgrid(x,y)

ZI = interpolate.griddata(data[:,0:2],data[:,2],
    (XI,YI),method='linear')
```

The results are plotted as contours. The plot also includes the locations of the control points:

```
v = np.linspace(-40,20,7)

plt.figure()
plt.contourf(XI,YI,ZI,v)
for i in range(0,len(data)):

    plt.plot(data[i,0],data[i,1],
        marker='o',
        markeredgecolor=(0,0,0),
        markerfacecolor='none',)
    plt.text(data[i,0]+1,data[i,1],labels[i],
        fontsize=8)
plt.colorbar()
plt.show()
```

The new surface is restricted to the area that contains control points: By default, bilinear interpolation does not extrapolate beyond this region. Furthermore, the contours are rather angular compared with the smooth shape of the contours from the bicubic spline interpolation. However, the most important character of the bilinear gridding technique is illustrated by projecting the data into a vertical plane:

```
plt.figure()
plt.plot(XI,ZI,
    linewidth=0.3,
    color='k')
for i in range(0,len(data)):
    plt.plot(data[i,0],data[i,2],
        marker='o',
        markeredgecolor=(0,0,0),
        markerfacecolor='none',)
    plt.text(data[i,0]+1,data[i,2],labels[i],
        fontsize=8)
```

This plot shows the projection of the estimated surface (vertical lines) and the labeled control points. The z-values at the grid points never exceed the z-values of the control points. As with the linear interpolation of time series (Sect. 5.5), bilinear interpolation causes both substantial smoothing of the data and a reduction in high-frequency variations.

Cubic spline interpolations represent the other extreme in many ways and are often used both with extremely unevenly spaced data and with noisy data:

```
x = np.linspace(420,470,201)
y = np.linspace(70,120,201)
XI,YI = np.meshgrid(x,y)

ZI = interpolate.griddata(data[:,0:2],data[:,2],
    (XI,YI),method='cubic')

v = np.linspace(-40,20,7)

plt.figure()
plt.contourf(XI,YI,ZI)
for i in range(0,len(data)):
    plt.plot(data[i,0],data[i,1],
        marker='o',
        markeredgecolor=(0,0,0),
        markerfacecolor='none',)
    plt.text(data[i,0]+1,data[i,1],labels[i],
        fontsize=8)
```

```
plt.colorbar()
plt.show()
```

The contours suggest an extremely smooth surface. This solution has proven highly useful in many applications, but the method also produces a number of artifacts. As we can see from the next plot, the estimated values at the grid points often lie beyond the range of the measured z-values:

```
plt.figure()
plt.plot(XI,ZI,
    linewidth=0.3,
    color='k')
for i in range(0,len(data)):
    plt.plot(data[i,0],data[i,2],
        marker='o',
        markeredgecolor=(0,0,0),
        markerfacecolor='none',)
    plt.text(data[i,0]+1,data[i,2],labels[i],
        fontsize=8)
```

This can sometimes be appropriate and does not smooth the data the same way that bilinear gridding does. However, introducing very close control points with different z-values can result in serious artifacts. As an example, we introduce one reference point with a z-value of $+5$ close to a reference point with a negative z-value of around -26:

```
data = np.loadtxt('normalfault.txt')
data = np.append(data,[[450,105,5]],axis=0)

labels = np.zeros(len(data))
for i in range(0,len(data)):
    labels[i] = round(data[i,2],
        -int(floor(log10(abs(0.1*data[i,2])))))

ZI = interpolate.griddata(data[:,0:2],data[:,2],
    (XI,YI),method='cubic')

v = np.linspace(-40,20,7)

plt.figure()
plt.contourf(XI,YI,ZI)
for i in range(0,len(data)):
    plt.plot(data[i,0],data[i,1],
        marker='o',
        markeredgecolor=(0,0,0),
```

```
            markerfacecolor='none',)
        plt.text(data[i,0]+1,data[i,1],labels[i],
            fontsize=8)
    plt.colorbar()
    plt.show()
```

The extreme gradient at the location (450, 105) results in a paired *low* and *high* (Fig. 7.8). In such cases, it is recommended that one of the two control points be deleted and that the z-value of the remaining control point be replaced by the arithmetic mean of both z-values.

The third method available with `griddata()` – in addition to *bilinear* (`linear`) and *bicubic* (`cubic`) interpolation – involves the use of *nearest neighbor* (`nearest`) interpolation methods. We compare all of these methods in the next example. We first clear the workspace, reload the data from *normalfault.txt*, and again add the outlier:

```
data = np.loadtxt('normalfault.txt')
data = np.append(data,[[450,105,5]],axis=0)

labels = np.zeros(len(data))
for i in range(0,len(data)):
    labels[i] = round(data[i,2],
        -int(floor(log10(abs(0.1*data[i,2])))))
```

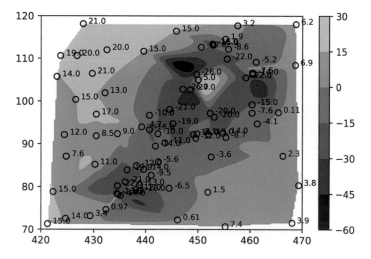

Fig. 7.8 Contour plot of a data set that was gridded using a bicubic spline interpolation. At the location (450,105), very close control points with different z-values were introduced. Interpolation causes a paired low and high, which is a common artifact in the spline interpolation of noisy data.

Fig. 7.9 This figure from the MATLAB-based book MATLAB Recipes for Earth Sciences (Trauth 2021) could not be translated to Python.

We then create titles for the results from the different interpolation methods:

```
titles = ['Linear interpolation',
    'Cubic Spline interpolation',
    'Nearest neighbor interpolation']
```

We again define the axis limits for gridding and contouring as xlim=[420 470] and ylim=[70 120]. The function meshgrid() transforms the domain specified by vectors x and y into arrays XI and YI. The rows of output array XI are copies of vector x, and the columns of output array YI are copies of vector y. We choose 1.0 as the grid interval:

```
x = np.linspace(420,470,51)
y = np.linspace(70,120,51)
XI,YI = np.meshgrid(x,y)
```

We then use griddata() with all available options and store the results in a three-dimensional array ZI:

```
ZILIN = interpolate.griddata(data[:,0:2],data[:,2],
    (XI,YI),method='linear')
ZICUB = interpolate.griddata(data[:,0:2],data[:,2],
    (XI,YI),method='cubic')
ZINEA = interpolate.griddata(data[:,0:2],data[:,2],
    (XI,YI),method='nearest')

ZI = np.dstack((ZILIN,ZICUB,ZINEA))
```

We compare the results in five different graphics in separate figure windows (which are slightly offset on the computer display) using figure(). The data are displayed as filled contours at values specified in a vector v:

```
v = np.linspace(-40,20,7)

for j in range(0,3):
    plt.figure()
    plt.contourf(XI,YI,ZI[:,:,j],v)
    for i in range(0,len(data)):
        plt.plot(data[i,0],data[i,1],
            marker='o',
            markeredgecolor=(0,0,0),
            markerfacecolor='none',)
        plt.text(data[i,0]+1,data[i,1],labels[i],
            fontsize=8)
        plt.title(titles[j])
    plt.colorbar()
    plt.show()
```

7.9   Statistics of Point Distributions

This section deals with the statistical distribution of objects within an area, which may help to explain the relationship between these objects and the properties of the area (Fig. 7.10). For instance, the spatial concentration of hand axes in an archaeological site may suggest that a larger population of hominins once lived in that area. As another example, the clustered distribution of fossils may document environmental conditions that were favorable to particular organisms that have been preserved as fossils. Furthermore, the alignment of volcanoes may help in mapping tectonic structures concealed beneath the Earth's surface.

Various methods for statistically analyzing point distributions are introduced below. We first consider a test for a uniform spatial distribution of objects followed by a test for a random spatial distribution and finally by a simple test for a clustered distribution of objects.

Test for Uniform Distribution

In order to illustrate the test for uniform distribution, we first need to compute some synthetic data (Fig. 7.10b). The function `random.rand()` computes uniformly distributed pseudorandom numbers drawn from a uniform distribution within the interval [0,1]. We compute xy data using `random.rand()` and multiply the data by ten in order to obtain data within the interval [0,10]:

```
%reset -f

import numpy as np
from scipy import stats
import matplotlib.pyplot as plt
```

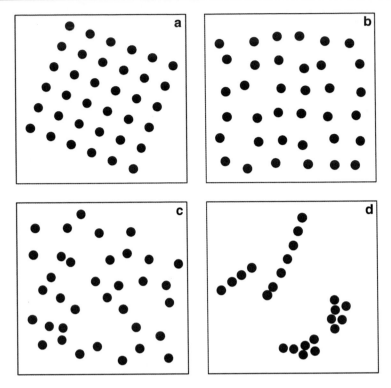

Fig. 7.10 Spatial distribution of objects. **a** Regular point distribution, **b** uniform point distribution, **c** random point distribution, and **d** anisotropic and clustered point distribution.

```
from scipy.special import factorial
import scipy.spatial.distance as distance

np.random.seed(0)
data = 10 * np.random.rand(100,2)
```

We can use the χ^2-test introduced in Sect. 3.8 to test the null hypothesis that the data have a uniform distribution. The *xy* data are now organized into 25 classes that are square subareas with dimensions of 2-by-2. This definition of classes ignores the rule of thumb that the number of classes should be close to the square root of the number of observations (see Sect. 3.3). However, our choice of classes does not result in any empty classes, which should be avoided when applying the χ^2-test. Furthermore, 25 classes produce integer values for the expected number of observations, and these values are easier to work with. We display the data as blue circles in a plot of y versus x. The rectangular areas are outlined with red lines (Fig. 7.11):

```
x = np.linspace(0,10,11)
y = np.ones(np.shape(x))

plt.figure()
plt.plot(data[:,0],data[:,1],'o')
for i in range(1,5):
    plt.plot(x,2*i*y,
        color='r',
        linewidth=0.5)
for i in range(1,5):
    plt.plot(2*i*y,x,
        color='r',
        linewidth=0.5)
plt.axis(np.array([0,10,0,10]))
plt.show()
```

A three-dimensional version of a histogram is used to display the spatial data, which is organized into classes (Fig. 7.12):

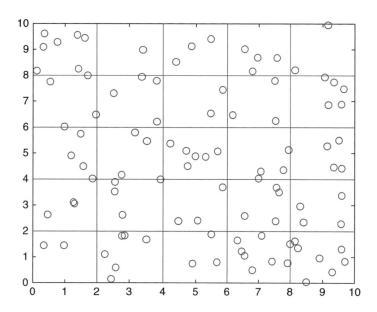

Fig. 7.11 Two-dimensional plot of a point distribution. The distribution of objects in the field is tested for uniform distribution using the χ^2-test. The xy data are organized into 25 classes, which are subareas with 2-by-2 dimensions.

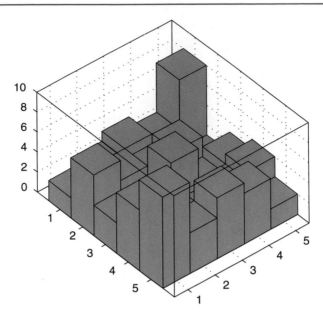

Fig. 7.12 Three-dimensional histogram displaying the numbers of objects for each subarea. The histogram was created using `histogram2d()`.

```
e = np.linspace(0,10,6)
H,xedges,yedges = np.histogram2d(data[:,0],
    data[:,1],bins=(e,e))

x = np.linspace(1,9,5)
y = np.linspace(1,9,5)
z = 0

X,Y = np.meshgrid(x,y)
Z = np.ones((5,5))

X = X.ravel()
Y = Y.ravel()
Z = 0

dx = 2
dy = 2
dz = H
dz = H.ravel()

fig = plt.figure()
ax1 = fig.add_subplot(121,projection='3d')
```

```
ax1.bar3d(X,Y,Z,dx,dy,dz)
plt.show()
```

As with the equivalent two-dimensional function, the output object H of histo-gram2D() can be used to compute the frequency distribution n_obs of the data,

```
n_obs = H.ravel()
```

where ravel() returns a contiguous flattened version of the 5-by-5 array H. The theoretical frequencies of the different classes in a uniform distribution are all identical. The expected number of objects in each square area is the size of the total area $(10 \cdot 10 = 100)$ divided by the 25 subareas (or classes), which equals four. To compare the theoretical frequency distribution with the actual spatial distribution of objects, we generate an array with an identical number of objects:

```
n_exp = 4*np.ones(len(n_obs))
```

The χ^2-test explores the squared differences between the observed and expected frequencies (Sect. 3.9). The quantity χ^2 is defined as the sum of the squared differences divided by the expected frequencies,

```
chi2_data = np.sum((n_obs-n_exp)**2/n_exp)
print(chi2_data)
```

which yields

```
18.5
```

The critical χ^2 can be calculated using chi2.ppf(). The χ^2-test requires the degrees of freedom Φ. In our example, we test the hypothesis that the data are uniformly distributed, that is, we estimate only one parameter (Sect. 3.5). The number of degrees of freedom is therefore $\Phi = 25 - (1 + 1) = 23$. We test the hypothesis at a $p = 5\%$ significance level. The function chi2.ppf() computes the inverse of the χ^2 CDF with parameters specified by Φ for the corresponding probabilities in p,

```
chi2_theo = stats.chi2.ppf(1-0.05,25-1-1)
print(chi2_theo)
```

which yields

```
35.17246162690806
```

Since the critical χ^2 of 35.1725 is well above the measured χ^2 of 22.5000, we cannot reject the null hypothesis and must therefore conclude that our data follow a uniform distribution.

Test for Random Distribution

The following example illustrates a test for the random distribution of objects within an area (Fig. 7.10c). We use the uniformly distributed data generated in the previous example and display the point distribution:

```python
np.random.seed(0)
data = 10 * np.random.rand(100,2)

x = np.linspace(0,10,11)
y = np.ones(np.shape(x))

plt.figure()
plt.plot(data[:,0],data[:,1],'o')
for i in range(1,10):
    plt.plot(x,i*y,
        color='r',
        linewidth=0.5)
for i in range(1,10):
    plt.plot(i*y,x,
        color='r',
        linewidth=0.5)
plt.axis(np.array([0,10,0,10]))
plt.show()
```

We then generate a three-dimensional histogram and use histogram2D() to count the objects per class. In contrast to the previous test, we now count the subareas that contain a certain number of observations. The number of subareas is larger than would normally be used for the previous test. In our example, we use 49 subareas (or classes):

```python
e = np.linspace(0,10,8)
H,xedges,yedges = np.histogram2d(data[:,0],data[:,1],
    bins=(e,e))

x = np.linspace(1,9,7)
y = np.linspace(1,9,7)
z = 0

X,Y = np.meshgrid(x,y)
Z = np.ones((7,7))

X = X.ravel()
Y = Y.ravel()
Z = 0
```

```
dx = 2
dy = 2
dz = H
dz = H.ravel()

fig = plt.figure()
ax1 = fig.add_subplot(121,projection='3d')
ax1.bar3d(X,Y,Z,dx,dy,dz)
plt.show()

counts = H.ravel()
```

The frequency distribution of the subareas that contain a specific number of objects follows a Poisson distribution (Sect. 3.5) if the objects are randomly distributed. First, we compute the frequency distribution of the subareas containing N objects. In our example, we count subareas with 0, …, 5 objects. We also display the histogram of the frequency distribution two-dimensionally using histogram() after first calculating the bin edges E from the bin centers N (Sect. 3.3) (Fig. 7.13):

```
N = np.linspace(0,5,6)
E = np.linspace(-0.5,5.5,7)

plt.hist(counts,bins=7)
n,e = np.histogram(counts,bins=7)
v = np.diff(e[0:2]) * 0.5 + e[0:-1]

n_obs = n
```

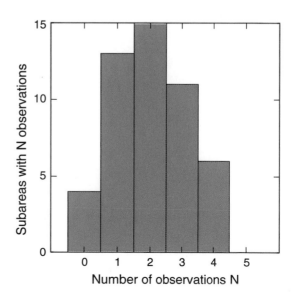

Fig. 7.13 Frequency distribution of subareas with N objects. In our example, the subareas with 0, …, 5 objects are counted. The histogram of the frequency distribution is displayed as a two-dimensional histogram using histogram().

The midpoints of the histogram intervals v correspond to the $N = 0, \ldots, 5$ objects contained within the subareas. The expected number of subareas E_j with a certain number of objects j can be computed using

$$E_j = Te^{-n/T}\frac{(n/T)^j}{j!}$$

where n is the total number of objects and T is the number of subareas. For $j = 0, j!$ is taken to be 1. We compute the expected number of subareas (i.e., the theoretical frequency distribution n_exp) using the equation shown above

```
n_exp = np.ones(7)
for i in range(0,6):
    n_exp[i] = 49*np.exp(-100/49)*(100/49) \
        **N[i]/factorial(N[i])

n_exp = np.sum(n_obs)*n_exp/np.sum(n_exp)
```

and display both the empirical and theoretical frequency distributions in a single plot:

```
plt.figure()
plt.bar(v,n_obs,
    edgecolor='b',
    color='none')
plt.bar(v,n_exp,
    edgecolor='r',
    color='none')
plt.show()
```

The χ^2-test is again used to compare the empirical and theoretical distributions, that is, to test the null hypothesis that the frequency distribution of the subareas that contain a specific number of objects follows a Poisson distribution. The test is performed at a $p = 5\%$ significance level. Since the Poisson distribution is defined by only one parameter (Sect. 3.4), the number of degrees of freedom is $\Phi = 6 - (1 + 1) = 4$. The measured χ^2 of

```
chi2_data = np.sum((n_obs-n_exp)**2/n_exp)
print(chi2_data)
```

which yields

```
3.546695554153232
```

is well below the critical χ^2, which is

```
chi2_theo = stats.chi2.ppf(1-0.05,6-1-1)
print(chi2_theo)
```

which yields

```
9.487729036781154
```

We therefore cannot reject the null hypothesis and must conclude that our data follow a Poisson distribution and that the point distribution is random.

Test for Clustering

Point distributions in the earth sciences are often clustered (Fig. 7.10d). We use a *nearest neighbor criterion* to test a spatial distribution for clustering. Davis (2002) published an excellent summary of the nearest neighbor analysis that summarizes the work of a number of other authors. Moreover, Swan and Sandilands (1996) presented a simplified description of this analysis. The test for clustering computes the distances d_i that separate all possible pairs of nearest points in the field. The *observed mean nearest neighbor distance* is

$$\overline{d} = \frac{1}{n} \sum_{i=1}^{n} d_i$$

where n is the total number of points (or objects) in the field. The arithmetic mean of all distances between possible pairs is related to the area covered by the map. This relationship is expressed by the *expected mean nearest neighbor distance*, which is

$$\overline{\delta} = \frac{1}{2} \sqrt{A/n}$$

where A is the area covered by the map. Small values for this ratio then suggest clustering, whereas large values indicate regularity (or uniformity). The test uses a Z-statistic (Sect. 3.5), which is

$$Z = \frac{\overline{d} - \overline{\delta}}{s_e}$$

where s_e is the standard error of the mean nearest neighbor distance, which is defined as

$$s_e = \frac{0.26136}{\sqrt{n^2/A}}$$

The null hypothesis of *randomness* is tested against two alternative hypotheses: *clustering* on one hand and *uniformity* or *regularity* on the other hand (Fig. 7.10). The Z-statistic has critical values of 1.96 and −1.96 at a significance level of 5%. If $-1.96 < Z < +1.96$, we cannot reject the null hypothesis that the data are randomly distributed. If $Z < -1.96$, we can reject the null hypothesis and accept the first alternative hypothesis of clustering. If $Z > +1.96$, we can also reject the null hypothesis but accept the second alternative hypothesis of uniformity or regularity.

As an example, we again use the synthetic data analyzed in the previous examples:

```
np.random.seed(0)
data = 10 * np.random.rand(100,2)

plt.figure()
plt.plot(data[:,0],data[:,1],'o')
plt.show()
```

We first compute the pairwise Euclidean distance between all pairs of observations using pdist() (Sect. 9.5). The resulting distance array distances is then converted to a symmetric square format distarray using squareform():

```
distances = distance.pdist(data)
distarray = distance.squareform(distances)
```

The following for loop finds the nearest neighbors, stores the nearest neighbor distances, and computes the mean distance,

```
nearest = np.zeros(100)
for i in range(0,100):
    distarray[i,i] = np.Inf
    k = np.where(distarray[i,:]==np.min(distarray[i,:]))
    nearest[i]=distarray[i,k]

observednearest = np.mean(nearest)
print(observednearest)
```

which yields

```
0.5284579351787346
```

In our example, the mean nearest distance observednearest equals 0.5285. We next calculate the area of the map. The expected mean nearest neighbor distance is half the square root of the map area divided by the number of observations,

```
maparea = (np.max(data[:,0])-np.min(data[:,0])) * \
    (np.max(data[:,1])-np.min(data[:,1]))
expectednearest = 0.5 * np.sqrt(maparea/len(data))
print(expectednearest)
```

which yields

```
0.49268341482387124
```

In our example, the expected mean nearest neighbor distance expectednearest is 0.4927. Finally, we compute the standard error of the mean nearest neighbor distance se,

```
se = 0.26136/sqrt(length(data).^2/maparea)
```

which yields

```
0.025753547459673395
```

and we also compute the test statistic z,

```
Z = (observednearest - expectednearest)/se
```

which yields

```
1.3891103899717698
```

Since $-1.96 < z < +1.96$, we cannot reject the null hypothesis and must conclude that the data are randomly distributed, but not clustered.

7.10   Analysis of Digital Elevation Models (by R. Gebbers)

Digital elevation models (DEMs) and their derivatives (e.g., slope and aspect) can indicate surface processes, such as lateral water flow, solar irradiation, or erosion. The simplest derivatives of a DEM are the slope and the aspect. The *slope* (or *gradient*) is a measure of the steepness, the incline, or the grade of a surface in either percentages or degrees. The *aspect* (or *exposure*) refers to the direction a slope faces.

We use the SRTM data set introduced in Sect. 7.5 to illustrate the analysis of a digital elevation model for slope, aspect, and other derivatives. The data are loaded by typing

```
%reset -f
```

```
import numpy as np
from scipy import signal
from matplotlib import cm
import matplotlib.pyplot as plt
from skimage.segmentation import watershed

with open('S01E036.hgt') as fid:
    rectype = np.dtype('>i2')
    SRTM = np.fromfile(fid,dtype=rectype)

SRTM = np.reshape(SRTM,(1201,1201))
SRTM = np.flipud(SRTM)

SRTM = SRTM.astype(float)
SRTM[SRTM==-32768] = np.mean(SRTM)
```

These data are elevation values in meters above sea level that were sampled on a 3 arc-second (or a 90 m grid). The SRTM data contain small-scale spatial disturbances and noise that could cause problems when computing a drainage pattern. We therefore filter the data with a two-dimensional moving average filter using convolve2d(). The filter calculates a spatial running mean of 3-by-3 elements. We use only the SRTM(750:850,700:800) subset of the original data set in order to reduce computation time. We also remove the data at the edges of the DEM in order to eliminate filter artifacts:

```
F = 1/9 * np.ones((3,3))
SRTM = signal.convolve2d(SRTM[749:850,699:800],
    F,mode='same')
SRTM = SRTM[1:99,1:99]
```

The DEM is displayed as a pseudocolor plot using pcolor() and the colormap cm. terrain, which assigns a colormap that is appropriate for elevation data since it relates land and sea colors to hypsometry and bathymetry, respectively:

```
plt.figure()
plt.axes(aspect='equal')
plt.pcolor(SRTM,
    cmap=cm.terrain)
plt.colorbar()
plt.title('Elevation [m]')
plt.show()
```

The DEM indicates a horseshoe-shaped mountain range surrounding a valley that slopes down toward the southeast (Fig. 7.16a).

Fig. 7.14 Local
neighborhood showing the
Python indexing convention.

Z(1)	Z(4)	Z(7)
Z(2)	Z(5)	Z(8)
Z(3)	Z(6)	Z(9)

The SRTM subset is now analyzed for slope and aspect. When working with DEMs on a regular grid, the slope and aspect can be estimated using centered finite differences in a local 3-by-3 neighborhood. Figure 7.14 shows a local neighborhood using the Python indexing convention. To calculate the slope and aspect, we need two finite differences in the DEM elements z in the x and y directions:

$$z_x = \frac{z_{r,c-1} - z_{r,c+1}}{2h}$$

and

$$z_y = \frac{z_{r-1,c} - z_{r+1,c}}{2h}$$

where h is the cell size and has the same units as the elevation. Using the finite differences, the slope is then calculated by

$$slp_{DF} = \sqrt{z_x^2 + z_y^2}$$

Other primary relief attributes (e.g., the *aspect*, the *plan*, the *profile*, and the *tangential curvature*) can be derived in a similar way using finite differences (Wilson and Gallant 2000). The function `gradient()` calculates the slope and aspect of a data grid z in degrees above the horizontal and degrees clockwise from north. Since the SRTM digital elevation model is sampled on a 3 arc-second grid, $60 \cdot 60/3 = 1{,}200$ elements of the DEM correspond to one degree of longitude or latitude. For simplicity, we ignore the actual coordinates of the SRTM subset in this example and instead use the indices of the DEM elements:

```
x,y = np.gradient(SRTM)

slp = np.sqrt(x**2+y**2)/90
slp = np.arctan(slp)*180./np.pi
asp = np.arctan2(x,y)
asp = asp*180./np.pi
asp[asp<0] = asp[asp<0]+360
```

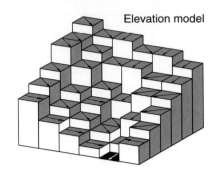

Elevation model

Flow direction					
↘	↘	↓	↘	↓	↓
↘	↘	↘	↓	↙	↓
↘	↘	↘	↓	↙	↙
↘	↘	→	↘	↓	↙
↘	↘	↘	↘	↓	↙
→	→	→	→	↓	←

Flow accumulation					
1	1	1	1	1	1
1	2	3	1	3	2
1	2	3	8	1	3
1	2	3	16	4	1
1	2	3	1	22	1
1	3	6	10	36	1

Fig. 7.15 Schematic of calculation of flow accumulation using the D8 method.

We then display a pseudocolor map of the DEM slope in degrees (Fig. 7.16b):

```
h = pcolor(slp);
colormap(jet), colorbar
plt.figure()
plt.axes(aspect='equal')
plt.pcolor(slp,
    cmap=cm.jet)
plt.colorbar()
plt.title('Slope')
plt.show()
```

Flat areas are common on the summits and on the valley floors. The southeastern and south–southwestern sectors are also relatively flat. The steepest slopes are concentrated in the center of the area and in the southwestern sector. Next, a pseudocolor map of the aspect is generated (Fig. 7.16c):

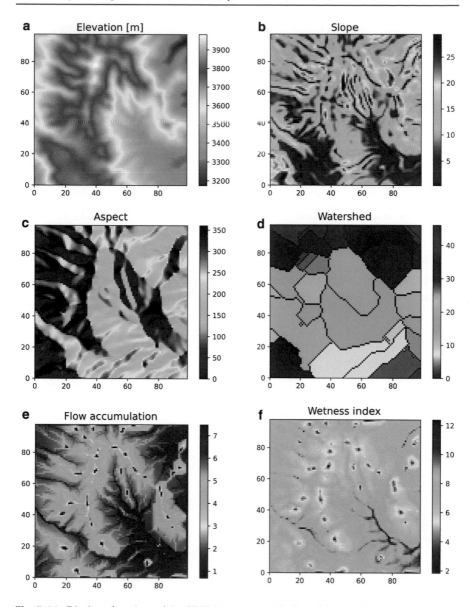

Fig. 7.16 Display of a subset of the SRTM data set used in Sect. 7.5 as well as primary and secondary attributes of the digital elevation model: **a** elevation, **b** slope, **c** aspect, **d** watershed, **e** flow accumulation, and **f** wetness index (data from Farr and Kobrick 2000, Farr et al. 2007).

```
plt.figure()
plt.axes(aspect='equal')
plt.pcolor(asp,
    cmap=cm.jet)
```

```
plt.colorbar()
plt.title('Aspect')
plt.show()
```

This plot displays the aspect in degrees clockwise from north. For instance, mountain slopes facing north are displayed in red, and east-facing slopes are displayed in green.

The aspect changes abruptly along the ridges of the mountain ranges where neighboring drainage basins are separated by *watersheds*. The *scikit-image* package includes watershed() for detecting these drainage divides and for ascribing numerical labels to each catchment area, starting with one:

```
watersh = watershed(SRTM,
    markers=None,
    watershed_line=True)
```

The catchment areas are displayed in a pseudocolor plot in which each area is assigned a color from the colormap hsv (Fig. 7.16d) according to its numerical label:

```
plt.figure()
plt.axes(aspect='equal')
plt.pcolor(watersh,
    cmap=cm.hsv)
plt.colorbar()
plt.title('Watershed')
plt.show()
```

The watersheds are represented by a series of red pixels. As in this example, watershed() often generates unrealistic results since watershed algorithms are sensitive to local minima, which act as spurious sinks and are potential locations of spurious areas of internal drainage. These minima should thus be borne in mind during any subsequent computation of hydrological characteristics from the DEM.

Flow accumulation (also called the *specific catchment area* or *upslope contributing area*) is defined as the number of cells (that define an uphill area) that contribute runoff to a particular cell (Fig. 7.15). In contrast to the local variables of slope and aspect, flow accumulation can only be determined from the global neighborhood. The principal operation is to add cell inflows from topographically higher neighboring cells starting from the specified cell and working up to the watersheds. Before adding together the outflows from each cell, we need to determine the gradient of each individual cell toward each neighboring cell, which is indexed by N. The array N contains indices for the eight adjacent cells according to the Python convention, as shown in Fig. 7.14. We make use of roll() to access the neighboring cells. For a two-dimensional array Z, roll() circularly shifts the values in the array Z by r rows and c columns. The individual gradients are calculated by

$$grad = \frac{z_{r+y,c+x} - z_{r,c}}{h}$$

for the eastern, southern, western, and northern neighbors (known as the *rook's case*) and by

$$grad = \frac{z_{r+y,c+x} - z_{r,c}}{\sqrt{2h}}$$

for the diagonal neighbors (known as the *bishop's case*). In these formulas, h is the cell size, $z_{r,c}$ is the elevation of the central cell, and $z_{r+y,c+x}$ is the elevation of the neighboring cells. Cell indices x and y are obtained from array N. The gradients are stored in a three-dimensional array grads, where grads[:,:,0] contains the gradient toward the neighboring cells to the east, grads[:,:,1] contains the gradient toward the neighboring cells to the southeast, and so on. Negative gradients indicate outflow from the central cell toward the relevant neighboring cell. In order to obtain the surface flow between cells, gradients are transformed using the inverse tangent of grads divided by π and multiplied by 180°:

```
N = np.array([[0,-1],[-1,-1],[-1,0],
    [+1,-1],[0,+1],[+1,+1],[+1,0],[-1,+1]]])
a,b = np.shape(SRTM)
grads = np.zeros((a,b,8))

for c in range(1,9,2):
    grads[:,:,c] = np.roll(SRTM,N[c,0],axis=0)
    grads[:,:,c] = np.roll(grads[:,:,c],N[c,1],axis=1)
    grads[:,:,c] = (grads[:,:,c]-SRTM)/np.sqrt(2*90**2)

for c in range(0,8,2):
    grads[:,:,c] = np.roll(SRTM,N[c,0],axis=0)
    grads[:,:,c] = np.roll(grads[:,:,c],N[c,1],axis=1)
    grads[:,:,c] = (grads[:,:,c]-SRTM)/90

grads = np.arctan(grads)/np.pi*180.
```

Since a central cell can have several downslope neighbors, water can flow in several directions. This phenomenon is called *divergent flow*. Early flow accumulation algorithms were based on the single flow direction method (known as the D8 method; Fig. 7.15), which allows flow to only one of the cell's eight neighboring cells. However, this method cannot model divergences in ridge areas and tends to produce parallel flow lines in some situations. In our example, we illustrate the use of a multiple flow direction method, which allows flow from a central cell to multiple neighboring cells. The proportion of the total outflow that is assigned to a neighboring cell is dependent on the gradient between the central cell and that particular neighboring cell. Even though multiple flow methods produce more realistic results in most situations, they tend to result in dispersion in valleys, where the flow should be more concentrated. A weighting factor w is therefore introduced, which controls the relationship between the outflows:

$$flow_i = \frac{grad_i^w}{\sum_{i=1}^{8} grad_i^w} \quad \text{for} \quad grad_i^w < 0$$

A recommended value for w is 1.1 because higher values would concentrate the flow in the direction of the steepest slope, while $w = 0$ would result in extreme dispersion. In the following sequence of commands, we first select the gradients that are less than zero and then multiply these gradients by the weighting:

```
w = 1.1
flow = (grads*(-1*grads<0.))**w
```

We then sum the upslope gradients along the third dimension of the `flow` array. Replacing all upslope gradient values of 0 with a value of 1 avoids the problems created by trying to divide by zero:

```
upssum = np.sum(flow,axis=2)
upssum[upssum==0] = 1
```

We divide the flows by `upssum` in order to obtain fractional weights that add to a total of one. This is achieved separately for each layer of the 3D `flow` array using a `for` loop:

```
for i in range(0,8):
    flow[:,:,i] = flow[:,:,i]*(flow[:,:,i]>0)/upssum
```

The 2D array `inflowsum` stores the intermediate inflow totals for each iteration. These intermediate totals are then summed to reach a figure for the total accumulated flow `flowac` at the end of each iteration. The initial values for `inflowsum` and `flowac` are obtained through `upssum`:

```
inflowsum = upssum
flowac = upssum
```

Another 3D array, `inflow`, is now needed in which to store the total intermediate inflow from all neighbors:

```
inflow = grads*0
```

Flow accumulation is terminated when there is no inflow. When translating this information to Python code, we use a conditional `while` loop that terminates if sum(inflowsum)===0. The number of non-zero entries in inflowsum decreases during each loop. This is achieved by alternately updating inflow and inflowsum. Here, inflowsum is updated with the intermediate inflow of the neighboring cells, which are weighted by flow under the condition that all neighboring cells are contributing cells (i.e., where grads are positive). Where not all neighboring cells are contributing cells, the intermediate inflowsum is reduced, as is inflow. The flow accumulation flowac increases through consecutive summations of the intermediate inflowsum:

```
while np.sum(inflowsum)>0:
    for i in range(0,8):
        inflow[:,:,i] = np.roll(inflowsum,N[i,0],axis=0)
        inflow[:,:,i] = np.roll(inflow[:,:,i],N[i,1],axis=1)
    inflowsum = np.sum(inflow*flow*grads>0,2)
    flowac = flowac + inflowsum
```

We display the result as a pseudocolor map with log-scaled values (Fig. 7.16e):

```
plt.figure()
plt.axes(aspect='equal')
plt.pcolor(np.log(1+flowac),
    cmap=cm.jet_r)
plt.colorbar()
plt.title('Flow accumulation')
plt.show()
```

The plot displays areas with high flow accumulation in shades of blue and areas with low flow accumulation (which usually correspond to ridges) in shades of red. We use a logarithmic scale to map the flow accumulation in order to obtain a better representation of the results. The simplified algorithm introduced here for calculating flow accumulation can be used to analyze sloping terrains in DEMs. In flat terrains, in which the slope approaches zero, no flow direction can be generated by our algorithm, and flow accumulation thus stops. Such situations require more advanced algorithms that can perform analyses on completely flat terrain. These more advanced algorithms also include sink-filling routines for avoiding spurious sinks that interrupt flow accumulation. Small depressions can be filled by smoothing, as we did at the beginning of this section.

The first part of this section dealt with primary relief attributes. Secondary attributes of a DEM are functions of two or more primary attributes. Examples of these secondary attributes include the wetness index and the stream power index. The *wetness index* for a cell is the log of the ratio between the area of the catchment for that particular cell and the tangent of its slope:

$$
weti = \log\left(\frac{1 + flowac}{\tan(slp)}\right)
$$

The term 1+*flowac* avoids the problems associated with calculating the logarithm of zero when `flowac = 0`. The wetness index is used to predict the soil water content (*saturation*) that results from lateral water movement. The potential for water logging is usually highest in the lower parts of catchments, where the slopes are gentler. Flat areas with a large upslope area have a high wetness index compared with steep areas with small catchments. The wetness index `weti` is computed and displayed by typing

```
weti = np.log((1+flowac)/np.tan(np.pi*slp/180))

plt.axes(aspect='equal')
plt.pcolor(weti,
    cmap=cm.jet_r)
plt.colorbar()
plt.title('Wetness index')
plt.show()
```

In this plot, blue colors represent high values for the wetness index, while red colors represent low values (Fig. 7.16f). In our example, soils in the southeast are the most likely to have a high water content due both to the runoff from the large central valley and to the flatness of the terrain.

The *stream power index* is another important secondary relief attribute that is frequently used in hillslope hydrology, geomorphology, soil science, and related disciplines. As a measure of stream power, it provides an indication of the potential for sediment transport and erosion by water. The stream power index is defined as the product of the catchment area for a specific cell and the tangent of the slope of that cell:

$$spi = flowac \cdot \tan(slp)$$

The potential for erosion is high when large quantities of water (calculated via flow accumulation) are fast flowing due to an extreme slope. The following series of commands compute and display the stream power index:

```
spi = flowac*np.tan(np.pi*slp/180)

plt.figure()
plt.axes(aspect='equal')
plt.pcolor(np.log(1+spi),
    cmap=cm.jet)
plt.colorbar()
plt.title('Stream power index')
plt.show()
axis([1 c 1 r])
set(gca,'TickDir','out');
```

The wetness and stream power indices are particularly useful in high-resolution terrain analysis, that is, in digital elevation models sampled at intervals of less than 30 m. In our terrain analysis example, we calculated `weti` and `spi` from a medium resolution DEM, and we must expect a degree of scale dependency in these attributes.

This section illustrated the use of basic tools for analyzing digital elevation models. A more detailed introduction to digital terrain modeling is given in the book by Wilson and Gallant (2000). Furthermore, the article by Freeman (1991) provides a comprehensive summary of digital terrain analysis, including an introduction to the use of advanced algorithms for flow accumulation.

7.11   Geostatistics and Kriging (by R. Gebbers)

Geostatistics describes the autocorrelation of one or more variables in 1D, 2D, or 3D space – or even in 4D space–time – in order to make predictions for unobserved locations, to obtain information on the accuracy of these predictions, or to reproduce spatial variability and uncertainty. The shape, range, and direction of the spatial autocorrelation are described by a *variogram*, which is the main tool used in linear geostatistics. The origins of geostatistics in its current form can be traced back to the early 1950s, when South African mining engineer Daniel G. Krige first published an interpolation method based on the spatial dependency of samples. In the 1960s and 1970s, French mathematician Georges Matheron developed the *theory of regionalized variables*, which provides the theoretical foundations for Krige's more practical methods. This theory forms the basis of several procedures for analyzing and estimating spatially dependent variables, which Matheron called geostatistics. Matheron also coined the term *kriging* for spatial interpolation via geostatistical methods.

Theoretical Background

A basic assumption in geostatistics is that a spatiotemporal process is composed of both deterministic and stochastic components (Fig. 7.17). The deterministic components can be *global* or *local trends* (sometimes called *drifts*). The stochastic component comprises both a purely random component and an autocorrelated component. The autocorrelated component suggests that on average, closer observations are more similar to one another than are more widely separated observations. This behavior is described by a variogram in which squared differences between observations are plotted against the separation distances of the observations. Krige's fundamental idea was to use the variogram for interpolation as a means of determining the amount of influence that neighboring observations have when predicting values for unobserved locations. Basic linear geostatistics includes two main procedures: variography for modeling a variogram, and kriging for interpolation.

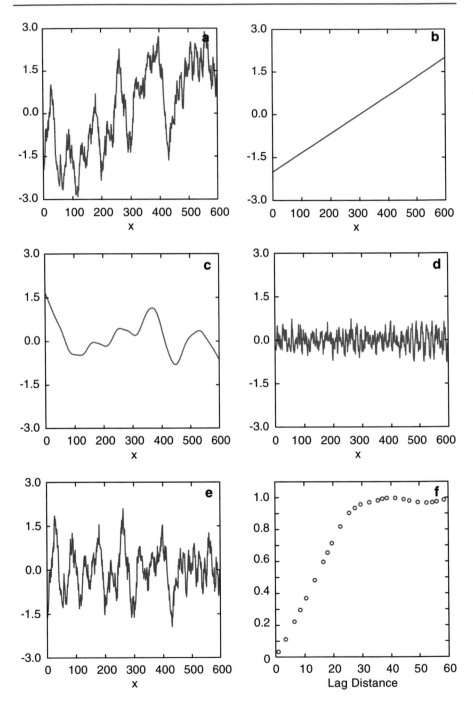

Fig. 7.17 Components of a spatiotemporal process and the variogram: **a** spatiotemporal process, **b** global trend component, **c** local trend component, **d** random component, **e** autocorrelation component, and **f** variogram. The variogram should only be derived from the autocorrelated component.

Preceding Analysis

Because linear geostatistics as presented in this book is a parametric method, its underlying assumptions need to be checked by a preceding analysis. As with other parametric methods, linear geostatistics is sensitive to outliers and to deviations from a normal distribution. We first open the data file *geost_dat.mat*, which contains *xyz* data triplets, and we then plot the sampling locations. This allows us to check the point distribution and to detect any major errors in the data coordinates *x* and *y*:

```
%reset -f

import numpy as np
from numpy import linalg
from scipy import stats
import matplotlib.pyplot as plt

data = np.loadtxt('geost_dat.txt')
x = data[:,0]
y = data[:,1]
z = data[:,2]

plt.figure()
plt.plot(data[:,0],data[:,1],
    marker='o',
    markeredgecolor=(0.3,0.5,0.8),
    markerfacecolor='none',
    linestyle='none')
plt.show()
```

The range of the observations *z* can be checked by typing

```
print(np.min(z))
print(np.max(z))
```

which yields

```
3.7198573
7.8459703
```

For linear geostatistics, the observations *z* should be Gaussian distributed. This is usually only tested by visually inspecting the histogram because statistical tests are often too sensitive if the number of samples exceeds about 100. It is also possible to calculate the skewness and kurtosis of the data by typing

```
plt.hist(z)
print(stats.skew(z))
print(stats.kurtosis(z,fisher=True))
```

which yields

```
0.2568129442682396
-0.4780014841875708
```

A flat-topped or multiple-peaked distribution suggests that there is more than one population present in the data set. If these populations can be related to particular areas, they should be treated separately. Another reason for the existence of multiple peaks can be due to the preferential sampling of areas with high and/or low values. This usually happens as a result of some a priori knowledge and is referred to as a cluster effect. Dealing with a cluster effect is described in Deutsch and Journel (1998) and in Isaaks and Srivastava (1990).

Most problems arise from positive skewness, that is, if the distribution has a long tail to the right. According to Webster and Oliver (2007), root transformation should be considered if the skewness is between 0.5 and 1, and logarithmic transformation should be considered if the skewness exceeds 1. A general transformation formula is

$$z^* = \begin{cases} \left((z+m)^k - 1\right)/k, & \text{for } k \neq 0 \\ \log(z+m), & \text{for } k = 0 \end{cases}$$

for $\min(z) + m > 0$. This is known as the Box–Cox transform, which has the special case $k = 0$ when a logarithm transformation is used. In the logarithm transformation, m should be added if z is zero or negative. Interpolation results of power-transformed values can be back-transformed directly after kriging. The back-transformation of log-transformed values is slightly more complicated and is explained later. The procedure is known as *lognormal kriging* and can be important because lognormal distributions are not uncommon in geology. Other transformations in geostatistics are discussed in Deutsch and Journel (1998) and in Remy et al. (2006). The transformation of compositional data is discussed in Chapter 9.7.

Variography with the Classic Variogram

A variogram describes the spatial dependency of referenced observations in a unidimensional or multidimensional space. Since the true variogram of the spatial process is usually unknown, it must be estimated from observations. This procedure is called variography and starts by calculating the *experimental variogram* from the raw data. In the next step, the experimental variogram is summarized by the *variogram estimator*. The variography then concludes by fitting a variogram model to the variogram estimator. The experimental variogram is calculated as the differences between pairs of observed values and is dependent on the *separation*

Fig. 7.18 Separation vector
h between two points.

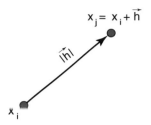

$$\vec{x_j} = \vec{x_i} + \vec{h}$$

$\vec{x_i}$

vector h (Fig. 7.18). The classic experimental variogram is defined by the *semi-variance*

$$\gamma(h) = 0.5 \cdot (z_x - z_{x+h})^2$$

where z_x is the observed value at location x and z_{x+h} is the observed value at another point at a distance h from x. The length of the separation vector h is called the *lag distance* (or simply the *lag*). The correct term for $\gamma(h)$ is the *semivariogram* (or *semivariance*), where *semi* refers to the fact that $\gamma(h)$ is half the variance of the differences between z_x and z_{x+h}. Nevertheless, $\gamma(h)$ is the variance per point when points are considered in pairs (Webster and Oliver 2007). However, by convention, $\gamma(h)$ is termed a *variogram* instead of a semivariogram – a convention that we follow for the rest of this section. To calculate the experimental variogram, we first need to group pairs of observations. This is achieved by typing

```
X1,X2 = np.meshgrid(x,x)
Y1,Y2 = np.meshgrid(y,y)
Z1,Z2 = np.meshgrid(z,z)
```

The array of separation distances D between the observation points is

```
D = np.sqrt((X1-X2)**2+(Y1-Y2)**2)
```

We then obtain the experimental variogram G as half the squared differences between the observed values:

```
G = 0.5*(Z1-Z2)**2
```

In order to speed up the processing, we use Python's ability to vectorize commands instead of using for loops. However, we compute n^2 pairs of observations even though only $n(n-1)/2$ pairs are required. For large data sets (e.g., more than 3000 data points), the software and physical memory of the computer may become limiting factors. In such a case, loops should be used, and aggregated results should be stored in a variogram grid instead of generating an array with individual pairs, as is done in the SURFER software (Golden Software, Inc.). The plot of the

experimental variogram is called the *variogram cloud* (Fig. 7.19), which we obtain
by extracting the lower triangular portions of the D and G arrays:

```
indx = np.linspace(1,len(z),len(z))
C,R = np.meshgrid(indx,indx)
I = R>C

plt.figure()
plt.plot(D[I],G[I],
    marker='.',
    linestyle='none')
plt.xlabel('lag distance')
plt.ylabel('variogram')
plt.show()
```

The variogram cloud provides a visual impression of the dispersion of values at
the different lags. It can be useful for detecting outliers or anomalies, but it is difficult to judge from this presentation whether there is any spatial correlation and, if
so, what form it might have or how we could model it (Webster and Oliver 2007).

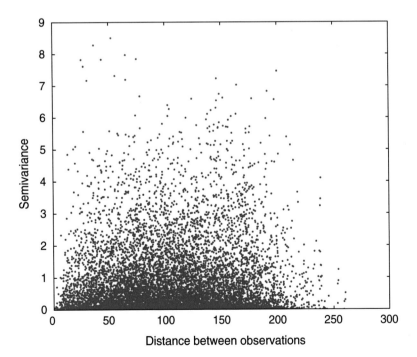

Fig. 7.19 Variogram cloud: plot of the experimental variogram (half the squared difference
between pairs of observations) versus the lag distance (separation distance between the two components of a pair).

In order to obtain a clearer view and prepare a variogram model, the experimental variogram is now replaced by the variogram estimator.

The variogram estimator is derived from the experimental variograms in order to summarize their central tendency (similar to the descriptive statistics derived from univariate observations in Sect. 3.2). The classic variogram estimator is the averaged empirical variogram within certain distance classes (or bins), which are defined by multiples of the lag interval. The classification of separation distances is illustrated in Fig. 7.20. The variogram estimator is calculated by

$$\gamma_E(h) = \frac{1}{2 * N(h)} \cdot \sum_{i=1}^{N(h)} (z_{x_i} - z_{x_{i+h}})^2$$

where $N(h)$ is the number of pairs within the lag interval h.

We first need to decide on a suitable lag interval h. If sampling has been carried out on a regular grid, the length of a grid cell can be used. If the samples are unevenly spaced (as in our case), the mean minimum separation distance of pairs is a good starting point for the lag interval (Webster and Oliver 2007). In order to calculate the mean minimum separation distance between pairs, we need to replace the zeros in the diagonal of the lag array D with NaNs; otherwise, the mean minimum separation distance will be zero. Typing

```
D2 = D*(np.diag(x*np.nan)+1)
D2m = np.zeros(len(D2))
for i in range(0,len(D2)):
    D2m[i] = np.nanmin(D2[i])
lag = np.mean(D2m)
print(lag)
```

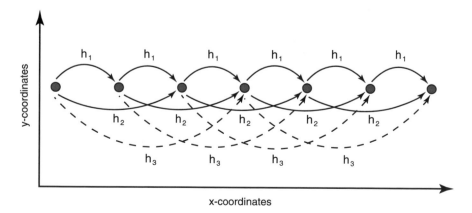

Fig. 7.20 Classification of separation distances for observations that are equally spaced. The lag interval is h_1, while h_2, h_3, etc. are multiples of the lag interval.

yields

```
8.010698943460342
```

Since the estimated variogram values tend to become more erratic with increasing separation distances, it is important to place a maximum separation distance limit on the calculation. As a rule of thumb, half of the maximum separation distance is a suitable limit for variogram analysis. We obtain the half-maximum distance and the maximum number of lags by typing

```
hmd = np.max(D)/2
max_lags = int(np.floor(hmd/lag))
print(hmd)
print(max_lags)
```

which yields

```
130.19008189950569
16
```

The separation distances are then classified – and the classic variogram estimator is calculated – by typing

```
LAGS = np.ceil(D2/lag)

DE = np.zeros(max_lags)
PN = np.zeros(max_lags)
GE = np.zeros(max_lags)

for i in range(1,max_lags+1):
    SEL = (LAGS == i)
    DE[i-1] = np.mean(D[SEL])
    PN[i-1] = np.sum(SEL)/2
    GE[i-1] = np.mean(G[SEL])
```

where SEL is the selection array defined by the lag classes in LAG, DE is the mean lag, PN is the number of pairs, and GE is the variogram estimator. We can now plot the classic variogram estimator (variogram versus mean separation distance) together with the population variance by typing

```
plt.figure()
plt.plot(DE,GE,
    marker='o',
    markerfacecolor = [0.6,0.6,0.6],
    linestyle='none')
```

```
var_z = np.var(z,ddof=1)
xl = np.max(DE)*1.1
yl = 1.1*max(GE)
b = np.array([0,xl])
c = np.array([var_z,var_z])

plt.plot(b,c,'r--')
plt.axis(np.array([0,xl,0,yl]))
plt.xlabel('Averaged distance between observations')
plt.ylabel('Averaged semivariance')
plt.show()
```

The variogram in Fig. 7.21 exhibits a typical pattern. Values are low at small separation distances (near the origin) and then increase with increasing distance until
reaching a plateau (*sill*) that is close to the population variance. This indicates that
the spatial process is correlated over short distances but not over longer distances.
The extent of the spatial dependency is called the *range* and is defined as the separation distance at which the variogram reaches the sill.

The *variogram model* is a parametric curve fitted to the variogram estimator. This is similar to frequency distribution fitting (see Sect. 3.5), in which the
frequency distribution is modeled by a distribution type and by its parameters

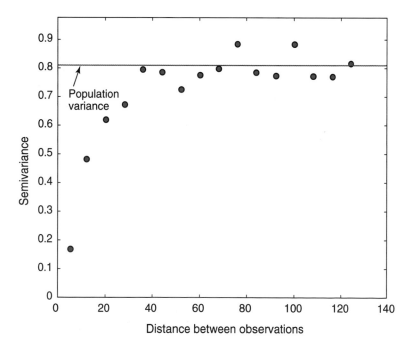

Fig. 7.21   The classic variogram estimator (blue) and the population variance (red line).

(e.g., a normal distribution with its mean and variance). For theoretical reasons, only functions with certain properties should be used as variogram models. Common *authorized* *models* are the spherical model, the exponential model, and the linear model (more models can be found in the relevant published literature),

Spherical model:

$$\gamma_{sph}(h) = \begin{cases} c \cdot \left(1.5 \cdot \frac{h}{a} - 0.5 \left(\frac{h}{a}\right)^3\right), & \text{for } 0 \le h \le a \\ c, & \text{for } h > a \end{cases}$$

Exponential model:

$$\gamma_{exp}(h) = c \cdot \left(1 - \exp\left(-3 \cdot \frac{h}{a}\right)\right)$$

Linear model:

$$\gamma_{lin}(h) = b \cdot h$$

where c is the sill, a is the range, and b is the slope (for a linear model). The parameters c and either a or b must be modified if a variogram model is fitted to the variogram estimator.

The *nugget* *effect* is a special type of variogram model. In practice, when extrapolating the variogram toward a separation distance of zero, we often observe a positive intercept on the y-axis. This is called the nugget effect and is explained by measurement errors as well as by small-scale fluctuations (*nuggets*) that are not captured due to overly large sampling intervals. We sometimes have expectations about the minimum nugget effect from the variance of repeated measurements in the laboratory or from other prior knowledge. More details concerning the nugget effect can be found in Cressie (1990, 1993) and in Kitanidis (1997). If there is a nugget effect, it can be added to the variogram model. An exponential model with a nugget effect looks like this:

$$\gamma_{exp+nug}(h) = c_0 + c \cdot \left(1 - \exp\left(-3 \cdot \frac{h}{a}\right)\right)$$

where c_0 is the nugget effect.

We can even combine variogram models (e.g., two spherical models with different ranges and sills). These combinations are called *nested* *models*. During variogram modeling, the components of a nested model are regarded as spatial structures that should be interpreted as the result of geological processes. Before we discuss further aspects of variogram modeling, we first fit some models to our data. We begin with a spherical model with no nugget effect and then add an exponential model and a linear model, both with nugget variances:

```
plt.figure()
plt.plot(DE,GE,
    marker='o',
    markerfacecolor = [0.6,0.6,0.6],
    linestyle='none')
var_z = np.var(z,ddof=1)
xl - np.max(DE)*1.1
yl = 1.1*max(GE)
b = np.array([0, xl])
c = np.array([var_z,var_z])
plt.plot(b,c,'r--')
plt.axis(np.array([0,xl,0,yl]))
plt.xlabel('Averaged distance between observations')
plt.ylabel('Averaged semivariance')

nugget = 0
sill = 0.803
rang = 45.9
lags = np.arange(0,max(DE),1)
Gsph = nugget + (sill*(1.5*lags/rang - \
    0.5*(lags/rang)**3)*(lags<=rang) + \
    sill*(lags>rang))

plt.plot(lags,Gsph,'g:')

nugget = 0.0239
sill = 0.78
rang = 45
Gexp = nugget+sill*(1-np.exp(-3*lags/rang))
plt.plot(lags,Gexp,'b:')

nugget = 0.153
slope = 0.0203
Glin = nugget + slope*lags

plt.plot(lags,Glin,'m:')
plt.xlabel('Distance between observations')
plt.ylabel('Semivariance')
plt.legend(['Variogram estimator','Population variance',\
    'Spherical model','Exponential model','Linear model'])
plt.show()
```

The techniques of variogram modeling are still very much under discussion. Some researchers advocate for *objective* variogram modeling via automated curve fitting using a weighted least-squares method, a maximum likelihood method, or a maximum entropy method. In contrast, it is often argued that geological knowledge should be included in the modeling process, in which case visual fitting is therefore recommended. In many cases, the problem with variogram modeling is much less a question of whether the appropriate procedure has been used than a question of the quality of the experimental variogram. If the experimental variogram is good, both procedures will yield similar results.

Another important question in variogram modeling is the intended use of the model. In our case, the linear model does not at first appear to be appropriate (Fig. 7.22). Upon closer inspection, however, we can see that the linear model fits reasonably well over the first three lags. This can be sufficient if we use the variogram model only for kriging because in kriging, the nearby points are the most important points for the estimate (see discussion of kriging below). Different variogram models with similar fits close to the origin therefore yield similar kriging results if the sampling points are regularly distributed. However, if the objective is to describe the spatial structures, the situation is quite different. It then becomes important to find a model that is suitable over all lags and to accurately determine the sill and the range. A collection of geological case studies in Rendu and Readdy (1982) show how process knowledge and variography can be interlinked. Good guidelines for variogram modeling are provided by Gringarten and Deutsch (2001) and by Webster and Oliver (2007).

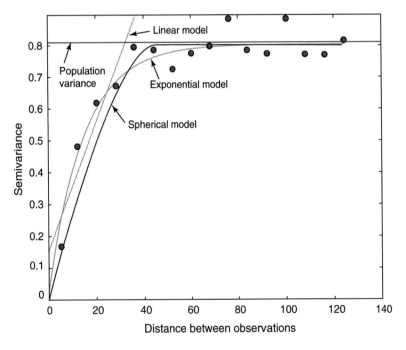

Fig. 7.22 Variogram estimator (blue circles), the population variance (red line), as well as spherical, exponential, and linear models (yellow, purple, and green lines, respectively).

We now briefly discuss a number of other aspects of variography:

- *Sample size* – As in any statistical procedure, as large a sample as possible is required in order to obtain a reliable estimate. For variography, the number of samples should be greater than 100 (and ideally, greater than 150) (Webster and Oliver 2007). For smaller sample numbers, a maximum likelihood variogram should be computed (Pardo-Igúzquiza and Dowd 1997).
- *Sampling design* – In order to obtain a good estimation close to the origin of the variogram, the sampling design should include observations over small distances. This can be achieved by means of a nested design (Webster and Oliver 2007). Other possible designs have been evaluated by Olea (1984).
- *Anisotropy* – Thus far, we have assumed that the structure of spatial correlation is independent of direction. We have calculated *omnidirectional variograms* while ignoring the direction of the separation vector *h*. In a more thorough analysis, the variogram should be discretized not only in distance, but also in direction (directional bins). By plotting *directional variograms* (usually in four directions), we are sometimes able to observe different ranges (*geometric anisotropy*), different scales (*zonal anisotropy*), and different shapes (which indicate a trend). Treating anisotropy requires a highly interactive graphical user interface, which lies beyond the scope of this book (see the VarioWin software by Panatier 1996).
- *Number of pairs and the lag interval* – When calculating the classic variogram estimator, more than 30 pairs of points (and ideally, more than 50) should be used per lag interval (Webster and Oliver 2007) due to the estimator's sensitivity to outliers. If there are fewer pairs, the lag interval should be increased. The lag spacing does not necessarily need to be uniform and can be chosen individually for each distance class. It is also possible to work with overlapping classes, in which case the *lag width* (*lag tolerance*) must be defined. However, increasing the lag width can cause unnecessary smoothing and a resulting loss of detail. The separation distance and the lag width must therefore be chosen with care. Another option is to use a more robust variogram estimator (Cressie 1990, 1993; Deutsch and Journel 1998).
- *Calculating the separation distance* – If the observations cover a large area (e.g., more than 1000 km^2),spherical distances instead of Pythagorean distances should be calculated from a planar Cartesian coordinate system.

Kriging

We now interpolate our observations to a regular grid using *ordinary point kriging*, which is the most popular kriging method. Ordinary point kriging uses a weighted average of the neighboring points to estimate the value of an unobserved point,

$$\widehat{z}_{x_0} = \sum_{i}^{n} \lambda_i \cdot z_{x_i}$$

where λ_i are the weights that must be estimated. The sum of the weights should be equal to one in order to guarantee that the estimates are unbiased:

$$\sum_{i}^{n} \lambda_i = 1$$

The expected (average) error for the estimation must be zero, that is,

$$E\left(\widehat{z}_{x_0} - z_{x_0}\right) = 0$$

where z_{x0} is the true (but unknown) value. We can use the equations above to algebraically compute the mean squared error in terms of the variogram,

$$E\left(\left(\widehat{z}_{x_0} - z_{x_0}\right)^2\right) = 2\sum_{i=1}^{n} \lambda_i \gamma(x_i, x_0) - \sum_{i=1}^{n}\sum_{j=1}^{n} \lambda_i \lambda_j \gamma(x_i, x_j)$$

where E is the estimation or *kriging variance* (which must be minimized), $\gamma(x_i, x_0)$ is the variogram (semivariance) between the data points and the unobserved points, $\gamma(x_i, x_j)$ is the variogram between the data points x_i and x_j, and λ_i and λ_j are the weights of the ith and jth data points.

For kriging, we must minimize this equation (a quadratic objective function) in order to satisfy the condition that the sum of the weights must be equal to one (linear constraint). This optimization problem can be solved using a Lagrange multiplier v, which results in a *linear kriging system* of $n+1$ equations and $n+1$ unknowns:

$$\sum_{i=1}^{n} \lambda_i \gamma(x_i, x_j) + v = \gamma(x_i, x_0)$$

After obtaining the weights λ_i, the kriging variance is given by

$$\sigma^2(x_0) = \sum_{i=1}^{n} \lambda_i \gamma(x_i, x_0) + v(x_0)$$

The kriging system can also be presented in array notation,

$$G\_mod \cdot E = G\_R$$

where

$$G\_mod = \begin{bmatrix} 0 & \gamma(x_1, x_2) & \dots & \gamma(x_1, x_n) & 1 \\ \gamma(x_2, x_1) & 0 & \dots & \gamma(x_2, x_n) & 1 \\ v & v & & v & v \\ \gamma(x_n, x_1) & \gamma(x_n, x_2) & \dots & 0 & 1 \\ 1 & 1 & \dots & 1 & 0 \end{bmatrix}$$

is the array of the coefficients (i.e., these are the modeled variogram values for the pairs of observations). Note that on the diagonal of the array (where the separation distance is zero), the value of γ disappears

$$E = \begin{bmatrix} \lambda_1 \\ \lambda_2 \\ \nu \\ \lambda_n \\ \nu \end{bmatrix}$$

is the vector of the unknown weights and the Lagrange multiplier

$$G\_R = \begin{bmatrix} \gamma(x_1, x_0) \\ \gamma(x_2, x_0) \\ \nu \\ \gamma(x_n, x_0) \\ 1 \end{bmatrix}$$

is the right-hand-side vector. To obtain the weights and the Lagrange multiplier, the array G_mod is inverted:

$$E = G\_mod^{-1} \cdot G\_R$$

The kriging variance is given by

$$\sigma^2 = G\_R^{-1} \cdot E$$

For our calculations using Python, we need the array of coefficients derived from the distance array D and a variogram model. The distance array D was calculated in the variography section above, and we use the exponential variogram model with a nugget, sill, and range from the previous section:

```
G_mod=(nugget+sill*(1-np.exp(-3*D/rang)))*(D>0)
```

We then take the number of observations and add a column and row vector for all values of one to the array G_mod and add a zero in the lower-left corner:

```
n = len(x)
o1 = np.ones((n,1))
o2 = np.ones((1,n+1))
G_mod = np.append(G_mod,o1,axis=1)
G_mod = np.append(G_mod,o2,axis=0)
G_mod[len(x):,len(x):] = 0.
```

The array G_mod must now be inverted:

```
G_inv = linalg.inv(G_mod)
```

A grid with the locations of the unknown values is needed. Here, we use a grid cell size of five within a quadratic area that ranges from 0 to 200 in the x and y directions. The coordinates are created in an array form by

```
R = np.arange(0,200+5,5)
Xg1,Xg2 = np.meshgrid(R,R)
```

and are converted to vectors by

```
Xg = np.reshape(Xg1,(len(Xg1)**2,1))
Yg = np.reshape(Xg2,(len(Xg2)**2,1))
```

We then allocate memory for the kriging estimates Zg and the kriging variance s2_k by typing

```
Zg = Xg * np.NaN
s2_k = Xg * np.NaN
```

We now krige the unknown values at each grid point:

```
for k in range(1,len(Xg)):
    DOR = ((x-Xg[k])**2+(y-Yg[k])**2)**0.5
    G_R = (nugget+sill*(1-np.exp(-3*DOR/rang)))*(DOR>0)
    G_R = np.append(G_R,1)
    E = np.matmul(G_inv,G_R)
    Zg[k] = np.sum(E[0:n]*z)
    s2_k[k] = np.sum(E[0:n]*G_R[0:n])+E[n]
```

The first command computes the distance between the grid points (Xg,Yg) and the observation points (x,y). We then create the right-hand-side vector of the kriging system using the variogram model G_R and by adding one to the last row. We next obtain the array E with the weights and the Lagrange multiplier. The estimate Zg at each point k is the weighted sum of the observations z. Finally, the kriging variance s2_k of the grid point is computed, and we can plot the results. We first create a grid of the kriging estimate and the kriging variance:

```
r = len(R)
Z = np.reshape(Zg,(r,r))
SK = np.reshape(s2_k,(r,r))
```

A subplot on the left presents the kriged values:

```
plt.figure()
fig,axs = plt.subplots(nrows=1,ncols=2,
```

```
        constrained_layout=True)
ax = axs[0]
im = axs[0].pcolor(Xg1,Xg2,Z,
        cmap='jet',
        shading='auto')
ax.set_aspect('equal','box')
plt.colorbar(im,ax=ax,orientation='horizontal')
ax.set_xlabel('x-coordinates')
ax.set_ylabel('y-coordinates')
```

The left-hand-side subplot presents the kriging variance,

```
ax = axs[1]
axs[1].pcolor(Xg1,Xg2,SK,
        cmap='jet',
        shading='auto')
```

and we overlay the sampling positions:

```
plt.plot(x,y,
        marker='o',
        markerfacecolor='none',
        markeredgecolor='k',
        linestyle='none')
ax.set_aspect('equal','box')
plt.colorbar(im,ax=ax,orientation='horizontal')
ax.set_xlabel('x-coordinates')
ax.set_ylabel('y-coordinates')
plt.show()
```

The kriged values are shown in Fig. 7.23a. The kriging variance depends only on the distance from the observations and not on the observed values (Fig. 7.23b). Kriging reproduces the population mean when observations lie beyond the range of the variogram, but the kriging variance also increases (lower-right corner of the maps in Fig. 7.23). The kriging variance can be used as a criterion to improve sampling design and is needed for back-transformation in lognormal kriging, which is achieved by

$$y(x_0) = \exp\Big(z(x_0) + 0.5 \cdot \sigma^2(x_0) - \nu\Big)$$

Discussion of Kriging

Point kriging as presented here is an exact interpolator. It reproduces the exact values at an observation point even though it uses a variogram with a nugget effect. Smoothing can be achieved by including the variance of the measurement errors

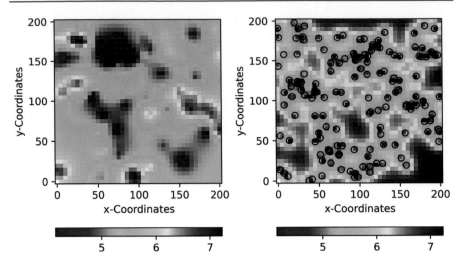

Fig. 7.23 Values interpolated to a regular grid by ordinary point kriging using **a** an exponential variogram model and **b** the kriging variance as a function of the distance from the observation (empty circles).

(see Kitanidis 1997) and by *block kriging*, which averages the observations within a certain neighborhood (or block). While the kriging variance depends only on the distance between the observed and unobserved locations, it is primarily a measure of the density of information (Wackernagel 2003). The accuracy of kriging is better evaluated by cross-validation using a resampling method or a surrogate test (Sects. 4.5 and 4.6). The influence of the neighboring observations on the estimation depends on their configuration, as summarized by Webster and Oliver (2007): "Near points carry more weight than more distant ones; the relative weight of a point decreases when the number of points in the neighborhood increases; clustered points carry less weight individually than isolated ones at the same distance; data points can be screened by ones lying between them and the target." The sampling design for kriging is different from the design that might be optimal for variography. A regular grid (be it triangular or quadratic) can be regarded as optimal.

The Python code presented here is a straightforward implementation of the formulas presented above. In professional programs, the number of data points that enter the array *G_mod* is restricted – and the inversion of *G_mod* is avoided – by working with the covariances instead of variograms (Kitanidis 1997; Webster and Oliver 2007). For those interested in programming or in gaining a deeper understanding of algorithms, the publication by Deutsch and Journel (1998) is essential reading. The best Internet source is the homepage for AI-GEOSTATISTICS:

https://wiki.52north.org/AI_GEOSTATS/WebHome

Recommended Reading

Amante C, Eakins BW (2009) ETOPO1 1 Arc-Minute Global Relief Model: Procedures, Data Sources and Analysis. NOAA Technical Memorandum NESDIS NGDC-24

Bohlander, J, Scambos T (2007) Antarctic coastlines and grounding line derived from MODIS Mosaic of Antarctica (MOA). National Snow and Ice Data Center, Boulder, Colorado

Cressie N (1990) The origins of kriging. Mat Geol 22:239–252

Cressie N (1993) Statistics for spatial data, Revised. John Wiley & Sons, New York

Davis JC (2002) Statistics and data analysis in geology, 3rd edn. John Wiley & Sons, New York

Deutsch CV, Journel AG (1998) GSLIB – Geostatistical software library and user's guide, 2nd edn. Oxford University Press, Oxford

Farr TG, Rosen P, Caro E, Crippen R, Duren R, Hensley S, Kobrick M, Paller M, Rodriguez E, Roth L, Seal D, Shaffer S, Shimada J, Umland J, Werner M, Oskin M, Burbank D, Alsdorf D (2007) The Shuttle Radar Topography Mission. Rev Geophys 45:RG2004

Farr TG, Kobrick M (2000) Shuttle radar topography mission produces a wealth of data. Am Geophys Union Eos 81:583–585

Freeman TG (1991) Calculating catchment area with divergent flow based on a regular grid. Comput Geosci 17:413–422

Gorny AJ (1977) World data bank II general user guide Rep. PB 271869. Central Intelligence Agency, Washington DC

Gringarten E, Deutsch CV (2001) Teacher's aide variogram interpretation and modeling. Math Geol 33:507–534

Isaaks E, Srivastava M (1990) An introduction to applied geostatistics. Oxford University Press, Oxford

Kitanidis P (1997) Introduction to geostatistics – applications in hydrogeology. Cambridge University Press, Cambridge

Olea RA (1984) Systematic sampling of spatial functions. Kansas series on spatial analysis 7, Kansas Geological Survey, Lawrence, KS

Pannatier Y (1996) VarioWin – software for spatial data analysis in 2D. Springer, Berlin Heidelberg New York

Pardo-Igúzquiza E, Dowd PA (1997) AMLE3D: a computer program for the inference of spatial covariance parameters by approximate maximum likelihood estimation. Comput Geosci 23:793–805

Remy N, Boucher A, Wu J (2006) Applied geostatistics with SGeMS: a user's guide. Cambridge University Press, Cambridge

Rendu JM, Readdy L (1982) Geology and semivariogram – a critical relationship. In: Johnson TB, Barns RJ (eds) Application of computer & operation research in the mineral industry. 17th Intern. Symp. American Institute of Mining. Metallurgical and Petroleum Engineers, New York, pp. 771–783

Schwanghart W, Scherler D (2014) TopoToolbox 2 – MATLAB-based software for topographic analysis and modeling in Earth surface sciences. Earth Surf Dyn 2:1–7

Schwanghart W, Kuhn NJ (2010) TopoToolbox: a set of Matlab functions for topographic analysis. Environ Model Softw 25:770–781

Soluri EA, Woodson VA (1990) World vector shoreline. Int Hydrogr Rev LXVI I(1):27–35

Swan ARH, Sandilands M (1995) Introduction to geological data analysis. Blackwell Sciences, Oxford

Trauth MH (2021) MATLAB Recipes for Earth Sciences – Fifth Edition. Springer, Berlin Heidelberg New York

Wackernagel H (2003) Multivariate geostatistics: an introduction with applications. Third, completely revised edition. Springer, Berlin Heidelberg New York

Webster R, Oliver MA (2007) Geostatistics for environmental scientists, 2nd edn. John Wiley & Sons, New York

Wessel P, Bercovici D (1998) Gridding with splines in tension: a green function approach. Math
    Geol 30:77–93
Wessel P, Smith WHF (1996) A global self-consistent, hierarchical, high-resolution shoreline
    database. J Geophys Res 101(B4):8741–8743
Wilson JP, Gallant JC (2000) Terrain analysis principles and applications. John Wiley & Sons,
    New York

Image Processing

8

8.1 Introduction

Computer graphics are stored and processed as either vector or raster data. Most of the data encountered in the previous chapter were vector data, including points, lines, and polygons. Examples of vector data include drainage networks, the outlines of geologic units, sampling locations, and topographic contours. In Chap. 7, coastlines were stored in a vector format, while bathymetric and topographic data were saved in a raster format. Vector and raster data are often combined in a single data set, for instance, when displaying the course of a river on a satellite image. Raster data are often converted to vector data by digitizing points, lines, or polygons. Conversely, vector data are sometimes transformed into raster data.

Images are generally represented as raster data, that is, as a 2D or 3D array of color intensities. Images are everywhere in the earth sciences. Field geologists use aerial photos and satellite images to identify lithologic units, tectonic structures, landslides, and other features within a study area. Geomorphologists use such images to analyze drainage networks, river catchments, and vegetation or soil types. Moreover, a great variety of image processing methods are used to analyze images from thin sections, to automatically identify objects, and to measure varve thicknesses.

Supplementary Information The online version contains supplementary material available at https://doi.org/10.1007/978-3-031-07719-7_8.

Sections 8.7–8.8 can be found in the MATLAB version of this book but could not be translated to Python. However, they are listed here with corresponding section numbers in order to create identical chapter numbering (and thereby also identical figure numbering and computer scripts) between the two versions of the book. The reader can thus use both books side by side to trace the translation of the scripts from MATLAB to Python (and back).

This chapter is concerned with analyzing and displaying image data. The various ways that raster data can be stored on the computer are first explained in Sect. 8.2. The main tools for importing, manipulating, and exporting image data are then presented in Sect. 8.3. This knowledge is next used to process and to geo-reference satellite images in Sects. 8.4 to 8.6. On-screen digitization techniques are discussed in Sect. 8.7, while Sects. 8.8 and 8.9 deal with importing, enhancing, and analyzing images from laminated lake sediments, including using color intensity measurements on transects across laminae. Sects. 8.10 to 8.12 deal with automated grain size analysis, charcoal quantification in microscopic images, and the detection of objects in microscopic images based on their shapes. Sect. 8.13 demonstrates how to calculate the normalized difference vegetation index (NDVI) both from handheld cameras and from satellite images. This chapter requires the *NumPy* (https://numpy.org), *Matplotlib* (https://matplotlib.org), *scikit-image* (https://scikit-image.org), *OpenCV* (https://pypi.org/project/opencv-python), and *pyhdf* (https://pypi.org/project/pyhdf), and *SciPy* packages (https://scipy.org).

8.2   Data Storage

Vector and raster graphics are the two fundamental formats for storing pictures. The typical format for storing *vector data* was introduced in the previous chapter. In the following example, the two columns in the file *coastline.txt* represent the longitudes and latitudes, respectively, of the points of a polygon:

```
NaN          NaN
42.892067    0.000000
42.893692    0.001760
NaN          NaN
42.891052    0.001467
42.898093    0.007921
42.904546    0.013201
42.907480    0.016721
42.910414    0.020828
42.913054    0.024642
(cont'd)
```

The NaNs help to identify break points in the data (Sect. 7.2).

Raster data are stored as 2D or 3D arrays. The elements of these arrays represent variables such as the altitude of a grid point above sea level, annual rainfall, or (in the case of an image) color intensity values:

```
174 177 180 182 182 182
165 169 170 168 168 170
171 174 173 168 167 170
184 186 183 177 174 176
191 192 190 185 181 181
189 190 190 188 186 183
```

Raster data can be visualized as 3D plots. The x-values and y-values are the indices of the 2D array or of any other reference frame, and z is the numerical value of the elements of the array (see also Chap. 7). The numerical values contained in the 2D array can be displayed as a pseudocolor plot, which is a rectangular array of cells with colors determined by a *colormap*. A colormap is an m-by-3 array of real numbers between 0.0 and 1.0. Each row defines a red, green, or blue (RGB) color. An example is the array above, which could be interpreted as grayscale intensities ranging from 0 (black) to 255 (white).

If we need more than one color or greyscale, we use the 3rd dimension of 3D arrays, for example, to store the three colors of red, green, and blue. More complex examples include satellite imagery acquired from different bands of the electromagnetic spectrum, such as the *visible and near-infrared* bands (VNIR; 400 to 1,400 nm), the *short-wave infrared* bands (SWIR; 1,400 to 3,000 nm), and the *thermal* (or *long-wave*) *infrared* bands (TIR; 3,000 to 15,000 nm). The imagery is then stored in 3D arrays, whose 3rd dimension corresponds to the number of spectral bands, which sometimes amount to several hundred (Sects. 8.4 to 8.6).

As previously discussed, a computer stores data as bits that have one of two states, which are represented by either a one or a zero (Chap. 2). If the elements of a 2D array represent the color intensity values of the *pixels* (short for *picture elements*) of an image, 1 bit arrays contain only ones and zeros:

```
0   0   1   1   1   1
1   1   0   0   1   1
1   1   1   1   0   0
1   1   1   1   0   1
0   0   0   0   0   0
0   0   0   0   0   0
```

This 2D array of ones and zeros can be simply interpreted as a black-and-white image in which the value of one corresponds to white and zero corresponds to black. Alternatively, the 1 bit array could be used to store an image consisting of any two different colors, such as red and blue.

In order to store more complex types of data, the bits are joined together to form larger groups, such as bytes consisting of eight bits. Since the earliest computers could only process eight bits at a time, early computer code was written in sets of eight bits, which came to be called bytes. Each element of the 2D array (i.e., each pixel) therefore contains a vector of eight ones and zeros:

```
1   0   1   0   0   0   0   1
```

These 8 bits (or 1 byte) allow $2^8 = 256$ possible combinations of the eight ones and zeros and are therefore able to represent 256 different intensities, such as grayscales. The 8 bits can be read in the following way (from right to left): A single bit represents two numbers, two bits represent four numbers, three bits represent eight numbers, and so forth up to one byte (or eight bits), which represents 256 numbers. Each added bit doubles the count of numbers. Below is a comparison of binary and decimal representations of the number 161:

```
128   64   32   16    8    4    2    1      (value of the bit)
  1    0    1    0    0    0    0    1      (binary)

128 +  0 + 32  + 0 +   0 +   0 +   0 +  1 = 161 (decimal)
```

The end members of the binary representation of grayscales are

```
0    0    0    0    0    0    0    0
```

which is pure black, and

```
1    1    1    1    1    1    1    1
```

which is pure white. In contrast to the 1 bit array above, the 1 byte array allows a grayscale image of 256 different levels to be stored. Alternatively, the 256 numbers could be interpreted as 256 discrete colors. In either case, displaying such an image requires an additional source of information concerning how the 256 intensity values are converted to colors. Numerous global colormaps for interpretating 8 bit color images exist that allow raster images to be exchanged across platforms, whereas local colormaps are often embedded in a graphics file.

The disadvantage of 8 bit color images is that the 256 discrete colorsteps are not enough to simulate smooth transitions for the human eye. A 24 bit system is therefore used in many applications, with 8 bits of data for each RGB channel creating a total of $256^3 = 16,777,216$ colors. Such a 24 bit image is stored in three 2D arrays (or in one 3D array) of intensity values between 0 and 255:

```
195   189   203   217   217   221
218   209   187   192   204   206
207   219   212   198   188   190
203   205   202   202   191   201
190   192   193   191   184   190
186   179   178   182   180   169


209   203   217   232   232   236
234   225   203   208   220   220
224   235   229   214   204   205
223   222   222   219   208   216
209   212   213   211   203   206
206   199   199   203   201   187


174   168   182   199   199   203
198   189   167   172   184   185
188   199   193   178   168   172
186   186   185   183   174   185
177   177   178   176   171   177
179   171   168   170   170   163
```

Compared with the 1 bit and 8 bit representations of raster data, 24 bit storage certainly requires much more computer memory. In the case of very large data sets, such as satellite images and digital elevation models, a user should therefore think carefully about the most suitable way of storing the data. The default floating-point data type in Python is the 64 bit array, which allows for storing the sign of a number (bit 63), the exponent (bits 62 to 52), and roughly 16 significant decimal digits between approximately 10^{-308} and 10^{+308} (bits 51 to 0). However, Python also works with other data types (e.g., 1 bit, 8 bit, and 24 bit raster data) in order to save memory space.

The amount of memory required for storing a raster image depends on the data type and the dimensions of the image. These image dimensions can be described by the number of pixels, which is the number of rows multiplied by the number of columns of the 2D array. Let us assume an image of 729-by-713 pixels, such as the one we use in the following section. If each pixel needs 8 bits to store a gray-scale value, the memory required by the data is $729 \cdot 713 \cdot 38 = 4,158,216$ its (or 519,777 bytes). This number is exactly what we obtain by typing whos in the command window. Common prefixes for bytes include kilo, mega, giga, and so forth:

```
bit = 1 or 0 (b)
8 bits = 1 byte (B)
1024 bytes = 1 kilobyte (kB)
1024 kilobytes = 1 megabyte (MB)
1024 megabytes = 1 gigabyte (GB)
1024 gigabytes = 1 terabyte (TB)
```

Note that in data communication, 1 kilobit = 1,000 bits, while in data storage, 1 kilobyte = 1,024 bytes. A 24 bit image (known as a *true-color image*) then requires three times the memory needed to store an 8 bit image, or 1,559,331 bytes = 1,559,331/1,024 kilobytes (kB) ≈ 1,523 kB ≈ $1,559,331/1,024^2 = 1.487$ megabytes (MB).

However, the dimensions of an image are often given not by the total number of pixels, but by the length and height of the image as well as by its resolution. The resolution of an image is the number of *pixels per inch* (ppi) or *dots per inch* (dpi) it contains. The standard resolution of a computer display remained at 72 dpi for a long time, but modern displays have a higher resolution, such as 96 dpi or even higher. In fact, the 5 K display (on which this very sentence is being written) has ~218 dpi. Thus, a 17 inch display with a 72 dpi resolution displays 1,024-by-768 pixels. If the display is used for images of a different (lower or higher) resolution, the image is resampled to match the display's resolution. For scanning and printing, a resolution of 300 or 600 dpi is enough in most applications. However, scanned images are often scaled for large printouts and therefore often have higher resolutions, such as 2,400 dpi. The image used in the next section has a width of 25.2 cm (or 9.92 inches) and a height of 25.7 cm (10.12 inches). The resolution of the image is 72 dpi. The total number of pixels is therefore $72 \cdot 9.92 \approx 713$ in a horizontal direction and $72 \cdot 10.12 \approx 729$ in a vertical direction.

Numerous formats are available for saving vector and raster data in a file, and each format has its own particular advantages and disadvantages. The choice of one format over another in an application depends on the way the images are to be used in a project and on whether or not the images are to be analyzed quantitatively. The most popular formats for storing vector and raster data are

- *CompuServe Graphics Interchange Format* (GIF) – This format was developed in 1987 for raster images that use a fixed 8 bit colormap of 256 colors. The GIF uses compression without losing data. It was designed for fast transfer rates over the Internet. The limited number of colors means that it is not the right format for the smooth color transitions used in aerial photos or satellite images. However, the format is often used for line art, maps, cartoons, and logos (http:// www.compuserve.com).
- *Portable Network Graphics* (PNG) – Developed in 1994, this image format is used as an alternative to the GIF. It is similar to the GIF in that it also uses a fixed 8 bit colormap of 256 colors. Alternatively, grayscale images of 1 to 16 bits can be stored, as can 24 bit and 48 bit color images. The PNG format uses compression without losing data, and the employed method is better than that used for GIF images.
- *Tagged Image File Format* (TIFF) – This format was jointly designed by the Aldus Corporation and Microsoft in 1986 with the goal of becoming an industry standard for exchanging image files. A TIFF file includes an image file header, a directory, and the data in all available graphics and image file formats. Some TIFF files even contain vector and raster versions of the same picture as well as images at different resolutions and with different colormaps. The main advantage of TIFF files was originally their portability. A TIFF should be able to perform on all computer platforms; unfortunately, however, numerous modifications to the TIFF have evolved in subsequent years, resulting in incompatibilities. The TIFF is therefore now often referred to as the *Thousands of Incompatible File Formats.*
- *PostScript* (PS) and *Encapsulated PostScript* (EPS) – The PS format was developed by John Warnock at PARC, the Xerox research institute. Warnock was also co-founder of Adobe Systems, where the EPS format was created. The PostScript vector format would never have become an industry standard without Apple Computers. In 1985, Apple needed a typesetter-quality controller for the new Apple LaserWriter printer and the Macintosh operating system, and the company thus adopted the PostScript format. The third partner in the history of PostScript was Aldus, which developed the PageMaker software and is now part of Adobe Systems. The combination of Aldus PageMaker software, the PS format, and the Apple LaserWriter printer led to the creation of desktop publishing. The EPS format was then developed by Adobe Systems as a standard file format for importing and exporting PS files. Whereas a PS file is generally a single-page format that contains either an illustration or a text, the purpose of an EPS file is to also allow other pages to be included (i.e., a file that can contain any combination of text, graphics and images) (http://www.adobe.com).

- In 1986, the *Joint Photographic Experts Group* (JPEG) was founded for the purpose of developing various standards for image compression. Although the name JPEG stands for the committee, it is now widely used as the name of an image compression format and a file format. This compression involves grouping pixel values into 8-by-8 blocks and transforming each block with a discrete cosine transform. As a result, all unnecessary high-frequency information is deleted, which makes this compression method irreversible. The advantage of the JPEG format is the availability of a three channel 24 bit true-color version, which allows images with smooth color transitions to be stored. The new JPEG 2000 format uses a wavelet transform instead of the cosine transform (Sect. 5.8) (http://www.jpeg.org).
- *Portable Document Format* (PDF) – The PDF, which was also designed by Adobe Systems, is now a true self-contained cross-platform document. PDF files contain different types of content, which include text as well as vector and raster graphics. The newer versions of PDFs can also contain interactive content, such as hyperlinks, animated objects, audio, and video. These files are highly compressed, thereby allowing for fast Internet download. Adobe Systems provides Acrobat Reader free of charge for all computer platforms so that PDF files can be read (http://www.adobe.com).

8.3   Importing, Processing, and Exporting Images

We first need to learn how to read an image from a graphics file in the workspace. As an example, we use a satellite image that shows a 10.5-km-by-11-km subarea in northern Chile:

```
https://asterweb.jpl.nasa.gov/gallery/images/unconform.jpg
```

The file *unconform.jpg* is a processed Terra ASTER satellite image that can be downloaded free of charge from the NASA website. We first save this image in the working directory. The command

```
%reset -f

import numpy as np
from matplotlib import cm
import matplotlib.pyplot as plt
from skimage import color
from skimage import exposure
from skimage import io
from matplotlib.colors import ListedColormap

I1 = io.imread('unconform.jpg')
```

reads and decompresses the JPEG file, imports the data as a 24 bit RGB image array, and stores it in a variable I1. The command

```
np.who()
```

shows how the RGB array is stored in the workspace:

```
Name              Shape              Bytes          Type
================================================================

I1                729 x 713 x 3      1559331        uint8

Upper bound on total bytes   =      1559331
```

The details indicate that the image is stored as a 729-by-713-by-3 array, which represents one 729-by-713 array for each of the colors red, green, and blue. The listing of the current variables in the workspace also includes the information array *uint8,* that is, each array element represents one pixel and contains 8 bit integers. These integers represent intensity values between 0 (minimum intensity) and 255 (maximum intensity). As an example, the following is a sector in the upper-left corner of the data array for red,

```
print(I1[49:55,49:55,0])
```

which yields

```
[[174 177 180 182 182 182]
 [165 169 170 168 168 170]
 [171 174 173 168 167 170]
 [184 186 183 177 174 176]
 [191 192 190 185 181 181]
 [189 190 190 188 186 183]]
```

We can now view the image using `imshow()`

```
plt.figure()
io.imshow(I1)
plt.axis('off')
io.show()
```

from the *scikit-image* package, which opens a new figure window that shows an RGB composite of the image (Fig. 8.1). In contrast to the RGB image, a grayscale image only needs a 2D array to store all the necessary information. We therefore convert the RGB image to a grayscale image using the command `rgb2gray()` (RGB to gray):

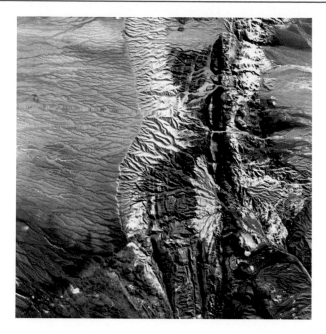

Fig. 8.1 RGB true-color image contained in the file *unconform.jpg*. After decompressing and reading the JPEG file in a 729-by-713-by-3 array, Python interprets and displays the RGB composite using `imshow()`. Original image courtesy of NASA/GSFC/METI/ERSDAC/JAROS and the U.S./Japan ASTER Science Team.

```
I2 = color.rgb2gray(I1)
```

The new workspace listing now reads

```
Name              Shape              Bytes          Type
================================================================

I1                729 x 713 x 3      1559331        uint8
I2                729 x 713          4158216        float64

Upper bound on total bytes  =      5717547
```

and the difference between the 24 bit RGB and the 8 bit grayscale arrays can be observed. The commands

```
plt.figure()
io.imshow(I2)
plt.axis('off')
io.show()
```

display the result. It is easy to see the difference between the two images in separate figure windows. Next, we process the grayscale image. To begin, we compute a histogram of the distribution of intensity values:

```
n,v = exposure.histogram(I2,nbins=256)

plt.figure()
plt.plot(v,n)
plt.show()
```

A simple technique for enhancing the contrast in such an image is to transform this histogram in order to obtain an equal distribution of grayscales:

```
I3 = exposure.equalize_hist(I2)

n,v = exposure.histogram(I3,nbins=256)

plt.figure()
plt.plot(v,n)
plt.show()
```

We can again view the difference using

```
plt.figure()
io.imshow(I3)
plt.axis('off')
io.show()
```

and save the results in a new file:

```
io.imsave('unconform_gray.jpg',I3)
```

We can also list the built-in colormaps by typing

```
plt.colormaps()
```

in the Console, which yields

```
['magma',
 'inferno',
 'plasma',
 'viridis',
 Z'cividis',
 'twilight',
```

```
'twilight_shifted',
'turbo',
'Blues',
'BrBG',
(cont'd)
```

As an example of colormaps, we can use hot from the list above by typing

```
print(cm.hot(range(16)))
newhot = cm.get_cmap(cm.hot,16)
newhot = cm.hot(np.linspace(0,1,16))
newhot = ListedColormap(newhot)

plt.figure()
plt.imshow(I3,cmap=newhot)
plt.axis('off')
plt.show()
```

in order to display the image with a black–red–yellow–white colormap. Typing

```
edit(cm.hot)
```

reveals that hot() is a function that creates an m-by-4 array containing floating-point values between 0 and 1. We can also design our own colormaps either by manually creating an m-by-3 array or by creating another function similar to hot(). As an example, we can use random numbers

```
np.random.seed(0)
newrnd = np.random.rand(16,3)
o1 = np.ones((len(newrnd),1))
newrnd = np.append(newrnd,o1,axis=1)
newrnd = ListedColormap(newrnd)

plt.figure()
plt.imshow(I3,cmap=newrnd)
plt.axis('off')
plt.show()
```

to display the image I3 with random colors. Finally, we can create an indexed color image of three different colors that are displayed with a simple colormap of full-intensity red, green, and blue:

```
newrgb = np.array([[1,0,0,1],[0,1,0,1],[0,0,1,1]])
newrgb = ListedColormap(newrgb)
```

```
plt.figure()
plt.imshow(I3,cmap=newrgb)
plt.axis('off')
plt.show()
```

8.4   Importing, Processing, and Exporting Landsat Images

The Landsat project is a satellite remote sensing program that is jointly managed by the U.S. National Aeronautics and Space Administration (NASA) and the U.S. Geological Survey (USGS). The project began with the launch of the Landsat 1 satellite (originally known as Earth Resources Technology Satellite 1) on 23 July 1972. According to the program's website, the latest in a series of successors is the Landsat 9 satellite, which was launched on 27 September 2021. The satellite has two sensors: the Operational Land Imager (OLI) and the Thermal Infrared Sensor (TIRS). These two sensors provide coverage of the global landmass at spatial resolutions of 30 m (visible, NIR, SWIR), 100 m (thermal), and 15 m (panchromatic) (Ochs et al. 2009; Irons et al. 2011). General information concerning the Landsat program can be obtained from the following website:

```
http://landsat.gsfc.nasa.gov/
```

Landsat data – together with data from other NASA satellites – can be obtained from the following website:

```
https://earthexplorer.usgs.gov
```

On this website, we first select the desired map section in the *Search Criteria* either by entering the coordinates of the four corners of the map or by zooming into the area of interest and selecting *Use Map*. As an example, we enter the coordinates 4°42′40.72″N 36°51′10.47″E, which locate the center of the Chew Bahir basin in the southern Ethiopian Rift. We then choose *L8 OLI/TIRS* from the *Landsat Archive* as the *Data Set* and click *Results,* which produces a list of records along with a toolbar for previewing and downloading data. By clicking the *Show Browse Overlay* button, we can examine the images for cloud cover. We then find the cloud-free image

```
Entity ID: LC81690572013358LGN00
Coordinates: 4.33915,36.76225
Acquisition Date: 24-DEC-13
Path: 169
Row: 57
```

taken on 24 December 2013. We need to register with the USGS website, log in, and download the Level 1 GeoTIFF Data Product (897.5 MB), which is then stored in the file *LC81690572013358LGN00.tar.gz.* The *.tar.gz* archive contains separate files for each spectral band as well as a metadata file that contains information about the data. We use Band 4 (red; 640 to 670 nm), Band 3 (green; 530 to 590 nm), and Band 2 (blue; 450 to 510 nm), each of which has a 30 m resolution. We can import the 118.4 MB TIFF files using

```
%reset -f

import numpy as np
import matplotlib.pyplot as plt
from skimage import exposure
from skimage import io
from skimage import filters

I1 = io.imread('LC81690572013358LGN00_B2.TIF')
I2 = io.imread('LC81690572013358LGN00_B3.TIF')
I3 = io.imread('LC81690572013358LGN00_B4.TIF')
```

Typing

```
np.who()
```

reveals that the data are in an unsigned 16 bit `uint16` format, that is, the maximum range of the data values lies between 0 and $2^{16}-1 = 65{,}535$:

```
Name            Shape               Bytes           Type
===============================================================

I1              7771 x 7611         118290162       uint16
I2              7771 x 7611         118290162       uint16
I3              7771 x 7611         118290162       uint16

Upper bound on total bytes   =      354870486
```

For quantitative analyses, these digital number (DN) values need to be converted to radiance and reflectance values, which is beyond the scope of this book. Radiance is the power density scattered from the Earth in a particular direction and is expressed in units of watts per square meter per steradian (Wm^{-2} sr^{-1}) (Richards 2013). The radiance values need to be corrected for atmospheric and topographic effects in order to obtain Earth surface reflectance percentages. The Landsat 8 Handbook provides the necessary information on these conversions:

```
https://www.usgs.gov/media/files/landsat-8-data-users-handbook
```

However, we use Landsat 8 data to create an RGB composite of Bands 4, 3, and 2 for use in fieldwork. Since the image has a relatively low level of contrast, we use equalize_adapthist() to perform *contrast-limited adaptive histogram equalization* (CLAHE) (Zuiderveld 1994). Unlike equalize_hist(), which we used before, equalize_adapthist() works on small regions (or tiles) of the image rather than on the entire image. The neighboring tiles are then combined using bilinear interpolation in order to eliminate edge effects:

```
I1 = exposure.equalize_adapthist(I1,
    clip_limit=0.1,nbins=256)
I2 = exposure.equalize_adapthist(I2,
    clip_limit=0.1,nbins=256)
I3 = exposure.equalize_adapthist(I3,
    clip_limit=0.1,nbins=256)
```

Using clip_limit with a real scalar between 0 and 1 limits the contrast enhancement. Higher numbers result in increased contrast, and the default value is 0.01. Using a clip_limit of 0.1 yields good results. The three bands are concatenated into a 24 bit RGB image using dstack():

```
I = np.dstack([I1,I2,I3])
```

We only display the section of the image that contains the Chew Bahir basin (using axes limits), and we hide the coordinate axes:

```
I8 = exposure.rescale_intensity(I,
    out_range='uint8')

plt.figure()
plt.imshow(I8)
plt.xlim([3000,5000])
plt.ylim([1000,4000])
plt.show()
```

The function rescale_intensity() from the *exposure* module of the *scikit-image* package does two things for us: First, it returns images after stretching or shrinking their intensity levels in order to improve the contrast. Second, it converts the data from any type to *uint8* without rolling the values around the value range [0, 255], as occurs when using uint8() from the *NumPy* package (see Sect. 2.5). To export the processed image, we use the command

```
io.imsave('chewbahirbasin.tif',
    I8[999:4000,2999:5000])
```

This command saves the RGB composite as the TIFF file *chewbahirbasin.tif* (about 18 MB) in the working directory, which can then be processed using other software, such as Adobe Photoshop.

According to the USGS Landsat website, Landsat data are among the most geometrically and radiometrically corrected data available. Data anomalies do occasionally occur, however, and the most common types are listed on the USGS website:

```
https://www.usgs.gov/land-resources/nli/landsat/landsat-known-issues
```

We explore one of these types of anomalies as an example: namely artifacts known as *Single Event Upsets* (SEUs), which cause anomalously high values in the image and are similar to *Impulse Noise* (IN), which is also described on the same website. These anomalies occur in some – but not all – Landsat images, and similarly anomalous high or low values can also occur in other satellite images. We therefore use part of a Landsat 7 image that covers an area in the southern Ethiopian Rift and that was acquired by the satellite's Enhanced Thematic Mapper (ETM+) instrument. We can load and display the image using

```
I1 = io.imread('ethiopianrift_blue.tif')
I1 = exposure.rescale_intensity(I1)

plt.figure()
io.imshow(I1,cmap='gray')
plt.title('Original image')
plt.axis('off')
io.show()
```

Image I1 shows numerous randomly distributed anomalously high or low values as well as a parallel track of paired anomalies in the right half. We first apply a 10-by-10-pixel median filter to the image (see Sect. 8.8):

```
I2 = filters.median(I1,
    np.ones((9,9)),mode='reflect')

plt.figure()
io.imshow(I2,cmap='gray')
plt.title('Median filtered image')
plt.axis('off')
io.show()
```

The median-filtered version of image I2 is, of course, very smooth compared with the original image I1; however, we would lose a lot of detail if we used this version of the image. We next subtract median-filtered image I2 from the original image I1, which yields image I3:

```
I3 = np.float64(I2)-np.float64(I1)
I3 = np.clip(I3,0,255)
I3 = np.uint8(I3)

plt.figure()
io.imshow(I3,cmap='gray')
plt.title('I2-I1')
plt.axis('off')
io.show()
```

We then subtract the original image I1 from the median-filtered image I2, which yields image I4:

```
I4 = np.float64(I1)-np.float64(I2)
I4 = np.clip(I4,0,255)
I4 = np.uint8(I4)

plt.figure()
io.imshow(I4,cmap='gray')
plt.title('I1-I2')
plt.axis('off')
io.show()
```

We next replace the original pixels with their median-filtered versions if the difference between the median-filtered image I2 and the original image I1 is greater than 10 in both directions (as it is in our example):

```
I5 = np.copy(I1)
I5[(I3>10) | (I4>10)] = I2[(I3>10) | (I4>10)]

plt.figure()
io.imshow(I5,cmap='gray')
plt.title('Despeckled image')
plt.axis('off')
io.show()
```

Image I5 obtained using this approach is the *despeckled* version of the image I1. We can also explore the pixel values of both versions of the image (i.e., I1 and I5) in a 3D surface plot using

```
b,a = np.shape(I1)
X,Y = np.meshgrid(np.linspace(0,a+1,a),
    np.linspace(0,b+1,b))
```

```
plt.figure()
fig,ax1 = plt.subplots(subplot_kw={"projection":"3d"})
ax1.plot_surface(X,Y,I1,
    rstride=1,
    cstride=1,
    linewidth=0,
    antialiased=False,
    shade=True)
ax1.view_init(33,120-90)
ax1.set_zlim(np.min(I1),np.max(I1))
ax1.axis('off')
plt.show()

plt.figure()
fig,ax2 = plt.subplots(subplot_kw={"projection":"3d"})
ax2.plot_surface(X,Y,I5,
    rstride=1,
    cstride=1,
    linewidth=0,
    antialiased=False,
    shade=True)
ax2.view_init(33, 120-90)
ax2.set_zlim(np.min(I1),np.max(I1))
ax2.axis('off')
plt.show()
```

We need convert the image data to the class double using im2double in order to be able to display the data using surface. Finally, we can display both images in the same figure window

```
plt.figure()
plt.subplot(1,2,1)
io.imshow(I1,cmap='gray')
plt.title('Original image')
plt.axis('off')
plt.subplot(1,2,2)
io.imshow(I5,cmap='gray')
plt.title('Despeckled image')
plt.axis('off')
io.show()
```

in order to see the results of *despeckling* image I1 (Fig. 8.2).

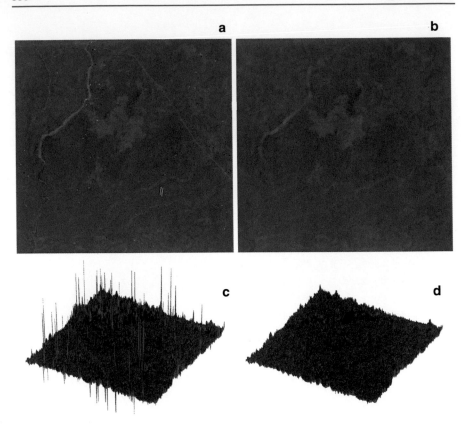

Fig. 8.2 Despeckled section of the blue band of a Landsat image covering the Chew Bahir catchment in southern Ethiopia: **a** original image, **b** despeckled image, **c** surface plots of the original image, and **d** surface plot of the image after despeckling. Original image courtesy of the Landsat Program of the U.S. National Aeronautics and Space Administration (NASA) and the U.S. Geological Survey (USGS).

8.5   Importing and Georeferencing Terra ASTER Images

In Sect. 8.3, we used a processed ASTER image that we had downloaded from the ASTER website. In this section, we use raw data from this sensor. The ASTER sensor is mounted on the Terra satellite, which was launched in 1999 as part of the Earth Observing System (EOS) series of multi-national NASA satellites (Abrams und Hook 2002). ASTER stands for *Advanced Spaceborne Thermal Emission and Reflection Radiometer* and provides high-resolution (15 to 90 m) images of the Earth in 14 bands, including three visible to near-infrared bands (VNIR Bands 1 to 3), six short-wave infrared bands (SWIR Bands 4 to 9), and five thermal (or long-wave) infrared bands (TIR Bands 10 to 14). ASTER images are used to map the temperature, emissivity, and reflectance of the Earth's surface. The 3rd VNIR band is recorded twice: once with the sensor pointing directly downward (Band 3 N, where *N* stands for *nadir,* from the Arabic word for *opposite*), as it does for all other channels, and a second time with the sensor angled backward at 27.6° (Band 3B, where *B* stands for *backward-looking*). These two bands are used to generate ASTER digital elevation models (DEMs).

The ASTER instrument produces two types of data: Level 1 A (L1A) data and Level 1B (L1B) data (Abrams und Hook 2002). Whereas L1A data are reconstructed, unprocessed instrument data, L1B data are radiometrically and geometrically corrected. Any data that ASTER has already acquired are available and can be obtained from the following website:

```
https://earthexplorer.usgs.gov
```

As an example, we now process an image of an area in Kenya that shows Lake Naivasha (0°46'31.38"S. 36°22'17.31"E). Level 1 A data are stored in two files:

```
AST_L1A_003_03082003080706_03242003202838.hdf
AST_L1A_003_03082003080706_03242003202838.hdf.met
```

The first file (116 MB) contains the actual raw data, while the second file (102 kB) contains the header together with all sorts of information about the data. We save both files in our working directory. Since the file name is very long, we first save it in `filename` and then use `filename` instead of the long file name. We then need to modify only this single line of Python code if we want to import and process other satellite images:

```
%reset -f

import numpy as np
import gdal as gdal
from pyhdf.SD import SD
import matplotlib.pyplot as plt
from skimage import io, exposure
from skimage import transform as tf

filename = 'AST_L1A_003_03082003080706_03242003202838.hdf'
```

The package *pyhdf* contains various tools for importing and processing files stored in the *hierarchical data format* (HDF). Information about the file can be obtained with `info()`,

```
file = SD(filename)
file.info()
```

which yields

```
(194, 10)
```

which suggests that we have 194 scientific data sets (SDS) within the HDF file. To acquire the SDS names from the Python *dict,* we use

```
datasets_dic = file.datasets()
print(datasets_dic)
datasets_dic
```

To print the SDS names (i.e., the IDs and the data fields), we use

```
for idx,sds in enumerate(datasets_dic.keys()):
    print(idx)
    print(sds)
```

To acquire a list of all 194 data sets, we use the commands shown below. Again, the first half of the name is the subdata set name, such as HDF4....:9, which is followed by the descriptor, such as [13x3] SatellitePosition VNIR_Band1 (64 bit floating-point):

```
asterfile = gdal.Open(filename)
aster_sds = asterfile.GetSubDatasets()
print(aster_sds)
```

We then choose Numbers 9, 20, and 31 as example data sets from the list above, which contains ImageData of *vnir_Band3n, vnir_Band2,* and *vnir_Band1,* each of which typically contains a lot of information about the lithology (including soils), vegetation, and water on the Earth's surface. These bands are therefore usually combined in 24 bit RGB images. We first read the data:

```
aster = gdal.Open('HDF4_SDS:UNKNOWN:"AST_L1A_003_'\
    '03082003080706_03242003202838.hdf":31')
I1 = aster.ReadAsArray()
aster = gdal.Open('HDF4_SDS:UNKNOWN:"AST_L1A_003_'\
    '03082003080706_03242003202838.hdf":20')
I2 = aster.ReadAsArray()
aster = gdal.Open('HDF4_SDS:UNKNOWN:"AST_L1A_003_'\
    '03082003080706_03242003202838.hdf":9')
I3 = aster.ReadAsArray()
```

These commands generate three 8 bit image arrays, each of which represents the intensity within a certain infrared (IR) frequency band of a 4,200-by-4,100-pixel image. We are not using the data for quantitative analyses and therefore do not need to convert the digital number (DN) values to radiance and reflectance values. The *ASTER User Handbook* provides the necessary information on these conversions (Abrams und Hook 2002). Instead, we are processing the ASTER image in order to create a georeferenced RGB composite of Bands 3 N, 2, and 1, which are used in fieldwork. We first use a contrast-limited adaptive histogram equalization method to enhance the contrast in the image by typing

```
I1 = exposure.equalize_adapthist(I1)
I2 = exposure.equalize_adapthist(I2)
I3 = exposure.equalize_adapthist(I3)
```

and then concatenate the result into a 24 bit RGB image using dstack():

```
naivasha_rgb = np.dstack([I1,I2,I3])
```

As with the previous examples, the 4,200-by-4,100-by-3 array can now be displayed using

```
naivasha_rgb = exposure.rescale_intensity(naivasha_rgb,
    out_range='uint8')
plt.figure()
io.imshow(naivasha_rgb)
io.imsave('naivasha.tif',naivasha_rgb)
plt.axis('off')
```

This command also saves the RGB composite as the TIFF file *naivasha.tif* (~52 MB) in the working directory, and the file can then be processed using other software, such as Adobe Photoshop. The processed ASTER image does not yet have a coordinate system and therefore needs to be tied to a geographical reference frame (*georeferencing*). The header of the HDF file can be used to extract the geodetic coordinates of the four corners of the image contained in *scenefourcorners*:

```
UL = [-0.319922,36.214332]
LL = [-0.878267,36.096003]
UR = [-0.400443,36.770406]
LR = [-0.958743,36.652213]
```

From these coordinates, we can calculate the `movingpoints` with respect to the original image by typing

```
movingpoints = np.array([
    [0.,4100.*(UL[1]-LL[1])/(UR[1]-LL[1])],
    [4200.*(UL[0]-LL[0])/(UL[0]-LR[0]),0.],
    [4200.*(UL[0]-UR[0])/(UL[0]-LR[0]),4100.],
    [4200.,4100.*(LR[1]-LL[1])/(UR[1]-LL[1])]])
```

The four corners of the image correspond to the pixels in the four corners of the original image, which we store in a variable named `fixedpoints`:

```
fixedpoints = np.array([
    [0.,0.],
    [4200.,0.],
    [0.,4100.],
    [4200.,4100.]])

fixedpoints = np.fliplr(fixedpoints)
movingpoints = np.fliplr(movingpoints)
```

The function `estimate_transform()` now takes the pairs of control points (i.e., `movingpoints` and `fixedpoints`) and uses them to infer a spatial transformation matrix `tform`. We use an affine transformation using `affine`, which preserves points, straight lines, and planes and also keeps parallel lines parallel after transforming the image:

```
tform = tf.estimate_transform('affine',movingpoints,fixedpoints)
```

Finally, the affine transformation can be applied to the original RGB composite `naivasha_rgb` using `warp()` in order to obtain a georeferenced version of the satellite image `newnaivasha_rgb` with the same size as `naivasha_rgb` using `tform`:

```
newnaivasha_rgb = tf.warp(naivasha_rgb,tform)
```

The georeferenced image is displayed with coordinates on the axes and with a superimposed grid (Figs. 8.3 and 8.4). By default, `imshow()` inverts the latitude axis when images are displayed. To invert the latitude axis direction back to normal, we need to use `invert_yaxis()` by typing

```
plt.figure()
io.imshow(np.flipud(newnaivasha_rgb),
    extent=[36.096003,36.770406,-0.319922,-0.958743])
ax=plt.gca()
ax.invert_yaxis()
```

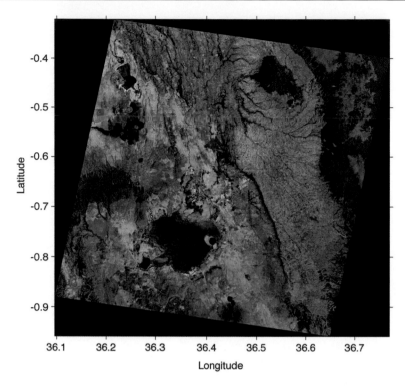

Fig. 8.3 Georeferenced RGB composite of a Terra ASTER image using the infrared Bands *vnir_Band3n, 2,* and *1.* The result is displayed using `imshow()`. Original image courtesy of NASA/GSFC/METI/ERSDAC/JAROS and the U.S./Japan ASTER Science Team.

Fig. 8.4 This figure from the MATLAB-based book MATLAB Recipes for Earth Sciences (Trauth 2021) could not be translated to Python.

```
plt.xlabel('Longitude')
plt.ylabel('Latitude')
plt.savefig('newnaivasha_rgb.png')
io.show()
```

This code also exports the image as the JPEG file *naivasha_georef.jpg.*

8.6   Processing and Exporting EO-1 Hyperion Images

The *Earth Observing-1* (EO-1) mission satellite was part of the New Millennium Program of the U.S. National Aeronautics and Space Administration (NASA) and the U.S. Geological Survey (USGS). The mission began with the launch of the satellite on 21 November 2000 and was deactivated on 30 March 2017. EO-1 has two sensors: the Advanced Land Image (ALI) (which has nine multispectral bands with a 30 m spatial resolution and a panchromatic band with a 10 m resolution) and the hyperspectral sensor (Hyperion) (which has 220 bands between 430 and 2,400 nm) (Mandl et al. 2002; Line 2012). General information about the EO-1 program can be obtained from the following website:

```
http://eo1.gsfc.nasa.gov
```

Hyperion data (together with data from of other NASA satellites) are freely available from the following website:

```
https://earthexplorer.usgs.gov
```

On this latter website, we first select the desired map section in the *Search Criteria* either by entering the coordinates of the four corners of the map or by zooming into the area of interest and selecting *Use Map.* As an example, we enter the coordinates 2°8'37.58"N 36°33'47.06"E and arrive at the center of the Suguta Valley in the northern Kenya Rift. We then choose *Hyperion* from the *EO-1* collection as the *Data Set* and click *Results,* which produces a list of records together with a toolbar for previewing and downloading data. Clicking the *Show Browse Overlay* button allows us to examine the images for cloud cover. In the overlay, we can find the cloud-free image

```
Entity ID: EO1H1690582013197110KF_PF2_01
Acquisition Date: 16-JUL-13
Target Path: 169
Target Row: 58
```

taken on 16 July 2013. As before, we need to register with the USGS website, log in, and then download the radiometrically (but not geometrically) corrected Level 1R (L1R) product (215.3 MB), which is then stored in the file *LEO1H1690582013197110KF_1R.ZIP.* The *.ZIP* archive consists of a metadata file (*.MET*), a Federal Geographic Data Committee (FGDC) metadata file (*.fgdc*), an HDF data set file (*.L1R*), and multiple auxiliary files. The EO-1 User's Guide provides some useful information on the data formats of these files (Barry 2001; Beck 2003). We can import the data from the file *EO1H1690582013197110KF. L1R* using

```
%reset -f
```

```
import numpy as np
import matplotlib.pyplot as plt
from skimage import io, exposure
from pyhdf.SD import SD

headerfile = 'EO1H1690582013197110KF.hdr'
imagefile = 'EO1H1690582013197110KF.L1R'

finfo = SD(imagefile)
file = finfo.select(0)
HYP = file.get()
```

Typing

```
np.who()
```

shows how the hyperspectral image is stored in the workspace:

```
Name              Shape                     Bytes             Type
===============================================================

HYP               3189 x 242 x 256          395129856         int16

Upper bound on total bytes  =        395129856
```

These details indicate that the image is stored as a 3,189-by-242-by-256 array, which represents a 3,189-by-256 array for each of the 242 spectral bands. The listing of the current variables in the workspace also includes the information *int16* (i.e., each array element represents one pixel and contains signed 16 bit integers). We need to permute the array in order to move the bands to the third dimension by typing

```
HYP = HYP.transpose(0,2,1)
```

We next need to determine the radiance values from the digital number (DN) values in HYP. The radiance is the power density scattered from the Earth in a particular direction and uses the units of watts per square meter per steradian ($Wm^{-2} sr^{-1}$) (Richards 2013). The EO-1 User Guide (v. 2.3) provides the necessary information on these conversions in its *Frequently Asked Questions* (FAQ) section (Beck 2003). According to this document, the radiance HYPR for the visible and near-infrared (VNIR) bands (Bands 1 to 70) is calculated by dividing the digital number in HYP by 40. The radiance for the shortwave infrared (SWIR) bands (Bands 71 to 242) is calculated by dividing HYP by 80:

```
HYPR = np.copy(HYP)
HYPR[:,:,0:70] = HYPR[:,:,0:70]/40
HYPR[:,:,70:242] = HYPR[:,:,70:242]/80
```

For quantitative analyses, the radiance values HYPR need to be corrected for atmospheric and topographic effects. This correction – which yields Earth surface reflectance values (in percentages) – is beyond the scope of the present book. The EO-1 User Guide (v. 2.3) again explains several methods of converting radiance values to reflectance values (Beck 2003).

We instead process the Hyperion image in order to create a georeferenced RGB composite of Bands 29, 23, and 16 for use in fieldwork. The header file *O1H1690582013197110KF.HDR* contains (*inter alia*) the wavelengths corresponding to the 242 spectral bands. We can read the wavelengths from the file using loadtext():

```
dtype = np.dtype("f8,f8,f8,f8,f8,f8,f8,f8")
C11 = np.loadtxt(headerfile,
    dtype = dtype,unpack=True,
    skiprows=257,max_rows=8,delimiter=",")
C11 = np.ravel(C11)
C11 = np.reshape(C11,(64,1))
dtype = np.dtype("f8,f8,f8,f8,f8,f8")
C12 = np.loadtxt(headerfile,
    dtype = dtype,unpack=True,skiprows=265,
    max_rows=1,delimiter=",",comments="}")
C12 = np.ravel(C12)
C12 = np.reshape(C12,(6,1))
C1 = np.vstack((C11,C12))
C1 = C1[C1[:,0].argsort()]
dtype = np.dtype("f8,f8,f8,f8,f8,f8,f8,f8")
C21 = np.loadtxt(headerfile,
    dtype = dtype,unpack=True,skiprows=265,
    max_rows=1,delimiter=",",comments="}")
C21 = np.ravel(C21)
C21 = np.reshape(C21,(8,1))
dtype = np.dtype("f8,f8,f8,f8,f8,f8,f8,f8")
C22 = np.loadtxt(headerfile,
    dtype = dtype,unpack=True,skiprows=266,
    max_rows=21,delimiter=",",comments="}")
C22 = np.ravel(C22)
C22 = np.reshape(C22,(168,1))
dtype = np.dtype("f8,f8")
C23 = np.loadtxt(headerfile,
    dtype = dtype,unpack=True,skiprows=287,
    max_rows=1,delimiter=",",comments="}")
C23 = np.ravel(C23)
C23 = np.reshape(C23,(2,1))
C2 = np.vstack((C21[6:8],C22,C23))
C2 = C2[C2[:,0].argsort()]
```

The character string f8,f8,f8,f8,f8,f8,f8,f8 enclosed in single quotation marks defines the conversion specifiers, where f8 stands for the double-precision floating-point 64 bit output class. We can easily obtain the wavelengths from C using

```
wavelengths = np.vstack((C1,C2))
```

Now, we can plot the radiance HYPR of the VNIR bands (in blue) and the SWIR bands (in red) in a single plot:

```
plt.figure()
plt.plot(wavelengths[0:70],HYPR[535,135,0:70],
    color=(0.3,0.5,0.8))
plt.plot(wavelengths[70:242],HYPR[535,135,70:242],
    color=(0.8,0.5,0.3))
plt.show()
```

According to v. 2.3 of the EO-1 User Guide (Beck 2003), Hyperion records 220 unique spectral channels that cover 357 to 2,576 nm. The L1R product has 242 bands, but only 198 bands are calibrated. Due to an overlap between the VNIR and SWIR focal planes, there are only 196 unique channels. The calibrated channels are 8 to 57 for the VNIR and 77 to 224 for the SWIR. The bands that are not calibrated are set to zero in these channels.

In order to create an RGB composite of Bands 29, 23, and 16, we can extract the bands from the radiance values data HYPR by typing

```
HYP1 = HYPR[:,:,28]
HYP2 = HYPR[:,:,22]
HYP3 = HYPR[:,:,15]
```

It is important to remember that Python indexing starts with zero and that we must therefore use *band*–1 as an index. To display the data with imshow, we need to convert the signed integer 16 bit (int16) data to unsigned integer 8 bit data (uint8). For this purpose, we first obtain an overview of the range of the data using a histogram plot with 100 classes:

```
n1,v1 = exposure.histogram(HYP1,nbins=100)
n2,v2 = exposure.histogram(HYP2,nbins=100)
n3,v3 = exposure.histogram(HYP3,nbins=100)

plt.figure()
plt.subplot(1,3,1)
plt.plot(v1,n1)
plt.title('Band 29')
plt.subplot(1,3,2)
plt.plot(v1,n1)
```

```
plt.title('Band 23')
plt.subplot(1,3,3)
plt.plot(v1,n1)
plt.title('Band 16')
plt.show()
```

As we can see, the radiance values of most pixels from spectral Bands 29, 23, and 16 lie between 0 and 200 Wm^{-2} sr^{-1}. Since most of our radiance values are within the range of [0,255], we use `rescale_intensity()` to convert our data to the `uint8` data type without losing much information:

```
HYP1 = exposure.rescale_intensity(HYP1,
    out_range='uint8')
HYP2 = exposure.rescale_intensity(HYP2,
    out_range='uint8')
HYP3 = exposure.rescale_intensity(HYP3,
    out_range='uint8')
```

Again, displaying the radiance values of the three bands in a histogram using

```
n1,v1 = exposure.histogram(HYP1,nbins=30)
n2,v2 = exposure.histogram(HYP2,nbins=30)
n3,v3 = exposure.histogram(HYP3,nbins=30)

plt.figure()
plt.subplot(1,3,1)
plt.plot(v1,n1)
plt.title('Band 29')
plt.subplot(1,3,2)
plt.plot(v1,n1)
plt.title('Band 23')
plt.subplot(1,3,3)
plt.plot(v1,n1)
plt.title('Band 16')
plt.show()
```

reveals that most radiance values are actually within the range of [20,80]. We next use `equalize_hist()` to enhance the contrast in the image and then concatenate the three bands into a 3,189-by-242-by-3 array:

```
HYP1 = exposure.equalize_hist(HYP1)
HYP2 = exposure.equalize_hist(HYP2)
HYP3 = exposure.equalize_hist(HYP3)
```

```
n1,v1 = exposure.histogram(HYP1,nbins=30)
n2,v2 = exposure.histogram(HYP2,nbins=30)
n3,v3 = exposure.histogram(HYP3,nbins=30)

plt.figure()
plt.subplot(1,3,1)
plt.plot(v1,n1)
plt.title('Band 29')
plt.subplot(1,3,2)
plt.plot(v1,n1)
plt.title('Band 23')
plt.subplot(1,3,3)
plt.plot(v1,n1)
plt.title('Band 16')
plt.show()

HYPC = np.dstack([HYP1,HYP2,HYP3])
HYPC = exposure.rescale_intensity(HYPC,
    out_range='uint8')
```

Finally, we can display the entire image using

```
plt.figure()
io.imshow(HYPC)
plt.axis('off')
io.show()
```

or (alternatively) by using only the part of the image that contains the Barrier Volcanic Complex in the northern Suguta Valley (Fig. 8.5):

```
plt.figure()
io.imshow(HYPC[899:1099,:,:])
plt.axis('off')
io.imsave('barrier_python.png',
    HYPC[899:1099,:,:])
io.show()
```

This code also exports the image as the TIFF file *barrier.tif* with a resolution of 600 dpi.

Fig. 8.5 RGB composite of an EO-1 Hyperion image using VNIR Bands 29, 23, and 16. The image shows the Barrier Volcanic Complex in the Suguta Valley of the northern Kenya Rift and was acquired on 16 July 2013. Original image courtesy of the NASA EO-1 Mission.

8.7   Digitizing from the Screen

This section from the MATLAB-based book MATLAB Recipes for Earth Sciences (Trauth 2021) could not be translated to Python.

8.8   Image Enhancement, Correction, and Rectification

This section introduces some fundamental tools for image enhancement, correction, and rectification. As an example, we use an image of varved sediments deposited around 33 kyrs ago in a landslide-dammed lake in the Quebrada de Cafayate in Argentina (25°58.900'S. 65°45.676'W) (Trauth et al. 1999, 2003). The diapositive was taken on 1 October 1996 with a film-based single-lens reflex (SLR) camera. A 30-by-20-cm print was made from the slide, which was scanned using a flatbed scanner and saved as a 394 kB JPEG file. We use this image as an example because it demonstrates some problems that we can solve with the help of image enhancement (Fig. 8.6). We then use the image to demonstrate how to measure color intensity transects for use in time series analysis (Sect. 8.9).

We can read and decompress the file *varves_original.jpg* by typing

```
%reset -f

import numpy as np
import skimage as ski
```

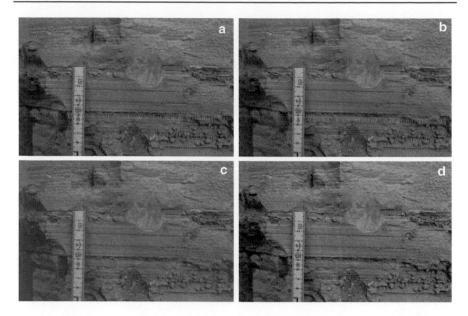

Fig. 8.6 Results of image enhancements: **a** original image, **b** image with intensity values adjusted using `rescale_intensity()`, **c** image with enhanced contrast using `equalize_adapthist()`, and **d** image after filtering with a 20-by-20-pixel filter with the shape of a Gaussian probability density function with a mean of zero and a standard deviation of 10 using `gaussian()`.

```
import matplotlib.pyplot as plt
from skimage import io, exposure, filters

I1 = io.imread('varves_original.jpg')
```

which yields a 24 bit RGB image array I1 in the Python workspace. Typing

```
np.who()
```

yields

Name	Shape	Bytes	Type
I1	1096 x 1674 x 3	5504112	uint8
Upper bound on total bytes =		5504112	

which reveals that the image is stored as a uint8 array of the size 1,096-by-1,674-by-3 (i.e., one 1,096-by-1,674 array each for the colors red, green, and blue). We can display the image using the command

```
plt.figure()
io.imshow(I1)
plt.axis('off')
io.show()
```

which opens a new figure window that shows an RGB composite of the image. As we can see, the image has a low level of contrast and very pale colors, and the sediment layers are not exactly horizontal. These are the characteristics of the image that we want to improve in the following steps.

First, we adjust the image intensity values. The function rescale_intensity() maps the values of image I1 to new values in I2,

```
I2 = exposure.rescale_intensity(I1)
```

which adjusts the ranges to the full range of [0,1] and then displays the result:

```
io.imshow(I2)
plt.axis('off')
io.show()
```

We can clearly see the difference between the very pale image I1 and the more saturated image I2.

The function equalize_hist() transforms the intensity of image I1 and returns an intensity image I5 with n discrete levels. A roughly equal number of pixels is ascribed to each of the n levels in I5 such that the histogram of I5 is approximately flat. Since equalize_hist() only works for two-dimensional images, histogram equalization must be carried out separately for each color. We use n=256 in our exercise:

```
I5 = np.zeros(np.shape(I1))
I5[:,:,0] = exposure.equalize_hist(I1[:,:,0],
    nbins=256)
I5[:,:,1] = exposure.equalize_hist(I1[:,:,1],
    nbins=256)
I5[:,:,2] = exposure.equalize_hist(I1[:,:,2],
    nbins=256)

plt.figure()
plt.subplot(2,2,1)
io.imshow(I1)
plt.axis('off')
plt.title('Original image')
plt.subplot(2,2,2)
n,v = exposure.histogram(I1,
    nbins=256)
```

```
plt.plot(v,n)
plt.title('Original image')
plt.subplot(2,2,3)
io.imshow(I5)
plt.axis('off')
plt.title('Enhanced image')
plt.subplot(2,2,4)
n,v = exposure.histogram(I5,
    nbins=256)
plt.plot(v,n)
plt.title('Enhanced image')
io.show()
```

The resulting image looks quite disappointing, and we therefore use the improved equalize_adapthist() instead of equalize_hist(). The function equalize_ adapthist() uses *contrast-limited adaptive histogram equalization* (CLAHE), which was created by Zuiderveld (1994). Unlike equalize_hist() and rescale_ intensity(), the equalize_adapthist() algorithm works on small regions (or tiles) of the image rather than on the entire image. The neighboring tiles are then combined using bilinear interpolation in order to eliminate edge effects:

```
I6 = np.zeros(np.shape(I1))
I6[:,:,0] = exposure.equalize_adapthist(I1[:,:,0],
    clip_limit=0.01,nbins=256)
I6[:,:,1] = exposure.equalize_adapthist(I1[:,:,1],
    clip_limit=0.01,nbins=256)
I6[:,:,2] = exposure.equalize_adapthist(I1[:,:,2],
    clip_limit=0.01,nbins=256)

plt.figure()
plt.subplot(2,2,1)
io.imshow(I1)
plt.axis('off')
plt.title('Original image')
plt.subplot(2,2,2)
n,v = exposure.histogram(I1,
    nbins=256)
plt.plot(v,n)
plt.title('Original image')
plt.subplot(2,2,3)
io.imshow(I6)
plt.axis('off')
plt.title('Enhanced image')
plt.subplot(2,2,4)
n,v = exposure.histogram(I6,
```

```
        nbins=256)
plt.plot(v,n)
plt.title('Enhanced image')
io.show()
```

The result looks slightly better than the result obtained using `equalize_hist()`. However, all three functions for image enhancement offer numerous ways of manipulating the final outcome. The excellent book by Gonzalez und Woods (2018) provides a more detailed introduction to using the various available parameters and the corresponding values of the image enhancement functions.

The *scikit-image* package also includes numerous functions for 2D image filtering. Many of the methods that we examined in Chap. 6 for one-dimensional data also work with two-dimensional data, as we saw in Chap. 7 when filtering digital terrain models. The most popular 2D filters for images are Gaussian filters and median filters as well as filters for image sharpening. Both Gaussian and median filters are used to smooth an image, mostly with the aim of reducing the amount of noise. In most examples, the signal-to-noise ratio is unknown, and adaptive filters (similar to those introduced in Sect. 6.10) are therefore used for noise reduction. A Gaussian filter can be designed using

```
I7 = filters.gaussian(I1)
```

which uses filter weights that follow the shape of a Gaussian probability density function. Next, we calculate `I8`, which is a median-filtered version of `I1`:

```
I8 = filters.median(I1)
```

The filter output pixels are the medians of the neighborhoods around the corresponding pixels in the input image. The third filter example demonstrates how to sharpen an image using `unsharp_mask()`:

```
I9 = filters.unsharp_mask(I1)
```

This function calculates the Gaussian lowpass filtered version of the image, which is used as an unsharp mask (i.e., the sharpened version of the image is calculated by subtracting the blurred filtered version from the original image). Comparing the results of the three filtering exercises with the original image

```
plt.figure()
plt.subplot(2,2,1)
io.imshow(I1)
plt.axis('off')
plt.title('Original image')
plt.subplot(2,2,2)
io.imshow(I7)
plt.axis('off')
```

```
plt.title('Gaussian filter')
plt.subplot(2,2,3)
io.imshow(I8)
plt.axis('off')
plt.title('Median filter')
plt.subplot(2,2,4)
io.imshow(I9)
plt.axis('off')
plt.title('Sharpening filter')
io.show()
```

clearly demonstrates the effect of the 2D filters. As an alternative to these space-domain filters, we can also design 2D filters with a specific frequency response, such as the 1D filters described in Sect. 6.9. Again, the book by Gonzalez und Woods (2018) provides an overview of 2D frequency-selective filtering for images, including the functions used to generate such filters. The authors also demonstrate the use of a 2D Butterworth lowpass filter in image processing applications.

8.9   Color Intensity Transects Across Varved Sediments

This Section from the MATLAB-based book MATLAB Recipes for Earth Sciences (Trauth 2021) could not be translated to Python (Figs. 8.7 and 8.8).

Fig. 8.7 This figure from the MATLAB-based book MATLAB Recipes for Earth Sciences (Trauth 2021) could not be translated to Python.

Fig. 8.8 This figure from the MATLAB-based book MATLAB Recipes for Earth Sciences (Trauth 2021) could not be translated to Python.

8.10   Grain Size Analysis from Microscopic images

Identifying, measuring, and counting particles in an image are the classic applications of image analysis. Examples from the geosciences include grain size analysis, counting pollen grains, and determining the mineral composition of rocks from thin sections. For grain size analysis, the task is to identify individual particles, to measure their sizes, and then to count the number of particles per size class. The reason for using image analysis lies in its ability to perform automated analyses of large sets of samples in a short period of time and at relatively low costs. Three different approaches are commonly used to identify and count objects in an image: (1) region-based segmentation using the watershed segmentation algorithm, (2) object detection using the Hough transform, and (3) thresholding using color differences to separate objects. Gonzalez und Woods (2018) describe these methods in great detail in the 3rd edition of their excellent book. We utilize two examples to demonstrate how image processing can be used to identify, measure, and count particles. In this section, we demonstrate an application of watershed segmentation in grain size analysis, and in Sect. 8.9, we introduce thresholding as a method for quantifying charcoal in microscopic images.

The following example covers segmenting, measuring, and counting objects using the watershed segmentation algorithm (Fig. 8.9). We first read an image of coarse lithic grains of different sizes and store it in the variable I1. The size of the image is $284 \cdot 367$ pixels, and since the width is 3 cm, the height is 3 cm $\cdot$ $284/67 = 2.32$ cm. We type

```
%reset -f

import cv2
import numpy as np
import matplotlib.pyplot as plt
from scipy import ndimage as ndi
from skimage import color
from skimage import exposure
from skimage import io
from skimage import feature
from skimage import measure
from skimage import segmentation
```

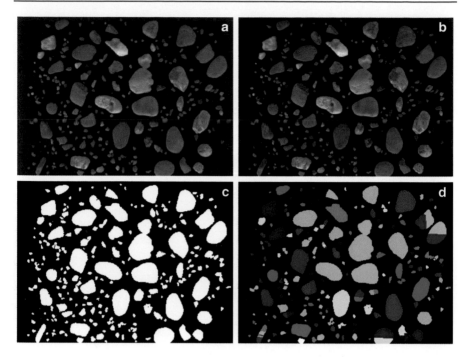

Fig. 8.9 Results from the automated grain size analysis of a microscopic image: **a** original gray-scale image, **b** image after adjusting intensity values, **c** image after conversion to a binary image, and **d** image with objects detected using a watershed segmentation algorithm.

```
I1 = io.imread('grainsize.tif')
ix = 3.
iy = 284*3/367

plt.figure()
io.imshow(I1,
    extent=[0,ix,0,iy])
plt.title('Original image')
plt.axis('off')
plt.show()
```

in which ix and iy denote the coordinate axes used to calibrate image I1 to a centimeter scale. The true number of objects counted in this image is 236 when including the three grains that overlap the borders of the image, and this number is therefore ignored in the following exercise. We can reject the color information of the image and convert image I1 to grayscale using rgb2gray():

```
I2 = color.rgb2gray(I1)
```

```
plt.figure()
io.imshow(I2,
    extent=[0,ix,0,iy])
plt.title('Grayscale image')
plt.axis('off')
plt.show()
```

This grayscale image I2 has a relatively low level of contrast. We therefore use rescale_intensity() to adjust the image intensity values, which increases the contrast in the new image I3:

```
I3 = exposure.rescale_intensity(I2,
    out_range='uint8')

plt.figure()
io.imshow(I3,
    extent=[0,ix,0,iy])
plt.title('Adjusted intensity values')
plt.axis('off')
plt.show()
```

The function threshold() converts the background-free image I3 to a binary image I4 via thresholding. If the threshold is 0, the image is all white, whereas if the threshold is 255, the image is all black. We manually change the threshold value until we get a reasonable result. We find 50 to be a suitable threshold:

```
thresh,I4 = cv2.threshold(I3,
    50,255,cv2.THRESH_BINARY)

plt.figure()
io.imshow(I4,
    extent=[0,ix,0,iy])
plt.title('Binary image')
plt.axis('off')
plt.show()
```

We next use the watershed segmentation algorithm to separate the grains in the image. Watersheds in geomorphology are ridges that divide areas that contribute to the hydrological budget of adjacent catchments (see Sect. 7.10). Watershed segmentation applies the same methods to grayscale images that are used to separate catchments in digital elevation models. In this application, the grayscale values are interpreted as elevations in a digital elevation model in which the watershed then separates the two objects of interest.

The criterion commonly used to identify pixels that belong to a particular object is the nearest neighbor distance. We use the distance transform performed by `distance_transform_edt()`, which assigns to each pixel a number that is the distance between the pixel and the nearest non-zero pixel in I4:

```
distance = ndi.distance_transform_edt(I4)
```

The function uses *Euclidean distances* to compute the nearest neighbor distances. The distance matrix now contains positive non-zero values in the object pixels and contains zeros elsewhere. We use this information to create a mask:

```
coords = feature.peak_local_max(distance,
    footprint=np.ones((15,15)),labels=I4)
mask = np.zeros(distance.shape,dtype=bool)
mask[tuple(coords.T)] = True
markers,_ = ndi.label(mask)
```

We then compute the watershed transform for the distance matrix and display the resulting label matrix:

```
L = segmentation.watershed(-distance,
    markers,mask=I4)

plt.figure()
io.imshow(color.label2rgb(L,bg_label=0),
    extent=[0,ix,0,iy])
plt.title('Watershed segmentation')
plt.axis('off')
plt.show()
```

After displaying the results from watershed segmentation, we determine the number of pixels for each object using the recipe described above, but index i runs from 2 to `max(objects)` since the value of 1 denotes the background and 0 denotes the boundaries of the objects. The first true object is therefore marked by the value of 2. Typing

```
print(np.max(L))

graindata = measure.regionprops(L)
grainareas = np.zeros(np.max(L))
for i in range(0,np.max(L)):
    grainareas[i] = graindata[i].area
```

yields

209

which suggests that we have now recognized 209 objects. However, visually inspecting the result reveals some oversegmentation (due to noise or other irregularities in the image) in which larger grains are divided into smaller pieces. We then scale the object sizes such that the area of one pixel is $(3 \text{ cm}/367)^2$:

```
objectareas = 3**2*grainareas*367**(-2)
```

We can now determine the areas for each of the grains and again find the maximum, minimum, and mean areas for all grains in the image in cm^2,

```
max_area = np.max(objectareas)
min_area = np.min(objectareas)
mean_area = np.mean(objectareas)
print(max_area)
print(min_area)
print(mean_area)
```

which yields

```
0.10156731433153413
6.682060153390403e-05
0.009231793632973584
```

The largest grain in the center of the image has a size of $\sim 0.10 \text{ cm}^2$, which represents the maximum size of all grains in the image. Finally, we plot the histogram of all the grain areas:

```
clf
e = 0 : 0.0005 : 0.15;
histogram(objectareas,e)
xlabel('Grain Size in Millimeters^2'),...
    ylabel('Number of Grains')
axis([0 0.1 0 70])
```

To check the final result, we digitize the outline of one of the larger grains and store the polygon in the variable data:

```
e = np.arange(0,0.15+0.0005,0.0005)

plt.figure()
plt.hist(objectareas,bins=e)
plt.xlabel('Grain Size in Millimeters^2')
plt.ylabel('Number of Grains')
plt.axis(np.array([0,0.1,0,30]))
```

If oversegmentation is a major problem when using segmentation to count objects in an image, it would be wise to refer to the book by Gonzalez und Woods (2018), which describes marker-controlled watershed segmentation as an alternative method that can be used to avoid oversegmentation.

8.11   Quantifying Charcoal in Microscopic images

Quantifying the composition of substances in geosciences (e.g., the mineral composition of a rock in thin sections or the amount of charcoal in sieved sediment samples) is facilitated by the use of image processing methods. *Thresholding* provides a simple solution to problems associated with segmenting objects within an image that have different coloration or grayscale values. During the thresholding process, pixels with an intensity value greater than a threshold value are marked as object pixels (e.g., pixels that represent charcoal in an image), and the rest are marked as background pixels (e.g., all other substances). The threshold value is usually defined manually by visually inspecting the image histogram, but numerous automated algorithms are also available.

As an example, we next analyze an image of a sieved lake sediment sample from Lake Nakuru, Kenya (Fig. 8.10). The image shows abundant light-gray oval ostracod shells and some mineral grains as well as gray plant remains and black charcoal fragments. We use thresholding to separate the dark charcoal particles and count the pixels of these particles after segmentation. After determining the number of pixels for all objects that can be distinguished from the background by thresholding, we use a lower threshold value to determine the ratio of the number of pixels that represent charcoal to the number of pixels that represent all particles in the sample (i.e., to determine the percentage of charcoal in the sample).

We read the image of size 1,500-by-1,500 pixels and assume that the width and height of the square image are both one centimeter:

```
%reset -f

import cv2
import numpy as np
import matplotlib.pyplot as plt
from skimage import color
from skimage import exposure
from skimage import io

I1 = io.imread('lakesediment.jpg')
ix = 1
iy = 1

plt.figure()
io.imshow(I1,
```

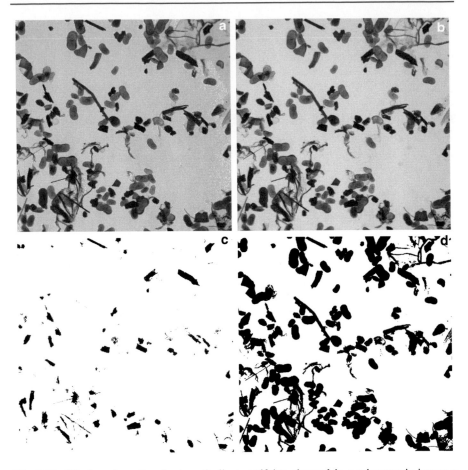

Fig. 8.10 Display of results of automatically quantifying charcoal in a microscopic image: **a** original grayscale image, **b** image after contrast enhancement, **c** image after thresholding in order to separate charcoal particles, and **d** image after thresholding in order to separate all objects.

```
        extent=[0,ix,0,iy])
    plt.xlabel('Centimeter')
    plt.ylabel('Centimeter')
    plt.title('Original image')
    io.show()
```

The RGB color image is then converted to a grayscale image using `rgb2gray`:

```
    I2 = color.rgb2gray(I1)

    plt.figure()
    io.imshow(I2,
        extent=[0,ix,0,iy])
```

```
plt.xlabel('Centimeter')
plt.ylabel('Centimeter')
plt.title('Grayscale')
io.show()
```

Since the image contrast is relatively low, we use `rescale_intensity()` to adjust the image intensity values. This increases the contrast in the new image I2:

```
I3 = exposure.rescale_intensity(I2,
    out_range='uint8')

plt.figure()
io.imshow(I3,
    extent=[0,ix,0,iy])
plt.xlabel('Centimeter')
plt.ylabel('Centimeter')
plt.title('Better contrast')
io.show()
```

The function `threshold()` converts the image I4 to a binary image I6 via thresholding. If the threshold is 0, the image is all white, whereas if the threshold is 255, the image is all black. We manually change the threshold value until we get a reasonable result. In our example, a threshold of 43 produces good results for identifying charcoal fragments:

```
thresh, I4 = cv2.threshold(I3,
    43,255,cv2.THRESH_BINARY)
I4 = exposure.rescale_intensity(I6,
    out_range='uint8')

plt.figure()
io.imshow(I4,
    extent=[0,ix,0,iy],
    cmap='gray')
plt.xlabel('Centimeter')
plt.ylabel('Centimeter')
plt.title('Only charcoal')
io.show()
```

Since we know the size of a pixel, we can now simply count the number of pixels in order to estimate the total amount of charcoal in the image. Finally, we compute the area of all objects, including the charcoal:

```
thresh, I5 = cv2.threshold(I3,
    150,255,cv2.THRESH_BINARY)
```

```
plt.figure()
io.imshow(I5,
    extent=[0,ix,0,iy])
plt.xlabel('Centimeter')
plt.ylabel('Centimeter')
plt.title('All objects')
io.show()
```

We are not interested in the absolute area of charcoal in the image, but rather in the percentage of charcoal in the sample. Typing

```
print(100*np.sum(np.sum(I4==0))/np.sum(np.sum(I5==0)))
```

therefore yields

```
13.672022985345837
```

This result suggests that approximately 13 % of the sieved sample is charcoal. As a next step, we could quantify the other components in the sample (e.g., ostracods or mineral grains) by choosing different threshold values.

8.12   Shape-Based Object Detection in Images

Counting objects within images based on their shapes is a highly time-consuming task. This is sometimes carried out for round objects, for example, when counting planktonic foraminifera shells in order to infer past sea surface temperatures, when counting diatom frustules in order to infer the past chemical composition of lake water, or when counting pollen grains in order to determine species assemblages, which can be used to reconstruct regional air temperature and precipitation. The linear objects that are determined include faults in aerial photos and satellite images (in order to derive the present-day stress field of an area) and annual layers (varves) in thin sections (in order to establish an annually resolved stratigraphy).

The *Hough transform* – which was named after Paul VC Hough's related 1962 patent – is a popular technique with which to detect objects within images based on their shapes. The Hough transform was originally used to detect linear features, but soon after being patented, it became generalized for use in identifying objects of any shape (Duda und Hart 1972; Ballard 1981). The book by Gonzalez und Woods (2018) contains a comprehensive introduction to the Hough transform and to its applications in detecting objects within images. According to the authors' introduction to the method, the Hough transform is performed in two steps: In the first step, an edge detector is used to extract edge features (e.g., distinct sediment layers or the outlines of pollen grains) from an image. In the second step, lines (or objects of any other shape) that trace these edge features are identified. The *scikit-*

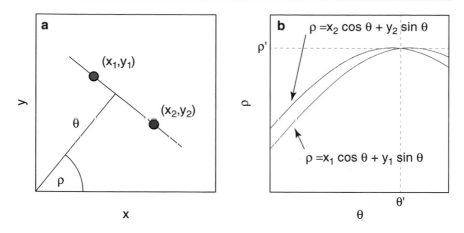

Fig. 8.11 Concept of the Hough transform: **a** parametrization of lines in the *xy*-plane and **b** sinusoidal curves in the (θ,ρ) parameter space, with the point of intersection corresponding to the line that passes through two different pixels of an edge feature. Modified from Gonzalez und Woods (2018).

image package contains functions that use the Hough transform to detect lines or circular objects.

The classic Hough transform is used to detect lines in images. After applying an edge detector of any kind, we end up with a binary image that has black pixels on the edges and white pixels in between. We next describe the lines through a given black pixel by using the Euclidean distance ρ between the line and the origin and by using the angle θ of the vector from the origin to the closest point on the line (Fig. 8.11a):

$$\rho = x \cos \theta + y \sin \theta$$

The family of all lines passing through this particular pixel (x_i, y_i) of an edge feature is displayed as a sinusoidal curve in the (θ, ρ) parameter space (Fig. 8.11b). The intersection point (θ', ρ') of two such sinusoidal curves corresponds to the line that passes through two different pixels (x_1, y_1) and (x_2, y_2) of an edge feature. Next, we search for n points (x_i, y_i) in the Hough transform in which many lines intersect since these points define the line that traces an edge feature. Detecting circles instead of lines works in a similar manner by using the coordinates of the center of the circle and its radius instead of ρ and θ.

For our first example, we use these functions to detect the thin layers of pure white diatomite within exposed varved sediments in the Quebrada de Cafayate in Argentina as these sediments were also used as examples in previous sections (Trauth et al. 1999, 2003) (Fig. 8.12). The quality of the image is not perfect, which is why we cannot expect optimal results. We first read the cropped version of the laminated sediment from Sect. 8.8 and store it in the variable I1. The size of the image is 1,047-by-1,691 pixels, which consist of three colors (red, green, and blue):

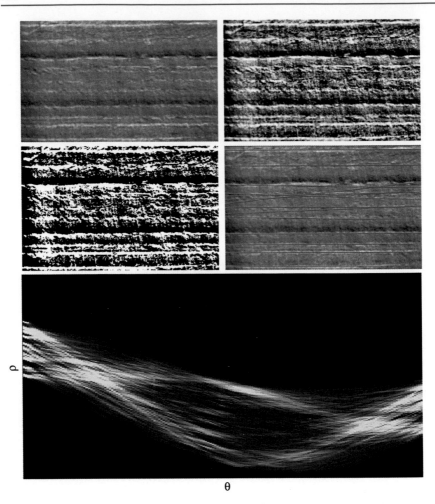

Fig. 8.12 Automated detection of thin layers of pure white diatomite within varved sediments exposed in the Quebrada de Cafayate in Argentina using `hough_line()` (Trauth et al. 1999, 2003): **a** grayscale image, **b** enhanced image, **c** binary image, **d** image with diatomite layers marked by red lines, and **e** Hough transform of the image.

```
%reset -f

import cv2
import numpy as np
import matplotlib.pyplot as plt
from skimage import color
from skimage import exposure
from skimage import io
from skimage import feature
from skimage import transform
from skimage import draw
```

```
I1 = io.imread('varves_cropped.tif')

plt.figure()
io.imshow(I1)
plt.axis('off')
io.show()
```

We can reject the color information of the image and convert I1 to grayscale using rgb2gray():

```
I2 = color.rgb2gray(I1)

plt.figure()
io.imshow(I2)
plt.axis('off')
io.show()
```

We then use equalize_adapthist() to perform *contrast-limited adaptive histogram equalization* (CLAHE) in order to enhance the contrast in the image (Zuiderveld 1994):

```
I3 = np.zeros(np.shape(I1))
I3 = exposure.equalize_adapthist(I2,
    clip_limit=0.1,nbins=256)

plt.figure()
io.imshow(I3)
plt.axis('off')
io.show()
```

In this case, clip_limit limits the contrast enhancement using a real scalar from 0 to 1, with higher numbers resulting in greater contrast (the default value is 0.01). Using a clip_limit of 0.1 yields good results. Using threshold() then converts image I3 to a binary image I4 via thresholding. We manually change the threshold value until we get a reasonable result and find 0.55 to be a suitable threshold:

```
thresh,I4 = cv2.threshold(I3,
    0.55,1.,cv2.THRESH_BINARY)
I4 = exposure.rescale_intensity(I4,
    out_range='uint8')

plt.figure()
io.imshow(I4)
plt.axis('off')
io.show()
```

The function `hough_line()` implements the Hough transform in order to return the Hough transform accumulator H, the angles theta at which the transform was computed (in radians), and the distance values dists:

```
H,theta,dists = transform.hough_line(I4)
```

We then display the Hough transform accumulator H using

```
plt.figure()
io.imshow(exposure.equalize_adapthist(H),
    cmap='hot',aspect='auto')
plt.xlabel('Theta')
plt.ylabel('Rho')
plt.title('Hough transform')
io.show()
```

We can determine the 15 maxima of the Hough transform, which correspond to the 15 most prominent lines in the image. We then calculate the *y*-intercepts yintcs and the slopes slopes of these lines in the image:

```
H,theta,dists = transform.hough_line_peaks(H,
    theta,dists,num_peaks=15,min_distance=20)

yintcs = dists*np.array([np.cos(theta),
    np.sin(theta)])
yintcs = np.transpose(yintcs)
slopes = np.tan(theta+np.pi/2)
```

This information can now be used to display the lines on the image:

```
s = np.shape(I1)

plt.figure()
ax = plt.axes()
io.imshow(I1)
plt.axis('off')
for i in range(0,len(dists)):
    ax.axline((yintcs[i,0],yintcs[i,1]),
        slope=slopes[i])
plt.axis(np.array([0,s[1],s[0],0]))
io.show()
```

The result reveals that the clearly recognizable white layers are well detected, whereas the less pronounced layers are not identified. The method also mistakenly marks non-existing lines on the image due to the low quality of the image. Using

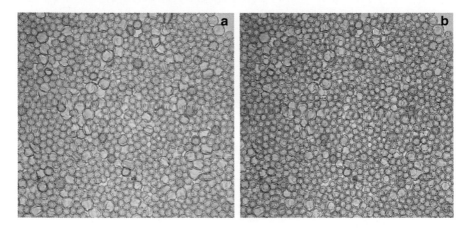

Fig. 8.13 Automated detection of pollen grains (mostly *Asteraceae,* with less abundant *Caesal-piniaceae* and *Lamiaceae* pollen) in a microscopic image of Argentine honey using `hough_cir-cle()` (original image courtesy of K. Schittek and F. Flores): **a** original RGB image and **b** pollen grains detected using the Hough transform.

a better-quality image and carefully adjusting the parameters used with the Hough transform would yield better results.

In the second example, the Hough transform is used to automatically count pollen grains in a microscopic image of Argentine honey (Fig. 8.13). The quality of the image is again not perfect, which is why we cannot expect optimum results. In particular, the image of three-dimensional objects was taken with very high magnification and is thus slightly blurred. We first read the pollen image and store it in the variable I1. The size of the image is 968-by-1,060 pixels of three colors (red, green, and blue). Since the image is relatively large, we reduce the image size by a factor of two:

```
I1 = io.imread('pollen.jpg')
I1 = I1[::2,::2,:]

plt.figure()
io.imshow(I1)
plt.axis('off')
io.show()
```

We can then reject the color information of the image and use the red color only:

```
I2 = I1[:,:,1]

plt.figure()
io.imshow(I2)
plt.axis('off')
io.show()
```

Next, we use `equalize_adapthist()` to perform contrast-limited adaptive histogram equalization (CLAHE) in order to enhance the contrast in the image (Zuiderveld 1994):

```
I3 = np.zeros(np.shape(I1))
I3 = exposure.equalize_adapthist(I2)

plt.figure()
io.imshow(I3)
plt.axis('off')
io.show()
```

The function `hough_circle()` implements the Hough transform and extracts circles from the original image:

```
I4 = feature.canny(I3,
    low_threshold=0.2,high_threshold=0.92)

hough_radii = np.arange(12,20)
hough_res = transform.hough_circle(I4,hough_radii)
H,cx,cy,radii = transform.hough_circle_peaks(
    hough_res,hough_radii,
    min_xdistance=20,min_ydistance=20,
    total_num_peaks=1000)
nstr = 'Number of pollen grains: '+str(len(radii))
```

We then display the circles on the grayscale image I2:

```
m,n,p = np.shape(I1)
plt.figure()
plt.axis('off')
I2 = color.gray2rgb(I2)
for i in range(0,len(H)):
    circy,circx = draw.circle_perimeter(cy[i],
        cx[i],radii[i])
    circyn = np.delete(circy,(circy>=m)|(circx>=n))
    circxn = np.delete(circx,(circy>=m)|(circx>=n))
    I2[circyn,circxn] = (255,0,0)
    io.imshow(I2)
io.imshow(I2,cmap='gray')
plt.title(nstr)
io.show()
```

The edge color of the circles in the graphics is set to b (for blue). The result indicates that we have counted 910 pollen grains using this method. The algorithm identifies most of the objects, but some are not recognized, and some of the larger objects are mistakenly identified as two or more pollen grains. Using a better-quality image and carefully adjusting the parameters used with the Hough transform would yield better results. Plotting the histogram of the pollen radii using `hist()`

```
plt.figure()
plt.hist(radii)
plt.show()
```

reveals that most of the grains have a radius of around 15 pixels.

8.13   The Normalized Difference Vegetation Index NDVI

Plant pigments reflect near-infrared light (700 to 1,100 nm), whereas they strongly absorb visible light (400 to 700 nm) for use in photosynthesis. The *normalized difference vegetation index* (NDVI) is an index used to map living green vegetation that is calculated with the formula

$$NDVI = \frac{(NIR - R)}{(NIR + R)}$$

using the spectral reflectance measurements acquired in the near-infrared (*NIR*) and red (R) spectral bands. The index can be measured with ground-based spectral cameras as well as by using drones, aircraft, or satellites. The cameras that are used are normally very expensive, but some rather inexpensive alternatives are now available, including the AgroCam (https://www.agrocam.eu) and the MAPIR Survey Camera (https://www.mapir.camera). These cameras are used in professional applications, but due to their low prices, they are also suitable for student exercises.

Since the MAPIR cameras (which are manufactured by Peau Productions, Inc., in San Diego, CA) have the same size and shape as the popular GoPro camera system, they can make use of all the accessories of the popular GoPro action camcorder, such as memory cards, camera mounts, and housings. We built an inexpensive multispectral array consisting of four cameras that record NDVI Red+NIR (650 and 850 nm), Blue (450 nm), Green (550 nm), and Visible Light RGB (~370 to 650 nm).

The following example explains how to acquire images with MAPIR cameras, how to import and process the images with Python, and how to calculate the NDVI in order to detect vegetation in an image. When editing images from the MAPIR Survey2 NDVI camera, we learn a lot about how cameras actually capture and store images. This is important in order to be able to correctly import the images into Python, to process them, and to compute the corresponding NDVI images. Failing to correctly complete any of these three steps yields an image that can be immediately seen to be unsatisfactory.

The following script uses information about the camera provided on the manufacturer's website (https://www.mapir.camera). The example does not include any calibration of the camera and uses an auto white balance rather than a true white

balance with a white reference. According to the MAPIR website's information
on *Calibrating Images in QGIS with MAPIR Plugin,* the Survey2 cameras capture
16 bit RAW photos, which means that there are 16 bit pixels with intensity values
ranging from 0 to 65,535. When the camera saves a JPEG file, it compresses the
image, thereby leaving a range of only 0 to 255. We first define a file name and the
size of the image:

```
%reset -f

import cv2
import numpy as np
import rasterio as rasterio
import matplotlib.pyplot as plt
from matplotlib.colors import ListedColormap
from skimage import color
from skimage import exposure
from skimage import io
from skimage import feature
from skimage import transform
from skimage import draw

filename = '2017_0328_110031_007.RAW';
row = 3456;
col = 4608;
```

According to the MAPIR specification page for the Survey2 NDVI camera, the
device uses a Sony Exmor IMX206 16MP (Bayer RGB), which is important
information required for correctly loading the images in Python. However, the
sensor specification page at Sony, Inc., says that the sensor actually acquires 12
bit images, which is in agreement with information received via email from the
software engineer at Peau Productions, Inc., in March 2017, but not with what is
stated on the company's website. With the support of this software engineer, our
own research, and some support from readers of Stack Overflow (http://stackover-
flow.com), we managed to solve the problem: If we read the data in a uint16 array
in Python, reshape it to (row,col), and display the image, we find that we have
arrived at the correct solution:

```
with open(filename) as fid:
    I1 = np.fromfile(fid,dtype='<u2')

I1 = np.reshape(I1,(row,col))
```

In this code, the data type <u2 stands for little-endian 2 byte (or 16 bit) unsigned
integers. The MAPIR Survey2 NDVI camera measures red (660 nm) and near-infrared (850 nm) radiances for the NDVI calculation. However, our red and infrared
data are nowhere to be found in the grayscale image. The Sony sensor is a single-

Fig. 8.14 Bayer RGGB
pattern helps to store RGB
images on the sensor of a
single-chip camera. The
pattern contains twice as
many green pixels as red
and blue pixels because
the human eye reacts more
sensitively to green than to
either of the other two colors.

chip sensor that stores the three colors (red, green, and blue) in a single array I1,
which is displayed as a grayscale image using the default gray colormap in Python.
Most cameras have single-chip sensors, as is even the case with single-lens reflex
cameras (SLRs), which can store images in the RAW format (in addition to in the
JPEG format) by using lossy compression. More expensive three-sensor cameras
also exist that use separate sensors for the filtered red, green, and blue color ranges.

Single-chip cameras store RGB images on their single chips using the Bayer
mosaic, which was named after Bryce Bayer (1929–2012), a research scientist at
Eastman Kodak (see http://en.wikipedia.org for more information on Bayer mosaics)
(Fig. 8.14). Bayer's idea was to use filters and to direct the individual colors of red,
green, and blue onto individual sensor pixels, which are usually in an RGGB arrange-
ment because the human eye reacts more sensitively to green than to red or blue. Due
to the Bayer mosaic, SLRs with 16 MP resolution (e.g., that of our Sony chip) do not
really take 16 MPs, but only 4 MPs each for red and blue (25 % each) and 8 MPs for
green (50 %), and they then interpolate each of the three colors to 16 MPs.

The RAW format provided by SLRs therefore does not mean that the colors are
stored in full resolution, but rather that the images are not compressed (e.g., using
the JPEG algorithm). Using Python instead, we can import and process the image
with its original resolution (i.e., 4 MPs each for red and blue and 8 MPs for green).
Alternatively, we can demosaic the image in the same way that an SLR does, for
example, by using the Python function cvtColor(). Demosaicing the image converts a Bayer pattern encoded image to a true-color (RGB) image using gradient-
corrected linear interpolation. According to the MAPIR website, the red (660 nm)
radiance is stored in the red image channel, the near-infrared (850 nm) radiance is
stored in the blue image channel, and no information is stored in the green channel.

When using imshow() to display I2 as a true-color RGB image, we find blue
hues in areas with high reflection in the NIR spectral band (and hence also denser
living green vegetation), whereas all other areas with low NIR values are more red
in color:

```
I2 = cv2.cvtColor(I1,cv2.COLOR_BAYER_BG2RGB)
I2 = exposure.rescale_intensity(I2,
    out_range='uint8')

plt.figure()
io.imshow(I2)
plt.axis('off')
plt.show()
```

To improve the contrast of the image, we can apply adaptive histogram equalization:

```
I3 = np.zeros(np.shape(I2))
I3[:,:,0] = exposure.equalize_adapthist(I2[:,:,0],
    clip_limit=0.01, nbins=256)
I3[:,:,1] = exposure.equalize_adapthist(I2[:,:,1],
    clip_limit=0.01, nbins=256)
I3[:,:,2] = exposure.equalize_adapthist(I2[:,:,2],
    clip_limit=0.01, nbins=256)

plt.figure()
io.imshow(I3)
plt.axis('off')
io.show()
```

We can now display the Red and NIR channels:

```
plt.figure()
plt.subplot(1,2,1)
io.imshow(I3[:,:,0])
plt.axis('off')
plt.title('Red')
plt.subplot(1,2,2)
io.imshow(I3[:,:,2])
plt.axis('off')
plt.title('NIR')
io.show()
```

We then perform a white balance to correct any unrealistic color casts. To do so, we need a reference card (usually gray), which is placed within the image and then used as a reference for the white balance. Alternatively, we can perform a fully automated white balance (as we do here) by dividing the values of each individual channel by the mean of the values of the individual colors of meanR, meanG, and meanB and multiplying the values of each channel by the mean value of the values of all channels meanGray:

```
I3G = color.rgb2gray(I3)
meanR = np.mean(I3[:,:,0])
meanG = np.mean(I3[:,:,1])
meanB = np.mean(I3[:,:,2])
meanGray = np.mean(I3G)
I3W = np.zeros(np.shape(I3))
I3W[:,:,0] = I3[:,:,0]*meanGray/meanR
I3W[:,:,1] = I3[:,:,1]*meanGray/meanG
I3W[:,:,2] = I3[:,:,2]*meanGray/meanB
```

Next, we calculate the NDVI using

```
I4 = np.copy(I3W)
ndvi = (I4[:,:,2]-I4[:,:,0])/(I4[:,:,2]+I4[:,:,0])

ndvi = exposure.rescale_intensity(ndvi,
    out_range='uint8')

plt.figure()
io.imshow(ndvi,
    cmap='gray')
plt.axis('off')
io.show()
```

Alternatively, we can display the NDVI as a pseudocolor plot using the colormap jet:

```
ndvi = exposure.rescale_intensity(ndvi,
    out_range='uint8')

plt.figure()
io.imshow(ndvi,
    cmap='jet')
plt.axis('off')
io.show()
```

Finally, we can define an arbitrary threshold th (e.g., 0.4) for separating vegetation (displayed as white pixels) from other materials (displayed as gray pixels). We then display NDVI>0.4 in green and the lower values in white (Fig. 8.15):

```
th = 0.4
ndvit = np.copy(ndvi)
for i in range(0,row):
        if ndvit[i,j] < th:
            ndvit[i,j] = 0
```

Fig. 8.15 RGB representation (left) of a demosaiced image captured with the MAPIR Survey2 NDVI camera and (right) of a threshold image from the normalized difference vegetation index (NDVI), with green pixels indicating the areas with high reflection values in the near-infrared frequency band. The red (660 nm) image is stored in the red image channel, and the near-infrared (850 nm) image is stored in the blue image channel. Since vegetation-rich areas reflect the near-infrared more strongly, they appear rather blue, while vegetation-poor areas appear red.

```
ndvicmap = np.array([[1,1,1,1],
    [0.5,0.6,0.3,1]])
newrgb = ListedColormap(ndvicmap)

thresh, ndvitth = cv2.threshold(ndvit,
    150,255,cv2.THRESH_BINARY)
plt.figure()
io.imshow(ndvitth,cmap=newrgb)
plt.axis('off')
io.show()
```

We can also calculate the NDVI from satellite data that include the red and near-infrared spectral bands. *Sentinel-2* is a system of four satellites designed to make terrestrial observations for a variety of tasks, including forest monitoring as well as mapping land surface changes and the impact of natural disasters. The data can be downloaded free of charge from the USGS EarthExplorer (https://earthexplorer. usgs.gov) website. We use two images (acquired on 22 and 29 May 2018, respectively) from which we load the 650 nm (B04) and 850 nm (B08) spectral bands required to calculate the NDVI:

```
filename1 = 'T32UQD_20180522T102031_B04_10m.jp2'
filename2 = 'T32UQD_20180522T102031_B08_10m.jp2'
filename3 = 'T32UQD_20180529T101031_B04_10m.jp2'
filename4 = 'T32UQD_20180529T101031_B08_10m.jp2'
img1 = rasterio.open(filename1)
img2 = rasterio.open(filename2)
img3 = rasterio.open(filename3)
img4 = rasterio.open(filename4)
B04_22 = img1.read(1)
B08_22 = img2.read(1)
B04_29 = img3.read(1)
B08_29 = img4.read(1)
```

We clip the image of the Berlin–Brandenburg area to the area of the city of Potsdam by typing

```
B04_22 = B04_22[8500:9500,6500:7500]
B08_22 = B08_22[8500:9500,6500:7500]
B04_29 = B04_29[8500:9500,6500:7500]
B08_29 = B08_29[8500:9500,6500:7500]
```

Next, we convert the uint16 data to float64 for the calculations:

```
B04_22 = np.float64(B04_22)
B08_22 = np.float64(B08_22)
B04_29 = np.float64(B04_29)
B08_29 = np.float64(B08_29)
```

To display the images, we first calculate the 5 % and 95 % percentiles

```
B04_22_05 = np.quantile(B04_22,0.05)
B04_22_95 = np.quantile(B04_22,0.95)
B08_22_05 = np.quantile(B08_22,0.05)
B08_22_95 = np.quantile(B08_22,0.95)
B04_29_05 = np.quantile(B04_29,0.05)
B04_29_95 = np.quantile(B04_29,0.95)
B08_29_05 = np.quantile(B08_29,0.05)
B08_29_95 = np.quantile(B08_29,0.95)
```

and then calculate the NDVI values for both images by typing

```
NDVI_22 = (B08_22-B04_22)/(B08_22+B04_22)
NDVI_29 = (B08_29-B04_29)/(B08_29+B04_29)
```

As an example, we display the 650 nm and 850 nm spectral bands of the image acquired on 22 May 2018 as grayscale images and display the NDVI map as a color image, both using the colormap jet,

```
plt.figure()
B04_22_X = exposure.rescale_intensity(B04_22,
    in_range=(B04_22_05,B04_22_95),
    out_range='uint16')
plt.imshow(B04_22_X,
    cmap='gray')
plt.colorbar()
plt.axis('off')
plt.title('650 nm')
plt.show()

plt.figure()
B08_22_X = exposure.rescale_intensity(B08_22,
    in_range=(B08_22_05,B08_22_95),
    out_range='uint16')
plt.imshow(B08_22_X,
    cmap='gray')
plt.colorbar()
plt.axis('off')
plt.title('850 nm')
plt.show()

plt.figure()
NDVI_22_X = exposure.rescale_intensity(NDVI_22,
    out_range=(0,1))
plt.imshow(NDVI_22_X,
    cmap='jet')
plt.colorbar()
plt.axis('off')
plt.title('NDVI 22 May 2018')
plt.show()
```

and we save the image in a PNG file:

```
print -dpng -r300 sentinel_ndiv_1_vs1.png
```

Then, we calculate the difference between the NDVI maps of the two images (acquired on 22 and 29 May 2018, respectively), and we again save the image in a PNG file:

```
DNDVI = NDVI_22 - NDVI_29

plt.figure()
B04_22_X = exposure.rescale_intensity(B04_22,
    in_range=(B04_22_05,B04_22_95),
    out_range='uint16')
plt.imshow(B04_22_X,
    cmap='gray')
plt.colorbar()
plt.axis('off')
plt.title('650 nm')
plt.show()

plt.figure()
B08_22_X = exposure.rescale_intensity(B08_22,
    in_range=(B08_22_05,B08_22_95),
    out_range='uint16')
plt.imshow(B08_22_X,
        cmap='gray')
plt.colorbar()
plt.axis('off')
plt.title('850 nm')
plt.show()

plt.figure()
DNDVI_X = exposure.rescale_intensity(DNDVI,
    in_range=(-0.2,0.2),
    out_range=(0,1))
plt.imshow(DNDVI_X,
    cmap='jet')
plt.colorbar()
plt.axis('off')
plt.title('Difference NDVI 22 - 29 May 2018')
plt.show()
```

The resulting image very nicely displays the progression in seasonal greening.

Recommended Reading

Abrams M, Hook S (2002) ASTER User Handbook—Version 2. Jet Propulsion Laboratory and EROS Data Center, Sioux Falls

Ballard DH (1981) Generalizing the Hough transform to detect arbitrary shapes. Pattern Recogn 13:111–122

Barry P (2001) EO-1/Hyperion Science Data User's Guide. TRW Space, Defense & Information Systems, Redondo Beach, CA

Beck R (2003) EO-1 User Guide. USGS earth resources observation Systems Data Center (EDC), Sioux Falls, SD

Duda RO, Hart PE (1972) Use of the Hough transform to detect lines and curves in pictures. Commun ACM 15:1–15

Gonzalez RC, Woods RE (2018) Digital image processing, Global. Pearson, London

Irons J, Riebeek H, Loveland T (2011) Landsat data continuity mission – continuously observing your world. NASA and USGS (available online)

Jensen JR (2013) Remote sensing of the environment: pearson new, International. Pearson, London

Lein JK (2012) Environmental sensing – analytical techniques for earth observation. Springer, Berlin Heidelberg New York

Mandl D, Cruz P, Frye S, Howard J (2002) A NASA/USGS collaboration to transform earth observing-1 Into a commercially viable mission to maximize technology infusion. SpaceOps 2002, Houston, TX, 9–12 October 2002

Ochs B., Hair D, Irons J, Loveland T (2009) Landsat data continuity mission, NASA and USGS (available online)

Richards JA (2013) Remote sensing digital image analysis –, 4th edn. Springer, Berlin Heidelberg New York

Trauth MH, Strecker MR (1999) Formation of landslide-dammed lakes during a wet period between 40,000 and 25,000 yr B.P. in northwestern Argentina. Palaeogeogr Palaeoclimatol Palaeoecol 153:277–287

Trauth MH, Bookhagen B, Marwan N, Strecker MR (2003) Multiple landslide clusters record Quaternary climate changes in the NW Argentine Andes. Palaeogeogr Palaeoclimatol Palaeoecol 194:109–121

Zuiderveld K (1994) Contrast limited adaptive histograph equalization. Academic Press Professional, Graphic Gems IV. San Diego, pp 474–485

Further Reading

Campbell JB (2002) Introduction to remote sensing. Taylor & Francis, London

Davies ER (2005) Machine vision: theory, algorithms, practicalities –, 3rd edn. Morgan Kaufman Publishers, Burlington MA

Francus P (2005) Image analysis, sediments and paleoenvironments – developments in paleoenvironmental research. Springer, Berlin Heidelberg New York

Hough PVC (1962) Method and means for recognizing complex patterns. US Patent No. 3069654

Marwan N, Trauth MH, Vuille M, Kurths J (2003) Nonlinear time-series analysis on present-day and Pleistocene precipitation data from the NW Argentine Andes. Clim Dyn 21:317–326

Seelos K, Sirocko F (2005) RADIUS - Rapid Particle Analysis of digital images by ultra-high-resolution scanning of thin sections. Sedimentology 52:669–681

Trauth MH, Alonso RA, Haselton KR, Hermanns RL, Strecker MR (2000) Climate change and mass movements in the northwest Argentine Andes. Earth Planet Sci Lett 179:243–256

Trauth MH (2021) MATLAB Recipes for Earth Sciences – Fifth Edition. Springer, Berlin Heidelberg New York

Yuen HK, Princen J, Illingworth J, Kittler J (1990) Comparative study of Hough transform methods for circle finding. Image Vis Comput 8:71–77

Multivariate Statistics

9

9.1   Introduction

Multivariate analysis is used to understand and describe the relationships between an arbitrary number of variables. Earth scientists often deal with multivariate data sets, such as microfossil assemblages, geochemical fingerprints of volcanic ash layers, or the clay mineral content of sedimentary sequences. Such multivariate data sets consist of measurements of p variables on n objects, which are usually stored in n-by-p arrays:

$$X = \begin{pmatrix} x_{11} & x_{12} & \cdots & x_{1p} \\ x_{21} & x_{22} & \cdots & x_{2p} \\ \vdots & \vdots & & \vdots \\ x_{n1} & x_{n2} & \cdots & x_{np} \end{pmatrix}$$

The rows of the array represent the n objects, and the columns represent the p variables. The characteristics of the 2nd object in the suite of samples are described by the vector in the second row of the data array:

$$X_2 = \begin{pmatrix} x_{21} & x_{22} & \cdots & x_{2p} \end{pmatrix}$$

As an example, consider a set of microprobe analyses on glass shards from volcanic ash layers in a tephrochronology project. The variables are the p chemical elements, and the objects are the n ash samples. The aim of the study is to correlate ash layers by means of their geochemical fingerprints.

Supplementary Information The online version contains supplementary material available at https://doi.org/10.1007/978-3-031-07719-7_9.

Most multivariate methods simply try to overcome the main difficulty associated with such data sets, which relates to data visualization. Whereas the character of univariate or bivariate data sets can easily be explored by visually inspecting a 2D histogram or an *xy* plot (Chap. 3), the graphical display of a three-variable data set requires projecting the 3D distribution of data points onto a two-dimensional display. It is impossible to imagine or display a larger number of variables. One solution to the problem of visualizing high-dimensional data sets is to reduce the number of dimensions. Several methods exist for grouping together highly correlated variables contained within a data set and for exploring the reduced number of groups.

The classic methods of reducing the number of dimensions are *principal component analysis* (PCA) and *factor analysis* (FA). These methods seek the directions of maximum variance in a data set and use these directions as new coordinate axes. The advantage of replacing existing variables with new groups of variables is that the groups are uncorrelated. These methods are therefore used for unmixing (or separating) mixed variables. Moreover, the new variables can often assist in interpreting a multivariate data set since they frequently contain valuable information on the processes responsible for the distribution of the data points. In a geochemical analysis of magmatic rocks, the groups defined by the method usually contain chemical elements with similarly sized ions in similar locations within the lattices of certain minerals. Examples include Si^{4+} and Al^{3+} as well as Fe^{2+} and Mg^{2+} in silicates.

A second important suite of multivariate methods aims at grouping objects by their similarity. As an example, *cluster analysis* (CA) is often used to correlate volcanic ash layers, such as those in the example above. Tephrochronology attempts to correlate tephra by means of their geochemical fingerprints. When combined with a few radiometric age determinations from key ash layers, this method allows for correlations between different sedimentary sequences that contain these ash layers (e.g., Westgate 1998; Hermanns et al. 2000). Cluster analysis is also used in the field of micropaleontology, for example, to compare different microfossil assemblages (e.g., Birks and Gordon 1985).

A third group of methods is concerned with classifying observations. Humans tend to want to classify the things around them even though nature rarely falls into discrete classes. Classification (or categorization) is useful because it can, for example, help decision-makers to take the necessary precautions to reduce risks when drilling an oil well or when assigning fossils to a particular genus or species. Most classification methods make decisions based on Boolean logic with two options, true or false, as in the use of a threshold value for identifying charcoal in microscope images (Sect. 8.11). Alternatively, fuzzy logic (which is not explained in this book) is a generalization of binary Boolean logic with respect to many real-world problems in decision-making, in which gradual transitions are reasonable (Zadeh 1965).

The following sections introduce the most important techniques of multivariate statistics. To begin, *principal component analysis* (PCA) and *cluster analysis* (CA) are introduced in Sects. 9.2 and 9.5, respectively, and *independent component analysis* (ICA) (which is a nonlinear extension of PCA) is outlined in Sect. 9.3. Section 9.4 introduces *discriminant analysis* (DA), which is a popular method of classification in the earth sciences. Section 9.6 introduces *multiple linear regression*, and Sect. 9.7 demonstrates the use of John Aitchison's *log-transformation* for overcoming the closed-sum problem—a problem that is very common in multivariate data sets (Aitchison 1984, 1986, 1999, 2003). For multivariate analysis, we use the *NumPy* (https://numpy.org), *Matplotlib* (https://matplotlib.org), *scikit-learn* (https://scikit-learn.org), and *SciPy* packages (https://scipy.org), which contain all the necessary routines.

9.2   Principal Component Analysis

Principal component analysis (PCA) detects linear dependencies between p variables and replaces groups of linearly correlated variables with p uncorrelated variables, which are referred to as the *principal components* (*PCs*) (Fig. 9.1). PCA—which was first introduced by Karl Pearson (1901) and was further

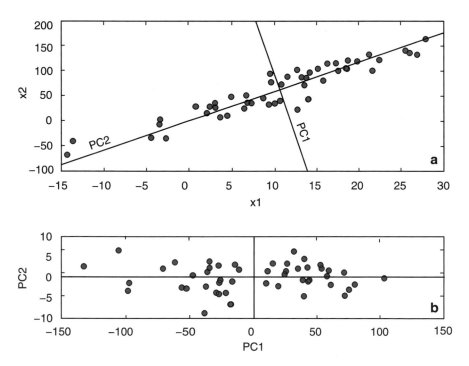

Fig. 9.1 Principal component analysis (PCA) illustrated for a bivariate scatter. The original x_1x_2 coordinate system is replaced with a new orthogonal system in which the 1st axis PC_1 passes through the long axis of the data scatter and the new origin is the bivariate mean. We can now reduce the number of dimensions by dropping the 2nd axis PC_2 with little loss of information. Modified from Swan and Sandilands (1995).

developed by Harold Hotelling (1931)—is the standard method used in the linear unmixing (or separating) of mixed variables. PCA is equivalent to fitting a p-dimensional ellipsoid to a data set of p variables, where the p eigenvectors of the p-by-p covariance matrix of the data set are the axes of the ellipsoid. PCA is therefore one of the eigenvector methods. The p eigenvalues represent the distribution of the variance between each of the eigenvectors.

The performance of PCA is best illustrated with a bivariate data set of the two variables x_1 and x_2 that exhibits a strong linear correlation by fitting a two-dimensional ellipse to the data with two eigenvectors PC_1 and PC_2 of the 2-by-2 covariance matrix of the data set (Fig. 9.1). Since the first principal component PC_1 passes through the long axis of the data scatter and the new origin is the bivariate mean, PC_1 can be used to describe most of the variance, while PC_2 contributes only a small amount of additional information. Prior to this transformation, two axes x_1 and x_2 had been required to describe the data set, but it is now possible to reduce the dimensions of the data by dropping the second axis PC_2 without losing very much information, as shown in Fig. 9.1.

Since this method produces signals PC_1 and PC_2, which are linearly uncorrelated, the method is also called *whitening* because this property is characteristic of white noise. An important prerequisite for successfully applying this method is that the data must be Gaussian distributed. Although the separated signals are uncorrelated, they can still be interdependent (i.e., they can retain a nonlinear correlation). This phenomenon arises, for example, if the data are not Gaussian distributed and PCA consequently does not yield useful results. Independent component analysis (ICA) was developed for this type of task as it separates p independent variables (which are then nonlinearly uncorrelated) from the p dependent variables (Sect. 9.3).

This process can now be expanded to an arbitrary number of variables and samples. Let us assume a data set that consists of measurements of p variables from n samples stored in an n-by-p array X:

$$X = \begin{pmatrix} x_{11} & x_{12} & \dots & x_{1p} \\ x_{21} & x_{22} & \dots & x_{2p} \\ \vdots & \vdots & & \vdots \\ x_{n1} & x_{n2} & \dots & x_{np} \end{pmatrix}$$

The columns of the array X represent the p variables, and the rows represent the n samples. After rotating the axis and moving the origin, the new coordinates Y_j can be computed by

$$Y_1 = a_{11}X_1 + a_{12}X_2 + \dots + a_{1p}X_p$$
$$Y_2 = a_{21}X_1 + a_{22}X_2 + \dots + a_{2p}X_p$$
$$\vdots$$
$$Y_p = a_{p1}X_1 + a_{p2}X_2 + \dots + a_{pp}X_p$$

The first principal component PC_1 (which is denoted by Y_1), contains the highest variance, while PC_2 contains the second-highest variance, and so forth. All the PCs together contain the full variance of the data set. However, this variance is largely concentrated in the first few PCs, which include most of the information content of the data set. The last PCs are therefore generally ignored in order to reduce the dimensions of the data. The factors a_{ij} in the equations above are the *principal component loadings*, and their values represent the relative contributions of the original variables to the new PCs. If the load a_{ij} of a variable X_j in PC_1 is close to zero, then the influence of this variable is low, whereas a high positive or negative a_{ij} suggests a strong influence. The new values Y_j of the variables computed from the linear combinations of the original variables X_j (weighted by the loadings) are called the *principal component scores*.

PCA is commonly used to unmix (or separate) variables X, which are a linear combination of independent source variables S,

$$X = A^T \cdot S$$

where A is the mixing matrix. PCA aims to determine (albeit not quantitatively) both the source variables S (represented by the principal components scores Y_1) and the mixing matrix A (represented by the principal component loadings a_{ij}). Unmixing such variables works best if the probability distribution of the original variables X is a Gaussian distribution, and only in such cases are the principal components completely linearly decorrelated. However, data in the earth sciences are often not Gaussian distributed, and alternative methods—such as independent component analysis (ICA)—should therefore be used instead (Sect. 9.3). For example, radiance and reflectance values from hyperspectral data are often not Gaussian distributed, and ICA rather than PCA is therefore widely used in remote sensing applications to decorrelate spectral bands. Examples in which PCA is used include assessing sediment provenance (as described in the example below), unmixing peridotite mantle sources of basalts, and multispectrally classifying satellite images.

The first example illustrates PCA without using PCA() and instead using cov() to calculate the p-by-p covariance matrix of a $p = 2$ dimensional data set and then using eig() to calculate the p eigenvectors and p eigenvalues of this matrix. We next assume a data set that consists of measurements of p variables from n samples that are stored in an n-by-p array called data. As an example, we can create a bivariate data set of 2-by-30 data points and with a strong linear correlation that is overlain by normally distributed noise:

```
%reset -f

import numpy as np
import numpy.linalg as linalg
from matplotlib import cm
import matplotlib.pyplot as plt
from numpy.random import default_rng
```

```
from sklearn.decomposition import PCA

rng = default_rng(0)
data = rng.standard_normal((30,2))
data[:,1] = 3.4 + 1.2*data[:,0]
data[:,1] = data[:,1] + \
    0.2*rng.standard_normal(np.shape(data[:,1]))

plt.figure()
plt.plot(data[:,0],data[:,1],
    marker='o',
    linestyle='none')
ax = plt.gca()
ax.spines['top'].set_color('none')
ax.spines['right'].set_color('none')
ax.spines['left'].set_position('zero')
ax.spines['bottom'].set_position('zero')
plt.axis(np.array([-3,2,0,6]))
plt.xlabel('x-value')
plt.ylabel('y-value')
plt.show()
```

We can display the data in a coordinate system in which the coordinate axes intersect at the origin. This is achieved using set_position('zero') (Fig. 9.2 a):

```
plt.figure()
plt.plot(data[:,0],data[:,1],
    marker='o',
    linestyle='none')
ax = plt.gca()
ax.spines['top'].set_color('none')
ax.spines['right'].set_color('none')
ax.spines['left'].set_position('zero')
ax.spines['bottom'].set_position('zero')
plt.axis(np.array([-3,2,0,6]))
plt.xlabel('x-value')
plt.ylabel('y-value')
plt.show()
```

The objective is to find the unit vector that points in the direction of the largest variance within the bivariate data set. The solution to this problem is to calculate the largest eigenvalue D of the covariance matrix C and the corresponding eigenvector V,

$$C \cdot V = \lambda \cdot V$$

Fig. 9.2 Principal
component analysis (PCA)
of a bivariate data set: **a**
the original data set of two
variables x_1 and x_2, **b** the
mean centered data set of two
variables x_1 and x_2 together
with the two principal
components PC_1 and PC_2 of
the covariance matrix, and **c**
the data set described by the
new coordinate system of
variables PC_1 and PC_2.

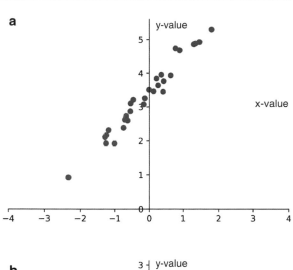

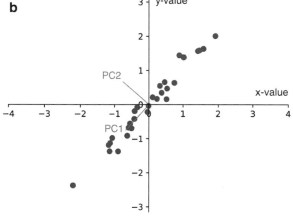

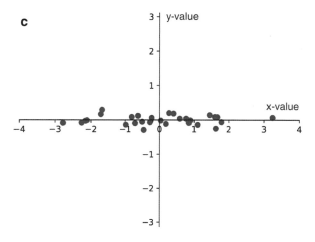

where

$$C = data' \cdot data$$

which is equivalent to

$$(C - D \cdot E) \cdot V = 0$$

where E is the identity matrix, which is a classic eigenvalue problem: It is a linear system of equations, and the eigenvectors V are the solution. We must first mean center the data, which involves subtracting the univariate means from the two columns (i.e., the two variables of data). In the following steps, we therefore study the deviations from the mean(s) only:

```
data[:,0] = data[:,0]-np.mean(data[:,0]);
data[:,1] = data[:,1]-np.mean(data[:,1]);
```

We next calculate the covariance matrix C of data. The covariance matrix contains all the information required to rotate the coordinate system,

```
C = np.cov(data[:,0],data[:,1])
print(C)
```

which yields

```
[[0.92061706 1.00801248]
 [1.00801248 1.14192816]]
```

The rotation helps to create new variables that are uncorrelated (i.e., the covariance is zero for all pairs of the new variables). The decorrelation is achieved by diagonalizing the covariance matrix C. The eigenvectors V that belong to the diagonalized covariance matrix are a linear combination of the old base vectors and thus express the correlation between the old and the new time series. We find the eigenvalues of the covariance matrix C by solving the equation

$$det(C - D \cdot E) = 0$$

The eigenvalues D of the covariance matrix (i.e., the diagonalized version of C) yield the variance within the new coordinate axes (i.e., the principal components). We now calculate the eigenvectors V and eigenvalues D of the covariance matrix C,

```
D,V = linalg.eig(C)
print(V)
print(D)
```

which yields

```
[[-0.744688    -0.66741275]
 [ 0.66741275 -0.744688   ]]
```

```
[0.01720466 2.04534057]
```

As we can see, the second eigenvalue of 2.0453 is larger than the first eigenvalue of 0.0172. We therefore need to change the order of the new variables after the transformation. We then display the data together with the eigenvectors that represent the new coordinate system (Fig. 9.2b):

```python
plt.figure()
plt.plot(data[:,0],data[:,1],
    marker='o',
    linestyle='none')
plt.plot((0,V[0,0]),(0,V[1,0]),
    color=(0.8,0.5,0.3),
    linewidth=0.75)
plt.text(V[0,0],V[1,0],'PC2',
    color=(0.8,0.5,0.3))
plt.plot((0,V[0,1]),(0,V[1,1]),
    color=(0.8,0.5,0.3),
    linewidth=0.75)
plt.text(V[0,1],V[1,1],'PC1',
    color=(0.8,0.5,0.3))
ax = plt.gca()
ax.spines['top'].set_color('none')
ax.spines['right'].set_color('none')
ax.spines['left'].set_position('zero')
ax.spines['bottom'].set_position('zero')
plt.axis(np.array([-6,6,-4,4]))
plt.xlabel('x-value')
plt.ylabel('y-value')
plt.show()
```

The eigenvectors are unit vectors and are orthogonal; therefore, the norm is one and the inner (scalar, dot) product is zero,

```python
print(linalg.norm(V[:,0]))
print(linalg.norm(V[:,1]))
print(np.dot(V[:,0],V[:,1]))
```

which yields

```
0.9999999999999999
0.9999999999999999
0.0
```

We now calculate the data set in the new coordinate system, newdata. We need to flip newdata left/right since the second column is the one with the largest eigenvalue (as stated previously),

```
newdata = np.matmul(data,V)
newdata = np.fliplr(newdata)
print(newdata)
```

which yields

```
newdata =

[[-0.28687857 -0.06796136]
 [-0.98110732 -0.13692354]
 [ 0.42153266  0.18529491]
 [-2.11734482 -0.00967294]
 [ 0.89395125 -0.01241517]
 [ 0.85585602 -0.08631932]
 [ 3.23801307  0.06391048]
 [ 1.59265288  0.08944404]
(cont'd)
```

The eigenvalues of the covariance matrix indicate the variance in this (new) coordinate direction. We can use this information to calculate the relative variance for each new variable by dividing the variances according to the eigenvectors by the sum of the variances. In our example, the 1st and 2nd principal components contain 99.57% and 0.43% of the total variance in the data, respectively,

```
print(np.var(newdata[:,0],ddof=1))
print(np.var(newdata[:,1],ddof=1))

print(np.var(newdata[:,0],ddof=1)/
    (np.var(newdata[:,0],ddof=1)+ \
    np.var(newdata[:,1],ddof=1)))
print(np.var(newdata[:,1],ddof=1)/
    (np.var(newdata[:,0],ddof=1)+ \
    np.var(newdata[:,1],ddof=1)))
```

which yields

```
2.045340567485195
0.017204658450711223
0.9916585303273026
0.008341469672697424
```

Alternatively, we can obtain the variances by normalizing the eigenvalues,

```
variance = D / np.sum(D)
print(variance)
```

which yields

```
[0.00834147 0.99165853]
```

and again indicates 99.17% and 0.83% of the variance, respectively, for the two principal components. We now display the data in newdata in the new coordinate system (Fig. 9.2c):

```
plt.figure()
plt.plot(newdata[:,0],newdata[:,1],
    marker='o',
    linestyle='none')
ax = plt.gca()
ax.spines['top'].set_color('none')
ax.spines['right'].set_color('none')
ax.spines['left'].set_position('zero')
ax.spines['bottom'].set_position('zero')
plt.axis(np.array([-6,6,-4,4]))
plt.xlabel('PC1',loc='right')
plt.ylabel('PC2',loc='top')
plt.show()
```

We can then carry out the same experiment using PCA(). The input for this function is data, and the function returns a pca object that contains PCA components (or loadings), the explained variance, and the transform needed to calculate newdata:

```
pca = PCA(n_components=2)
pca.fit(data)
print(pca.components_)
print(pca.explained_variance_ratio_)
newdata = pca.transform(data)
print(newdata)
```

As we can see, we obtain the same values as before:

```
[[-0.66741275 -0.744688  ]
 [-0.744688    0.66741275]]
```

```
[0.99165853 0.00834147]

[[-0.28687857 -0.06796136]
 [-0.98110732 -0.13692354]
 [ 0.42153266  0.18529491]
 [-2.11734482 -0.00967294]
 [ 0.89395125 -0.01241517]
 [ 0.85585602 -0.08631932]
 [ 3.23801307  0.06391048]
 [ 1.59265288  0.08944404]
(cont'd)
```

As expected, the data have been successfully decorrelated, which we can verify
with corrcoef(),

```
r = np.corrcoef(newdata[:,0],newdata[:,1])
print(r[0,1])
```

which yields a value close to zero

```
-5.385152103520556e-16
```

The second synthetic example demonstrates linearly unmixing a data set of sediment samples. One sample each was taken from thirty different levels in a sedimentary sequence that contained varying proportions of the three different
minerals stored in the columns of array x. The sediments were derived from three
distinct rock types (with unknown mineral compositions), whose relative contributions to each of the thirty sediment samples are represented by s1, s2, and s3,
respectively. Variations in these relative contributions (as represented by the thirty
values in s1, s2, and s3) could, for example, reflect climatic variability within the
catchment area of the sedimentary basin. It may therefore be possible to use the
sediment compositions in array x (from which we calculate s1, s2, and s3 using a
PCA) to derive information on past climate variations.

We need to create a synthetic data set consisting of three measurements that
represent the proportions of each of the three minerals in the each of the thirty
sediment samples. We first clear the workspace and reset the random number generator with default_rng(0). Then, we create s1, s2, and s3 using standard_nor-
mal(), which generates Gaussian-distributed random numbers with means of zero
and standard deviations of 10, 7, and 12, respectively:

```
rng = default_rng(0)
s1 = 10*rng.standard_normal(30)
s2 =  7*rng.standard_normal(30)
s3 = 12*rng.standard_normal(30)
```

We then calculate the varying proportions of each of the three minerals in the thirty sediment samples by summing the values in s1, s2, and s3 after first multiplying them by a weighting factor:

```
x = np.zeros((30,3))
x[:,0] =  15.4+ 7.2*s1+10.5*s2+2.5*s3
x[:,1] = 124.0 8.73*s1| 0.1*s2+2.6*s3
x[:,2] = 100.0+5.25*s1- 6.5*s2+3.5*s3
```

The weighting factors (which together represent the mixing matrix in our exercise) reflect differences not only in the mineral compositions of the source rocks, but also in the weathering, mobilization, and deposition of minerals within sedimentary basins. Hence, if two minerals have weighting factors with different signs, one could be (for example) the weathering product of the other mineral, which would explain why their proportions in the thirty sediment samples are anti-correlated. Alternatively, the different signs could indicate a dilution effect: If the proportions of one of the minerals in the sediment samples remain constant but the proportions of all other minerals vary in a similar way, they will be anti-correlated with the proportions of the first mineral. To complete the generation of the data set, we add some Gaussian noise with a standard deviation of 3.8 to the proportions of the minerals:

```
x = x + 3.8*rng.standard_normal(np.shape(x))
```

Now that we have seen how the sedimentary record was created and have examined the dependencies that exist within the data set, we next pretend that we do not know the relative contributions that the three source rocks made to the thirty sediment samples. The aim of PCA is now to decipher the statistically independent contribution of the three source rocks to the sediment compositions. We can display the histograms of the data and see that they are not perfectly Gaussian distributed, which means that we cannot expect a perfect unmixing result:

```
plt.figure()
plt.subplot(1,3,1)
plt.hist(x[:,0])
plt.subplot(1,3,2)
plt.hist(x[:,1])
plt.subplot(1,3,3)
plt.hist(x[:,2])
plt.show()
```

We display the proportions of the three minerals in the thirty samples along the sedimentary section. In this graphic, we can see weak correlations and anti-correlations between the proportions of the three minerals:

```
plt.figure()
plt.plot(x)
plt.legend(['Min1','Min2','Min3'])
plt.show()
```

Before running PCA, we define labels for the various graphics created during the exercise. We number the samples from 1 to 30, with the minerals being identified by the four-character abbreviations Min1, Min2, and Min3, respectively:

```
labels = []
for i in range(0,30):
    labels.append(str(i))
minerals = (['Min1','Min2','Min3'])
```

We can now explore the correlations between the minerals in pairwise bivariate scatter plots and observe a strong negative correlation between minerals 1 and 2, a weak positive correlation between minerals 1 and 3, and a moderate negative correlation between minerals 2 and 3:

```
plt.figure()
plt.subplot(1,5,1)
plt.plot(x[:,0],x[:,1],
    marker='o',
    linestyle='none')
plt.xlabel('Mineral 1')
plt.ylabel('Mineral 2')
plt.subplot(1,5,3)
plt.plot(x[:,0],x[:,2],
    marker='o',
    linestyle='none')
plt.xlabel('Mineral 1')
plt.ylabel('Mineral 3')
plt.subplot(1,5,5)
plt.plot(x[:,1],x[:,2],
    marker='o',
    linestyle='none')
plt.xlabel('Mineral 2')
plt.ylabel('Mineral 3')
plt.show()
```

The *correlation matrix* provides a technique for exploring such dependencies between the variables in the data set (i.e., the three minerals in our example). The elements of the correlation matrix are Pearson's correlation coefficients (Chap. 4) for each pair of variables, as shown in Fig. 9.3,

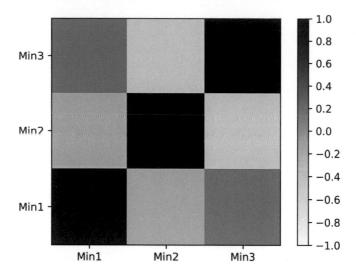

Fig. 9.3 Correlation matrix containing Pearson's correlation coefficients for each pair of variables (e.g., minerals in a sediment sample). Dark colors represent strong positive linear correlations, while light colors represent negative correlations. Intermediate colors suggest no correlation.

```
corrmatrix = np.corrcoef((x[:,0],x[:,1],x[:,2]))
corrmatrix = np.flipud(corrmatrix)
print(corrmatrix)
```

which yields

```
[[ 0.34831277 -0.5556862           1.  ]
 [-0.81838779  1.          -0.5556862 ]
 [ 1.          -0.81838779  0.34831277]]
```

We can now display the correlation matrix using a heatmap chart:

```
fig, ax = plt.subplots()
c = ax.imshow(corrmatrix,cmap=cm.Blues,clim = (-1,1))
ax.set_xticks(np.linspace(0,2,3))
ax.set_yticks(np.linspace(0,2,3))
ax.set_xticklabels(minerals)
ax.set_yticklabels(np.flipud(minerals))
plt.colorbar(c,ticks=np.linspace(-1,1,11))
plt.show()
```

This heatmap of the correlation coefficients confirms the correlations between the minerals that were revealed in the pairwise bivariate scatter plots, and these correlations again show a strong negative correlation between minerals 1 and 2 ($r=-0.8184$), a weak positive correlation between minerals 1 and 3 ($r=0.3483$), and a moderate negative correlation between minerals 2 and 3 ($r=-0.5557$). These observed dependencies would lead us to expect interesting results from the application of a PCA.

Various methods exist for scaling original data before applying a PCA. Such methods include *mean entering* (using a mean equal to zero) and *standardizing* (using a mean equal to zero and a standard deviation equal to one). However, we use the original data x for PCA. The output of this function is a pca object that contains PCA components (or loadings), the explained variance, and the transform needed to calculate PCA scores newx from the original data x. The loadings pcs are weights (or weighting factors) that indicate the extent to which the old variables (i.e., the minerals) contribute to the new variables (i.e., the principal components, or PCs). The principal component scores newx are the coordinates of the thirty samples in the new coordinate system, which is defined by the three principal components PC_1, PC_2, and PC_3 (stored in the three columns of pcs), which we interpret as the three source rocks:

```
pca = PCA(n_components=3)
pca.fit(x)

pcs = pca.components_
newx = pca.transform(x)
variances = pca.explained_variance_ratio_
```

The loadings of the three principal components can be shown by typing

```
print(pcs)
```

which yields

```
[[ 0.63417264 -0.62146955  0.46000072]
 [-0.50849     0.11295575  0.85362692]
 [ 0.58246286  0.7752526   0.24437761]]
```

We observe that PC_1 (first column) has high positive loadings in variables 1 and 3 (first and third rows) and a high negative load in variable 2 (second row). PC_2 (second column) has a high negative load in variable 1 and a high positive load in variable 3, while the load in variable 2 is close to zero. PC_3 (third column) has high loadings in variables 1 and 2, with the load in variable 3 being relatively low but also positive. We now create a number of plots to visualize the PCs:

```
a = np.array([0,2])
b = np.array([0,0])
plt.figure()
```

```
plt.subplot(1,3,1)
plt.plot(pcs[:,0],
    marker='o',
    linestyle='none')
plt.plot(a,b,
    color=(0.8,0.3,0.1),
    linewidth=0.75)
for i in range (0,3):
    plt.text(i+0.2,pcs[i,0],
        minerals[i])
plt.subplot(1,3,2)
plt.plot(pcs[:,1],
    marker='o',
    linestyle='none')
plt.plot(a,b,
    color=(0.8,0.3,0.1),
    linewidth=0.75)
for i in range (0,3):
    plt.text(i+0.2,pcs[i,1],
        minerals[i])
plt.subplot(1,3,3)
plt.plot(pcs[:,2],
    marker='o',
    linestyle='none')
plt.plot(a,b,
    color=(0.8,0.3,0.1),
    linewidth=0.75)
for i in range (0,3):
    plt.text(i+0.2,pcs[i,2],
        minerals[i])
plt.show()
```

The loadings of the minerals and their relationships to the PCs can be used to interpret the relative (but not absolute) influences of the different source rocks. PC_1 is characterized by strong positive contributions of minerals 1 and 3, which reflects a relatively strong influence of the first rock type as a source of the sediments. The opposite sign of the contribution of mineral 2 to the sediment could indicate that the process of mobilization of the mineral in the catchment is different from that of minerals 1 and 3; in other words, a certain climate could favor the weathering of minerals 1 and 3 but reduce the weathering of mineral 2. Alternatively, the negative sign could also indicate a dilution effect, that is, that there is no such process in which the minerals differ, but the negative sign is a statistical artifact. The second principal component PC_2 is also dominated by minerals 1 and 3, but with opposite signs, whereas mineral 2 has relatively little influence. The third principal component PC_3 is influenced by all three minerals, with the same sign. An alternative way of plotting the loadings is as a bivariate plot of two principal components. We ignore PC_3 at this point and concentrate on PC_1 and PC_2. It is important to remember to either close the figure window before plotting the loadings or to clear the figure window using clf in order to avoid integrating the new plot as a fourth subplot in the previous figure window:

```
a = np.array([0,0])
b = np.array([-1,1])
plt.figure()
plt.plot(pcs[:,0],pcs[:,1],
    marker='o',
    linestyle='none')
plt.plot(a,b,
    color=(0.8,0.3,0.1),
    linewidth=0.75,
    linestyle='--')
plt.plot(b,a,
    color=(0.8,0.3,0.1),
    linewidth=0.75,
    linestyle='--')
for i in range (0,3):
    plt.text(pcs[i,0]+0.05,pcs[i,1]+0.05,
        minerals[i])
plt.axis(np.array([-1,1,-1,1]))
plt.xlabel('First Principal Component Loadings')
plt.ylabel('Second Principal Component Loadings')
plt.show()
```

We can now observe in a single plot the same relationships that were previously shown in several graphics (Fig. 9.4). It is also possible to plot the data set as functions of the new variables (i.e., the source rocks). This requires the second output of PCA(), which contains the principal component scores newx:

```
a = np.array([0,0])
b = np.array([-150,200])
c = np.array([-300,300])
plt.figure()
plt.plot(newx[:,0],newx[:,1],
    marker='o',
    linestyle='none')
plt.plot(a,b,
    color=(0.8,0.3,0.1),
    linewidth=0.75,
    linestyle='--')
```

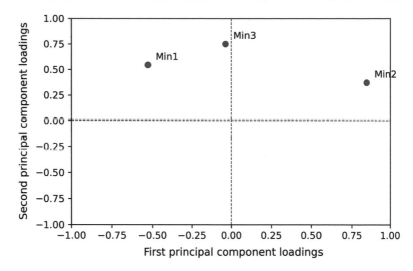

Fig. 9.4 Principal component loadings, which suggest that the *PC*s are influenced by different minerals. See the text for a detailed interpretation of the *PC*s.

```
plt.plot(c,a,
    color=(0.8,0.3,0.1),
    linewidth=0.75,
    linestyle='--')
for i in range (1,30):
    plt.text(newx[i,0]+5,newx[i,1]+5,
        labels[i])
plt.xlabel('First Principal Component Scores')
plt.ylabel('Second Principal Component Scores')
plt.show()
```

This plot clearly defines groups of samples with similar influences, such as samples 19, 20, and 26, and possibly also sample 18. We next use the third output from PCA() to compute the variances of the *PC*s,

```
percent_explained = 100*variances/np.sum(variances)
print(percent_explained)
```

which yields

```
[72.43620937  22.71744468      4.84634596]
```

We can see that more than 72% of the total variance is contained in PC_1 and that about 23% is contained in PC_2, while PC_3 contributes very little to the total variance of the data set (~5%). That means that most of the variability in the data set can be described by just two new variables. As would be expected, the two new variables do not correlate with each other. This is illustrated by a correlation coefficient between newx[:,0] and newx[:,1]

```
r = np.corrcoef(newx[:,0],newx[:,1])
print(r[0,1])
```

that is close to zero

```
-3.05521182351808e-17
```

We can therefore plot the time series of the thirty samples as two independent variables PC_1 and PC_2 in a single plot:

```
plt.figure()
plt.plot(newx[:,0])
plt.plot(newx[:,1])
plt.legend(['PC1','PC2'])
plt.show()
```

This plot displays approximately 94% (72%+22%) of the variance contained in the multivariate data set. According to our interpretation of PC_1 and PC_2, this plot shows the variability in the relative contributions from the two sources to the sedimentary column under investigation. Since we are working with a synthetic data set and the actual contribution of the three source rocks to the sediment is known, we can estimate the quality of the results by comparing the initial variations in s1, s2, and s3 with the (more or less) independent variables of PC_1, PC_2, and PC_3, which are stored in the three columns of newx:

```
plt.figure()
ax1 = plt.subplot(5,1,1)
ax2 = ax1.twinx()
ax1.plot(-newx[:,0])
ax2.plot(s1,
      color=(0.8,0.3,0.1))
ax1 = plt.subplot(5,1,3)
ax2 = ax1.twinx()
ax1.plot(newx[:,1])
ax2.plot(s2,
      color=(0.8,0.3,0.1))
ax1 = plt.subplot(5,1,5)
ax2 = ax1.twinx()
ax1.plot(newx[:,2])
ax2.plot(s3,
```

```
        color=(0.8,0.3,0.1))
    plt.show()
```

The sign and the amplitude cannot be determined quantitatively; therefore, in this case, we change the sign of the second *PC*, and we use `twinx()` to display the data on different axes in order to compare the results. As we can see, we have successfully unmixed the varying contributions of source rocks ʂ1, ʂ2, and ʂ3 to the mineral composition of the sedimentary sequence.

The approach described above has been used to study the provenance of the varved lake sediments described in the previous chapter (Sect. 8.9), which were deposited around 33 kyrs ago in a landslide-dammed lake in the Quebrada de Cafayate (Trauth et al. 2003). The provenance of the sediments contained in the varved layers can be traced using index minerals that are characteristic of the various possible source areas within the catchment. A comparison of the mineral assemblages in the sediments with those of potential source rocks within the catchment area indicates that exposed Fe-rich Tertiary sedimentary rocks in the Santa Maria Basin are the source of the red-colored basal portion of the varves. In contrast, metamorphic rocks in the mountainous parts of the catchment area are the most likely source of the relatively drab-colored upper part of the varves.

9.3   Independent Component Analysis (by N. Marwan)

Principal component analysis (PCA) is the standard method for unmixing (or separating) mixed variables (Sect. 9.2). Such analyses produce signals that are linearly uncorrelated, and the method is therefore also called *whitening* because this property is characteristic of *white noise*. Although the separated signals are uncorrelated, they can still be interdependent (i.e., they may retain a nonlinear correlation). This phenomenon arises, for example, when data are not Gaussian distributed and PCA consequently does not yield good results. *Independent component analysis* (ICA) was developed for this type of task; it separates the variables X into independent variables S, which are then nonlinearly uncorrelated. In order to perform an ICA (according to the central limit theorem), the mixture of standardized random numbers must be Gaussian distributed. ICA algorithms therefore use a criterion that estimates how Gaussian the combined distribution of the independent components is (Hyvärinen 1999). The less Gaussian this distribution is, the more independent the individual components are.

According to the linear mixing model, p independent variables $X = (x_1, x_2, ..., x_p)$ are linearly mixed in n measurements:

$$X = A^T \cdot S$$

We are interested in the source variables S and in the mixing matrix A (see Sect. 9.2). This task is perhaps analogous to being at a party in which many people are carrying on independent conversations: We can hear a mixture of the conversations but cannot distinguish them individually. We could perhaps install some microphones and use them to separate out the individual conversations. This dilemma is therefore sometimes known as the *cocktail party problem*. Officially, however, it is called *blind source separation*, which is defined by

$$S = W^T X$$

where W^T is the separation matrix required to reverse the mixing and obtain the original signals. In the earth sciences, we encounter similar problems, for example, if we want to determine the relative contributions of different source rocks to basin sediments (as we did for PCA), but this time with the possibility that there are nonlinear dependencies in the data and that the data are not Gaussian distributed (Sect. 9.2).

We again create a synthetic data set consisting of thirty total measurements (the proportions of each of the three minerals) from the thirty sediment samples. In contrast to the PCA example, however, the temporal variation in the source rocks is not Gaussian distributed; rather, it is uniformly distributed since we use rng. uniform() instead of standard_normal() to create the pseudorandom numbers:

```
%reset -f

import scipy
import numpy as np
import matplotlib.pyplot as plt
from sklearn.decomposition import PCA

rng = np.random.default_rng(0)
s1 = rng.uniform(0,10,30)
s2 = rng.uniform(0,7,30)
s3 = rng.uniform(0,12,30)
```

We use the same mixing equation as in the PCA example to create the three columns of x (which correspond to the three minerals) by linearly mixing source rocks s1, s2, and s3 (Fig. 9.5):

```
x = np.zeros((30,3))
x[:,0] =   15.4+ 7.2*s1+10.5*s2+2.5*s3
x[:,1] = 124.0-8.73*s1+ 0.1*s2+2.6*s3
x[:,2] = 100.0+5.25*s1- 6.5*s2+3.5*s3
x = x + 3.8*rng.standard_normal(np.shape(x))
```

By displaying histograms of the data, we see that the data are not Gaussian distributed:

```
fig, ax = plt.subplots(3)
ax[0].hist(x[:,0])
```

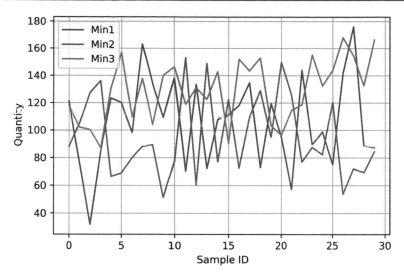

Fig. 9.5 Sample input for independent component analysis. The relative proportions of three minerals in 30 sediment samples reflect temporal variations in the contributions of the three source rocks within the catchment of a sedimentary basin.

```
ax[1].hist(x[:,1])
ax[2].hist(x[:,2])
plt.show()
```

We then display the proportions of the minerals in the thirty samples as a time series:

```
plt.figure()
plt.plot(x)
plt.grid(True)
plt.legend(('Min1','Min2','Min3'))
plt.xlabel('Sample ID')
plt.ylabel('Quantity')
plt.show()
```

We begin by unmixing the variables using PCA. We calculate the principal components pcs, the mixing matrix a_pca, and the whitening matrix w_pca using.

```
pca = PCA(n_components=3)
pca.fit(x)

newx = pca.transform(x)
newx = newx/np.std(newx, 0, ddof=1)
```

```
pcs = pca.components_
variances = pca.explained_variance_

a_pca = np.dot(pcs, np.sqrt(np.diag(variances)))
w_pca = np.dot(np.linalg.inv(np.sqrt(np.diag(variances))), pcs)
```

The pre-whitening reduces the process of unmixing into independent components (S) to a problem of finding a suitable rotation matrix B that can be applied to the variables X_{PC}:

$$S = B^T X_{PC}$$

We need to find a rotation matrix B such that the variables S have a completely non-Gaussian distribution. There are several possibilities for implementing such a non-Gaussian criterion, such as minimizing or maximizing the excess kurtosis

$$\gamma = E\left(x^4\right) - 3\left(E\left(x^2\right)\right)^2$$

because this excess kurtosis γ for normally distributed data is zero. E stands for the expected value (i.e., the mean value) of x. Please note that the excess kurtosis differs from the kurtosis: The excess kurtosis is the kurtosis minus three since the kurtosis of Gaussian-distributed data is three (Sect. 3.2) but the excess kurtosis of such a distribution is zero. To find a B that ensures a minimum or maximum excess kurtosis, a learning algorithm can be used that finds the fixed points of the learning rule:

$$E\left(x\left(B^T x\right)^3\right) - 3\|B\|^2 B + f\left(\|B\|^2\right)B = 0$$

The FastICA algorithm by Hyvärinen (1999)—which is based on a fixed-point iteration scheme—is an efficient algorithm for solving this problem. The learning rule is reduced in this example to the form

$$B(k) = E\left(x\left(B(k-1)^T x\right)^3\right) - 3B(k-1)$$

We begin with an initial rotation matrix B, which consists only of random numbers:

```
B = scipy.linalg.orth(rng.uniform(-.5,.5,(3,3)))
BOld = np.zeros(np.shape(B))
```

The iteration is continued until a divergence criterion div is reached. We start with a value of zero for this criterion:

```
div = 0
```

The *fixed-point iteration scheme* consists of two steps: a symmetric orthogonalization and the application of the learning rule. The divergence criterion `div` is updated with each step of the iteration and is checked in order to stop the iteration process as soon as the divergence criterion has been reached. Since Python works with floating point numbers, `div` does not actually reach zero, and we therefore allow (1-div) to fall below the floating-point relative accuracy eps(1) (or eps without input argument 1 since the default input for eps is 1), which is about 2.2204e-16 and is therefore close to zero (see Sect. 2.5):

```
while (1-div) > np.spacing(1):
    B = np.dot(B,np.real(scipy.linalg.sqrtm(
        np.linalg.inv(np.dot(B.T, B)))))

    div = np.min(np.abs(np.diag(np.dot(B.T,BOld))))
    BOld = B

    B = np.dot(newx.T, np.dot(newx,B)**3)/len(newx)-3*B
    sica = np.dot(newx,B)
```

Finally, we compare synthetic source rocks s1, s2, and s3 with unmixed variables IC1, IC2, and IC3 (Fig. 9.6):

```
fig, ax = plt.subplots(3)
ax[0].plot(sica[:,0])
ax[0].set_title('IC1')
ax0 = ax[0].twinx()
ax0.plot(s3,'C1')
ax[1].plot(-sica[:,1])
ax[1].set_title('IC2')
ax1 = ax[1].twinx()
ax1.plot(-s1,'C1')
ax[2].plot(-sica[:,2])
ax[2].set_title('IC3')
ax2 = ax[2].twinx()
ax2.plot(-s2,'C1')
plt.show()
```

ICA identifies the source signals almost perfectly. We can also notice that the descending order of the ICs is different from the initial order of s1, s2, and s3 due to the commutativity of addition. In real-world examples, of course, the order of the ICs is not relevant. Furthermore, the exact signs and amplitudes do not match the original values, and ICA therefore yields only semi-quantitative results, as was

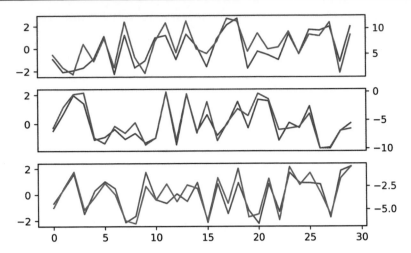

Fig. 9.6 Independent component analysis output. ICA identifies the source signals almost perfectly, as suggested by the pairwise similarity of the original inputs (orange) and the independent components (blue).

also the case for PCA. Finally, we can actually compute the mixing matrix a_ica and the separation matrix w_ica using

```
a_ica = np.dot(a_pca,B)
w_ica = np.dot(B.T,w_pca)
```

The mixing matrix a_ica can be used to estimate the proportions of the separated variables in our measurements. The components a_{ij} of the mixing matrix a_ica correspond to the principal component loadings, which were introduced in Sect. 9.2.

9.4   Discriminant Analysis

Discriminant analysis helps to assign objects to established categories or groups. Examples include assigning fossil specimens to established genera or species, identifying rock types following mineralogical (or chemical) analysis, and mapping vegetation types using satellite images. Discriminant analysis is different from simple classification, which does not define the number of groups or categories prior to the analysis.

The classic example of a discriminant analysis in petrography is the QAPF (or Streckeisen) diagram (Streckeisen 1974, 1976). This diagram categorizes igneous rocks (especially plutonic rocks) by their percentages of quartz (Q), alkali feldspar (including albite) (A), plagioclase (P), and feldspathoids (F), which are all normalized to a total of 100%. The QAPF diagram displays the percentages of the four minerals in a double-ternary plot, with QAP percentages in the upper half of the graphics and FAP percentages in the lower half. The QAPF diagram is commonly used by the International Union of Geological Sciences (IUGS) to classify plutonic rocks. As an example, a plutonic rock with 50% quartz, 30% alkali feldspar, 20% plagioclase, and 0% feldspathoids is termed a granite. Whereas Albert Streckeisen's definition of plutonic rocks represents compromises between established usages in different parts of the world (Streckeisen 1974, 1976), discriminant analysis is based solely on mathematical constraints for classifying objects. Furthermore, discriminant analysis assumes normality for the measured values within a class, which is probably not a valid assumption for the Streckeisen classification. Since normality is important to the success of the method, the user must be careful to ensure that this condition is actually met, especially when analyzing compositional (closed) data (see also Sect. 9.5).

Discriminant analysis was first introduced by Sir Ronald A. Fisher (1936) to discriminate between two or more populations of flowering plants. The first step involves determining the discriminant function Y_i that best separates two groups of objects described by the normally distributed variables X_i:

$$Y = a_0 + a_1 X_1 + a_2 X_2 + \ldots + a_p X_p$$

The parameters a_i are determined such that they maximize the distance between the multivariate means of the individual groups. In other words, we determine a_i such that the ratio of the distances between the means of the groups to the distances between group members is high. In the second step, once the discriminant function has been determined from a training set of objects, new objects can be assigned to one group or the other. Using the Streckeisen diagram, a rock sample is assigned to an established rock type (e.g., granite) on the basis of the percentages of Q, A, P, and F.

As an example, we first create a synthetic data set of granite rock samples that can be described by two variables x_1 and x_2. These two variables could represent the percentages of two chemical elements expressed as oxides (in weight percent). Let us assume that we know from preliminary studies that these rock samples come from three different granites that were formed at different times during three separate magmatic events. Apart from natural inhomogeneities within a granite intrusion, we can assume that the measured values from the granite samples are normally distributed. In this example, we first determine the discriminant functions that separate the three groups (or types of granite). We then use the discriminant functions to assign rock samples (which were collected during a subsequent field campaign) to one of the three types of granite.

We first clear the workspace

```
%reset -f

import numpy as np
import numpy.ma as ma
```

```
import matplotlib.pyplot as plt
from numpy.random import default_rng
from sklearn.discriminant_analysis import \
    LinearDiscriminantAnalysis
```

and then reset the random number generator:

```
rng = default_rng(100)
```

Next, we generate a data set from the chemical compositions of the three types of granite. Sixty granite samples were collected from each rock type and were chemically analyzed. The percentages of two chemical elements x1 and x2 are stored in the three variables data1, data2, and data3:

```
data1 = rng.standard_normal((60,2))
data1[:,0] = 8.5 + 1.8*data1[:,0]
data1[:,1] = 3.9 + 0.6*data1[:,1]

data2 = rng.standard_normal((60,2))
data2[:,0] = 6.4 + 1.2*data2[:,0]
data2[:,1] = 2.7 + 0.4*data2[:,1]

data3 = rng.standard_normal((60,2))
data3[:,0] = 5.5 + 1.8*data3[:,0]
data3[:,1] = 1.3 + 0.8*data3[:,1]
```

In order to define the established categories (or classes), we create an array containing labels 1, 2, and 3 for each pair of measurements and store them in tar:

```
tar = np.ones(60)
tar = np.append(tar,2*np.ones(60))
tar = np.append(tar,3*np.ones(60))
```

We then vertically concatenate the three variables data1, data2, and data3 into a single variable data:

```
data = ma.concatenate([data1,data2,data3])
```

We have thus generated a synthetic data set from three groups of normally distributed data. We then create a linear discriminant analysis classifier using LinearDiscriminantAnalysis():

```
cls = LinearDiscriminantAnalysis()
cls.fit(data,tar)
```

The function returns a discriminant analysis model for the predictors data and the class labels tar. The layers of cls can be listed by typing

```
coeffs = cls.coef_
means = cls.means_
interc = cls.intercept_
print(coeffs)
print(means)
print(interc)
```

which yields the output

```
[[ 0.80294382  3.66238131]
 [-0.133724   0.22444128]
 [-0.66921982 -3.88682258]]

[[8.97020713 3.84392049]
 [6.61472503 2.66054978]
 [5.22893157 1.25608319]]

[-19.26126129  -0.78131967  10.44095035]
```

Finally, we display the result in a graphic using

```
plt.figure()
plt.plot(data[0:60,0],data[0:60,1],
    marker='o',
    color=(0.8,0.3,0.1),
    linestyle='none')
plt.plot(data[60:120,0],data[60:120,1],
    marker='o',
    color=(0.3,0.8,0.1),
    linestyle='none')
plt.plot(data[120:180,0],data[120:180,1],
    marker='o',
    color=(0.1,0.3,0.8),
    linestyle='none')
plt.plot(means[:,0],means[:,1],
    marker='o',
    color='k',
    linestyle='none')
x = np.linspace(0,12,10)
y = -(interc[0] + coeffs[0,0]*x)/coeffs[0,1]
plt.plot(x,y,'k:')
y = -(interc[2] + coeffs[2,0]*x)/coeffs[2,1]
plt.plot(x,y,'k:')
```

```
plt.title('Discriminant Analysis')
plt.legend(('Granite 1','Granite 2','Granite 3'),
    loc='upper left')
plt.show()
```

This graphic shows the members of the three classes (or types of granite) in three different colors, the two lines separating the three classes, and the bivariate means of the classes (marked as black filled circles) (Fig. 9.7). A new sample with the composition x1=5.2 and x2=2.9 can be easily assigned to class Granite 2,

```
print(cls.predict([[5.2,3.5]]))
```

```
plt.figure()
plt.plot(5.2,3.5,
    marker='o',
    color='k',
    linestyle='none')
x = np.linspace(0,12,10)
y = -(interc[0] + coeffs[0,0]*x)/coeffs[0,1]
plt.plot(x,y,'k:')
y = -(interc[2] + coeffs[2,0]*x)/coeffs[2,1]
plt.plot(x,y,'k:')
plt.text(6.0,4.5,'Granite 1',color=(0.8,0.3,0.1))
plt.text(4.0,3,'Granite 2',color=(0.3,0.8,0.1))
plt.text(2.5,1.2,'Granite 3',color=(0.1,0.3,0.8))
plt.title('Discriminant Analysis')
plt.show()
```

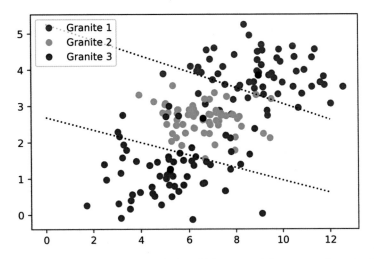

Fig. 9.7 Result of linear discriminant analysis, which separated three types of granite (1, 2, and 3).

which yields

[2.]

which suggests that the new sample is a member of the 2nd type of granite.

9.5 Cluster Analysis

Cluster analysis creates groups of objects that are very similar to one another compared with other individual objects or groups of objects. This method first computes the *similarity* or (alternatively) the *dissimilarity* (or *distance*) between all pairs of objects, It then ranks the groups according to their similarity (or distance from one another) before finally creating a hierarchical tree, which is visualized as a *dendrogram*. The grouping of objects can be useful in the earth sciences, for example, when making correlations within volcanic ash layers (Hermanns et al. 2000) or comparing different microfossil assemblages (Birks and Gordon 1985).

Numerous methods exist for calculating the similarity or (alternatively) the dissimilarity (or distance) between two data vectors. Let us define two data sets consisting of multiple measurements from the same object. These data can be described by vectors:

$$X_1 = (x_{11}, x_{12}, \ldots, x_{1p})$$

$$X_2 = (x_{21}, x_{22}, \ldots, x_{2p})$$

The most popular measures of dissimilarity (or distance) between any two sample vectors are

- the *Euclidian distance*—This is simply the shortest distance between the two points that describe two measurements in the multivariate space:

$$\Delta_{12} = \sqrt{(x_{11} - x_{21})^2 + (x_{12} - x_{22})^2 + \ldots + (x_{1p} - x_{2p})^2}$$

The Euclidian distance is certainly the most intuitive measure of similarity. However, in heterogeneous data sets consisting of a number of different types of variables, a better alternative would be

- the *Manhattan* (or *city block*) *distance*—In the city of Manhattan, it is necessary to follow perpendicular avenues rather than crossing blocks diagonally. The Manhattan distance is therefore the sum of all differences:

$$\Delta_{12} = |x_{11} - x_{21}| + |x_{12} - x_{22}| + \ldots + |x_{1p} - x_{2p}|$$

Measures of similarity include

- the *correlation similarity coefficient*—This measure uses Pearson's linear product-moment correlation coefficient to compute the similarity between two objects:

$$r_{X_1 X_2} = \frac{\sum\limits_{i=1}^{n} (x_{1i} - \bar{x}_1)(x_{2i} - \bar{x}_2)}{(n-1)s_{X_1} s_{X_2}}$$

  The measure is used if a researcher is interested in the ratios between the variables measured on the objects. However, Pearson's correlation coefficient is highly sensitive to outliers and should be used with care (see also Sect. 4.2).

- the *inner-product similarity index*—Normalizing the length of the data vectors to a value of one and computing their *inner product* yields another important similarity index that is often used in transfer function applications. In this example, a set of modern flora or fauna assemblages with known environmental preferences is compared with a fossil sample in order to reconstruct past environmental conditions:

$$s_{12} = \frac{1}{|X_1|}\frac{1}{|X_2|}\left(x_{11} x_{12} \ldots x_{1p}\right)\begin{pmatrix} x_{21} \\ x_{22} \\ \vdots \\ x_{2p} \end{pmatrix}$$

  The inner-product similarity varies between 0 and 1. A zero value suggests no similarity, and a value of one suggests maximum similarity.

The second step in performing a cluster analysis is to rank the groups by their similarity and to build a hierarchical tree, which is visualized as a dendrogram. Most clustering algorithms simply link the two objects with the highest level of similarity or dissimilarity (or distance). In the following steps, the most similar pairs of objects or clusters are linked iteratively. The difference between clusters (each of which is made up of groups of objects) is described in different ways depending on the type of data and the application:

- *K-means clustering* uses the Euclidean distance between the multivariate means of a number of *K* clusters as a measure of the difference between the groups of objects. This distance is used if the data suggest that there is a true mean value surrounded by random noise. Alternatively,
- *K-nearest neighbor clustering* uses the Euclidean distance of the nearest neighbors as a measure of this difference. This measure is used if there is natural heterogeneity in the data set that cannot be attributed to random noise.

It is important to evaluate the data properties prior to applying a clustering algorithm. The absolute values of the variables should first be considered. For example, a geochemical sample from volcanic ash might show an SiO_2 content of around 77% and an Na_2O content of only 3.5%, but the Na_2O content may be considered of greater importance. In such a case, the data need to be transformed such that they have means equal to zero (*mean centering*). Differences in both the variances and the means are corrected by *standardizing* (i.e., the data are standardized to means equal to zero and to variances equal to one). Artifacts arising from closed data (e.g., spurious negative correlations) are avoided by using *Aitchison's log-ratio transformation* (Aitchison 1984, 1986) (see Sect. 9.7). This transformation ensures data independence and avoids the constant sum normalization constraints. The log-ratio transformation is

$$x_{tr} = \log(x_i/x_d)$$

where x_{tr} denotes the transformed score ($i = 1, 2, 3, ..., d-1$) of some raw data x_i. The procedure is invariant under the group of permutations of the variables, and any variable can be used as the divisor x_d.

As an exercise in performing a cluster analysis, the sediment data stored in *sediment_3.txt* are loaded. This data set includes the percentages of various minerals contained in sediment samples. The sediments are sourced from three rock types: 1) a magmatic rock containing amphibole, pyroxene, and plagioclase; 2) a hydrothermal vein characterized by the presence of fluorite, sphalerite, and galena as well as some feldspars (plagioclase and potassium feldspars) and quartz; and 3) a sandstone unit containing feldspars, quartz, and clay minerals. Ten samples were taken from various levels in this sedimentary sequence, and each sample contains varying proportions of these minerals. First, the distances between pairs of samples can be computed. The function `pdist()` provides many different measures of distance, such as the Euclidian or Manhattan (or city block) distance. We use the default setting, which is the Euclidian distance:

```
%reset -f

import numpy as np
from matplotlib import cm
import matplotlib.pyplot as plt
import scipy.spatial.distance as distance
from scipy.cluster.hierarchy import dendrogram, linkage
from scipy.cluster.hierarchy import single, cophenet

data = np.loadtxt('sediments_3.txt')

Y = distance.pdist(data,'euclidean')
```

The function `pdist()` returns a vector Y containing the distances between each pair of observations in the original data matrix. We can visualize the distances in a heatmap:

```
S = distance.squareform(Y)

labels = []
for i in range(1,11):
    labels.append(str(i))

fig, ax = plt.subplots()
c = ax.imshow(S,cmap=cm.Blues)
ax.set_xticks(np.linspace(0,9,10))
ax.set_yticks(np.linspace(0,9,10))
ax.set_xticklabels(labels,
    fontsize=8)
ax.set_yticklabels(labels,
    fontsize=8)
plt.title('Euclidean distance between pairs of samples',
    fontsize=8)
plt.xlabel('First Sample No.',
    fontsize=8)
plt.ylabel('Second Sample No.',
    fontsize=8)
cbar = plt.colorbar(c)
for cax in cbar.ax.get_yticklabels():
    cax.set_fontsize(8)
plt.show()
```

The function squareform() converts Y to a symmetric, square format such that the elements (i,j) of the matrix denote the distance between the i and j objects in the original data. We next rank and link the samples with respect to the inverse of their separation distances using linkage(),

```
Z = linkage(Y,'single')
print(Z)
```

which yields

```
[[ 1.          8.          0.05639246  2.        ]
 [ 7.          9.          0.07304444  2.        ]
 [ 0.         11.          0.09233992  3.        ]
 [ 5.          6.          0.10218992  2.        ]
 [10.         12.          0.11291417  5.        ]
 [ 2.          3.          0.16042391  2.        ]
 [14.         15.          0.17373155  7.        ]
 [ 4.         16.          0.17635918  8.        ]
 [13.         17.          0.21460694 10.        ]]
```

In this 3-column array Z, each row identifies a link. The first two columns identify the objects (or samples) that have been linked, while the third column contains the separation distance between these two objects. The first row (link) between objects (or samples) 1 and 2 has the smallest distance, which corresponds to the greatest similarity. In our example, samples 2 and 9 have the smallest separation distance (i.e., 0.0564) and are therefore grouped together and given the label 11 (i.e., the next-available index higher than the highest sample index of 10.) Next, samples 8 and 10 are grouped into 12 since they have the second-smallest separation difference (i.e., 0.0730). The next row shows that the new group 12 is then grouped with sample 1 (the two items have a separation difference of 0.0923), and so forth. Finally, we visualize the hierarchical clusters as a dendrogram (as shown in Fig. 9.8):

```
plt.figure()
dendrogram(Z)
plt.xlabel('Sample ID - 1')
plt.ylabel('Distance')

plt.show()
```

Clustering finds the same groups as the principal component analysis. Please note that due to Python indexing, the sample numbers are reduced by one and thus start with zero (i.e., sample 1 is represented as sample 0). We observe clear groups consisting of (a) samples 1, 2, 8, 9, and 10 (the magmatic source rocks); (b) samples

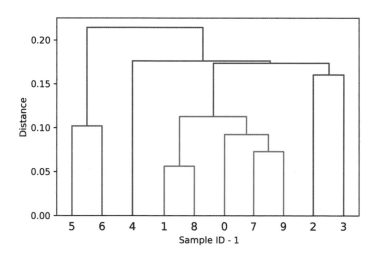

Fig. 9.8 Cluster analysis output. Please note that due to Python indexing, the sample numbers are reduced by one and thus start with zero (i.e., sample 1 is represented as sample 0). The dendrogram shows clear groups consisting of **a** samples 1, 2, 8, 9, and 10 (the magmatic source rocks); **b** samples 3, 4, and 5 (the magmatic dyke containing ore minerals); and **c** samples 6 and 7 (the sandstone unit).

3, 4, and 5 (the hydrothermal vein); and (c) samples 6 and 7 (the sandstone). One way to test the validity of our clustering result is to use the *cophenetic correlation coefficient*,

```
cophenet(Z,Y)
```

which yields

```
0.7579381416951301
```

The result is convincing since the closer this coefficient is to one, the better the cluster solution is.

9.6   Multiple Linear Regression

In Chap. 4, linear regression models were introduced as a way of describing the relationship between a dependent variable y and an independent variable x. The dependent variable is also known as the *response variable*, and the independent variable is also known as the *predictor variable*. A multiple linear regression model describes the relationship between a dependent (or response) variable y and n independent (or predictor) variables x_i,

$$y = b_0 + b_1 x_1 + b_2 x_2 + \ldots + b_p x_p$$

where b_i are the $n+1$ regression coefficients of the linear model. The linearity in the term multiple linear regression refers to the fact that the response variable is a linear function of the predictor variables. The regression coefficients are estimated by minimizing the mean squared difference between the predicted and true values of the response variable y. An example that is commonly used in the earth sciences is the quality of crude oil, which is assumed to be linearly dependent on the age of the sediment, the burial depth, and the temperature. In practice, the plausibility of the assumption of linearity must first be examined. If this assumption is probably true, several methods of multiple linear regression are available.

As a first example, we create a noise-free synthetic data set with three variables Var1, Var2, and Var3. We wish to uncover the influence of variables Var1 and Var2 on variable Var3. Variables Var1 and Var2 are therefore the predictor variables, and variable Var3 is the response variable. The linear relationship between the response variable and the predictor variables is Var3=0.2-52.0*Var1+276.0*Var2. The three variables Var1, Var2, and Var3 are stored as columns 1, 2, and 3, respectively, in a single array data:

```
%reset -f

import numpy as np
from numpy.random import default_rng
import matplotlib.pyplot as plt
from sklearn import linear_model
from sklearn metrics import mean_squared_error, r2_score

rng = default_rng(0)
data = np.zeros((50,3))
data[:,0] = 0.3 + 0.03*rng.standard_normal(50)
data[:,1] = 0.2 + 0.01*rng.standard_normal(50)
data[:,2] =      0.2 \
            -  52.0*data[:,0] \
            + 276.0*data[:,1]
```

We create labels for the names of the samples and the names of the variables (as we did in Sect. 9.2):

```
labels = []
for i in range(1,len(data)+1):
    labels.append(str(i))

variables = (['Var1','Var2','Var3'])
```

We then calculate the coefficients beta of the multiple linear regression model using LinearRegression.fit(). Typing

```
beta = linear_model.LinearRegression(). \
    fit(data[:,0:2],data[:,2])
beta.score(data[:,0:2],data[:,2])
coeffs = beta.coef_
interc = beta.intercept_
```

yields

```
[-52. 276.]
0.20000000000001705
```

The function LinearRegression.fit() uses a least-mean-squares criterion to calculate beta, which contains the coefficients coeffs=coef_ and interc=intercept_. Both coefficients of course match (almost) exactly with the original coefficients of the synthetic data since we have no noise. Because we have only three variables, we can display the results in a three-dimensional plot. We begin by creating a rectangular grid for the first two variables var1 and var2 using meshgrid(), and we next calculate the predicted values for the second variable var3 using `coeffs` and `interc`. We then use `plot_surface()` to display the linear regression plane of the model, and we use `plot()` to plot the measurements as red points with a marker size of 30 points:

```
var1 = np.linspace(0.20,0.45,26)
var2 = np.linspace(0.17,0.23,7)
Var1,Var2 = np.meshgrid(var1,var2)
Var3 = interc + coeffs[0]*Var1 + coeffs[1]*Var2

fig, ax = plt.subplots(subplot_kw={"projection": "3d"})
ax.plot_surface(Var1,Var2,Var3,
    rstride=1, cstride=1,
    linewidth=0,
    alpha=0.3,
    antialiased=True,
    shade=True)
for i in range(0,len(data)):
    plt.plot(data[i,0],data[i,1],data[i,2],
        marker='o',
        color=(0.1,0.3,0.8))
ax.view_init(30,70-90)
plt.show()
```

Since the data set is noise-free, the data points all lie on the linear regression plane. This changes if we introduce normally distributed noise with a standard deviation of Sect. 2.4:

```
rng = default_rng(0)
data = np.zeros((50,3))
data[:,0] = 0.3 + 0.03*rng.standard_normal(50)
data[:,1] = 0.2 + 0.01*rng.standard_normal(50)
data[:,2] =     0.2 \
            -  52.0*data[:,0] \
            + 276.0*data[:,1] \
            +   2.2*rng.standard_normal(50)

labels = []
for i in range(1,len(data)+1):
    labels.append(str(i))

variables = (['Var1','Var2','Var3'])
```

Typing

```
beta = linear_model.LinearRegression(). \
    fit(data[:,0:2],data[:,2])
```

```
beta.score(data[:,0:2],data[:,2])
coeffs = beta.coef_
interc = beta.intercept_
print(coeffs)
print(interc)
```

yields

```
[-50.21197011 303.02828779]
-5.6988484229812
```

which suggests that the estimates of the regression coefficients (−5.6988, −50.2120, and 303.0283) do not match exactly those that were used to create the synthetic data (0.2, −52.0, and 276.0). We can again display the results in a three-dimensional plot (Fig. 9.9).

```
var1 = np.linspace(0.20,0.45,26)
var2 = np.linspace(0.17,0.23,7)
Var1,Var2 = np.meshgrid(var1,var2)
Var3 = interc + coeffs[0]*Var1 + coeffs[1]*Var2

fig, ax = plt.subplots(subplot_kw={"projection":"3d"})
ax.plot_surface(Var1,Var2,Var3,
    rstride=1,cstride=1,
    linewidth=0,
    alpha=0.3,
    antialiased=True,
    shade=True)
```

Fig. 9.9 Linear regression model of a synthetic data set with three variables and random noise.

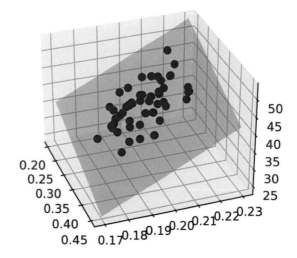

```
for i in range(0,len(data)):
    plt.plot(data[i,0],data[i,1],data[i,2],
        marker='o',
        color=(0.1,0.3,0.8))
ax.view_init(30,70-90)
plt.show()
```

and observe some differences between the data and the regression plane if we rotate the graph in three dimensions using the tool *Rotate 3D* on the toolbar of the figure window.

As a second example, we use laboratory data from the Sherwood Sandstone Group in England that suggest that intergranular permeability (as the response variable) is mainly influenced by porosity and matrix conductivity (as predictor variables) and that other variables have less influence. This example was discussed in detail in an article by Olorunfemi (1985) and was used as an example in the textbook by Swan and Sandilands (1995). The data for 40 rock samples from the Sherwood Sandstone Group were taken from M.O. Olorunfemi's publication and are stored in the file *sherwoodsandstone.txt*. We load the data from the file by typing

```
data = np.loadtxt('sherwoodsandstone.txt')
```

The five columns of the array data contain the numerical values of the petrophysical measures of permeability (in m/s), porosity (without a physical unit), matrix conductivity (in S/m), true formation factor (without a physical unit), and induced polarization (in %). We also load the sample IDs (from Olorunfemi 1985) from the file *sherwoodsandstone_samples.txt* using readlines() and save them in the list samples:

```
with open('sherwoodsandstone_samples.txt') as f:
    samples = f.readlines()[8:]
```

We next create another list that contains the variable names: Perm for permeability, Poro for porosity, MaCo for matrix conductivity, TrFF for true formation factor, and InPo for induced polarization:

```
variables = (['Perm',
    'Poro',
    'MaCo',
    'TrFF',
    'InPo'])
```

We then display the variables in a plot that includes a legend. Since the values of the variables are very small, we plot the logarithm of the values instead of the true values:

```
depth = np.linspace(1,40,40)
plt.figure()
for i in range(5):
    plt.plot(np.log(data[:,i]),depth,
        linestyle = 'none',
        marker = 's')
plt.legend(variables)
plt.show()
```

We then calculate the coefficients `beta` of the multiple linear regression model using `LinearRegression.fit()`,

```
beta = linear_model.LinearRegression(). \
    fit(data[:,1:6],data[:,0])
beta.score(data[:,1:6],data[:,0])
coeffs = beta.coef_
interc = beta.intercept_
print(coeffs)
print(interc)
```

which yields

```
[ 5.06357746e+01 -3.30900095e+02
 -1.78566908e-01 -4.52634500e-01]
0.1586815190858486
```

which indicates that only porosity and matrix conductivity influence the permeability of the Sherwood Sandstone Group, as the two larger coefficients (50.6358 and −330.9000) suggest.

There are multiple ways of examining the quality of the result. For example, we can examine the residuals of the model using

```
res = data[:,0] - beta.predict(data[:,1:6])

plt.figure()
plt.plot(res,depth,
    linestyle='none',
    marker='s')
plt.show()
```

which suggests that measurements above 9 are outliers, as is also shown in the residual histogram:

```
e = np.linspace(-6,12,7)

fig,ax = plt.subplots()
plt.hist(res,bins=e)
plt.axis(np.array([-6,12,0,25]))
ax.set_xticks(np.linspace(-6,12,7))
plt.show()
```

The linear regression model can now be used to predict the permeability of a new
sample from the values of the other petrophysical variables. The effect that each
predictor variable has on the model can be examined in many different ways. Let
us, for example, calculate the linear regression model based on the first 34 values
only using

```
beta = linear_model.LinearRegression(). \
    fit(data[0:35,1:6],data[0:35,0])
beta.score(data[:,1:6],data[:,0])
coeffs = beta.coef_
interc = beta.intercept_
print(coeffs)
print(interc)
```

and while keeping the half-open interval [0,35) (or [0:35]) in mind. We can then
we predict the remaining five values using

```
data_pred = beta.predict(data[35:,1:6])
print(data_pred)
```

which yields

```
[5.27453296 4.88171822 2.33299313 4.88200648 5.79731973]
```

The model now enables us to calculate the mean squared error and the coefficient
of determination using

```
print("Mean squared error: %.2f" \
    % mean_squared_error(data[35:,0],data_pred))

print("Coefficient of determination: %.2f" \
    % r2_score(data[35:,0],data_pred))
```

which yields

```
Mean squared error: 4.92
Coefficient of determination: 0.18
```

Finally, we can plot the predicted vs. real value for all data points using

```
plt.figure()
for i in range(5):
    plt.plot(data[35:,0],data_pred,
        linestyle='none',
        marker='s',
        color=(0.8,0.3,0.1))
plt.plot([0,7],[0,7],
    linestyle=':',
    color='k')
plt.axis(np.array([0,7,0,7]))
plt.xlabel('Real Value')
plt.ylabel('Predicted Value')
plt.axis('square')
plt.show()
```

9.7   Aitchison's Log-Ratio Transformation

Anyone who has ever dealt with the statistical analysis of compositional data must have stumbled across John Aitchison's (1926–2016) log-ratio transformation. The Scottish statistician spent much of his career dealing with the statistics of such data. Indeed, he wrote the famous book *The Statistical Analysis of Compositional Data* (Aitchison 1984, 1986, 1999, 2003) and published multiple papers on the topic. Aitchison's log-ratio transformation overcomes the closed-sum problem of closed data (i.e., data that are expressed as proportions and added to a fixed total, such as 100%). The closed-sum problem causes spurious negative correlations between pairs of variables in multivariate data sets, which are avoided by logarithmizing ratios of these variables. As an example, multiproxy data in paleoclimate studies are often closed data (i.e., they are very much influenced by dilution effects). In the simplest example, this could be a mixture of three components that together make up 100% of the sediment. Imagine that one of the components is now constant (i.e., the same amount of this component is always delivered per unit time). Since the total of all three components is always 100%, the other two components are necessarily anticorrelated: If one goes up, the other two must go down by the same total amount.

Using ratios helps to overcome the problem that the magnitudes of compositional data are actually ratios whose denominators are the sums of the concentrations of all constituents (see Davis 2003 for a short summary of Aitchison's log-ratio transformation). However, the product-moment covariances of ratios are awkward to manipulate and result in complicated products and quotients. John Aitchison therefore suggested taking the logarithms of these ratios and thereby also taking advantage of the simple properties of the logarithmic function, for example, $\log(x/y) = \log(x)-\log(y)$ and hence also $\log(x/y) = -\log(y/x)$. This process allows for reducing the number of the covariances that are necessary to describe the data set (Aitchison 2003, pages 65 to 66).

Here is a simple example that illustrates the closed-sum problem of a system with three variables and the use of ratios as well as log-ratios to overcome the problem of spurious correlations between pairs of variables. First, we clear the workspace and choose colors for the plots:

```
%reset -f

import numpy as np
import matplotlib.pyplot as plt

colors = np.array([
    [0,114,189],
    [217,83,25],
    [237,177,32],
    [126,47,142],
    ])/255
```

We are interested in `element1a` and `element1b`, which contribute to a sediment and are diluted by `element1c`. We simply create three variables with magnitudes measured in milligrams and with a sinusoidal variation of 200 samples down-core while making sure that all absolute values are > 0:

```
t = np.linspace(0.1,20,200)
t = np.transpose(t)

element1a = np.sin(2*np.pi*t/2) + 5
element1b = np.sin(2*np.pi*t/5) + 5
element1c = 2*np.sin(2*np.pi*t/20) + 5

plt.figure()
plt.plot(t,element1a,
    color=colors[0,:])
plt.plot(t,element1b,
    color=colors[1,:])
plt.plot(t,element1c,
    color=colors[3,:])
plt.legend(['1a','1b','1c'])
plt.title('Absolute amount of 1a-c (mg)')
plt.show()
```

We calculate the percentages of `element1a`, `element1b`, and `element1c` (i.e., we create ratios of the concentrations of the individual elements and the sum of the concentrations of all elements). This process creates closed data (i.e., the data are now expressed as proportions and add to a fixed total of 100 percent): `element1a+element1b+element1c` = 100%. Both `element1a` and `element1b` are now affected by the dilution due to `element1c`. These elements show a major sinusoidal long-term trend that is not real (as is shown by the first figure):

```
element1a_perc = element1a/(element1a+ \
    element1b+element1c)
element1b_perc = element1b/(element1a+ \
    element1b+element1c)
element1c_perc = element1c/(element1a+ \
    element1b+element1c)

plt.figure()
plt.plot(t,element1a_perc,
    color=colors[0,:])
plt.plot(t,element1b_perc,
    color=colors[1,:])
plt.plot(t,element1c_perc,
    color=colors[3,:])
plt.legend(['1a','1b','1c'])
plt.title('Relative concentration 1a-c (%)')
plt.show()
```

We have now established ratios of `element1a/element1b` and of `element1b/element1a`, both of which are independent of `element1c`. Note the change of sign and the difference in amplitude. The ratios of `element1a/element1b` and `element1b/element1a` do not show any evidence of a trend caused by the dilution effect of `element1c`. However, the two curves are not identical (i.e., `element1a/element1b` and `element1b/element1a` are not symmetric) (see Weltje and Tjallingii 2008, page 426):

```
ratio12 = element1a_perc/element1b_perc
ratio21 = element1b_perc/element1a_perc

plt.figure()
ax1 = plt.subplot()
ax2 = ax1.twinx()
line1, = ax1.plot(t,ratio12,
    color=colors[0,:],
    label='1a/1b')
line2, = ax2.plot(t,ratio21,
```

```
        color=colors[1,:],
        label='1b/1a')
ax2.invert_yaxis()
plt.legend(handles=[line1,line2])
plt.title('Ratios elements 1a/1b and 1b/1a')
plt.show()
```

We now display log-ratios according to Aitchison (1986, 2003). The log-ratios `log(element1a/element1b)` and `log(element1b/element1a)` are identical except for the sign. It therefore makes no difference whether we use `log(element1a/element1b)` or `log(element1b/element1a)`, for instance, when running further statistical analyses on the data:

```
ratio12log = np.log10(ratio12)
ratio21log = np.log10(ratio21)

plt.figure()
ax1 = plt.subplot()
ax2 = ax1.twinx()
line1, = ax1.plot(t,ratio12log,
        color=colors[0,:],
        label='log(1a/1b)')
line2, = ax2.plot(t,ratio21log,
        color=colors[1,:],
        linestyle='--',
        label='log(1b/1a)')
ax2.invert_yaxis()
plt.legend(handles=[line1,line2])
plt.title('Log-ratio elements log(1a/1b) and log(1b/1a)')
plt.show()
```

Aitchison's log-ratio transformation must be used in all multivariate statistical methods that analyze closed data (e.g., Sects. 9.2 to 9.6) (Fig. 9.10).

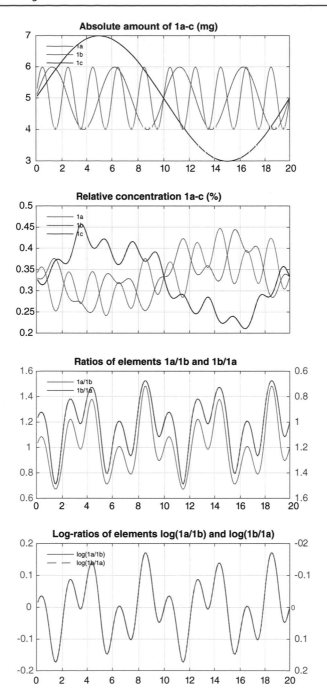

Fig. 9.10 Demonstration of Aitchison's log-ratio transformation being used to overcome the closed-sum problem of closed data: **a** absolute values of element1a, element1b, and element1c, **b** relative amounts of element1a, element1b, and element1c, which shows the closed-sum problem, **c** ratios of element1a/element1b and element1b/element1a, which are both independent of element1c, and **d** log-ratios according to Aitchison (1986, 2003), which are identical except for their sign and thus overcome the closed-sum problem.

Recommended Reading

Aitchison J (1984) The statistical analysis of geochemical composition. Math Geol 16(6):531–564

Aitchison J (1986 reprinted 2003) The statistical analysis of compositional data. The Blackburn Press, Caldwell

Aitchison J (1999) Logratios and natural laws in compositional data analysis. Math Geol 31(5):563–580

Birks HJB, Gordon AD (1985) Numerical methods in quaternary pollen analysis. Academic Press, London

Brown CE (1998) Applied multivariate statistics in geohydrology and related sciences. Springer, Berlin Heidelberg New York

Davis JC (2002) Statistics and data analysis in geology, 3rd edn. John Wiley & Sons, New York

Fisher RA (1936) The use of multiple measurements in taxonomic problems. Ann Eugen 7:179–188

Härdle WK, Simar L (2012) Applied multivariate statistical analysis. Springer, Berlin Heidelberg New York

Hermanns R, Trauth MH, McWilliams M, Strecker M (2000) Tephrochronologic constraints on temporal distribution of large landslides in NW-Argentina. J Geol 108:35–52

Hotelling H (1931) Analysis of a complex of statistical variables with principal components. J Educ Psychol 24(6):417–441

Hyvärinen A (1999) Fast and robust fixed-point algorithms for independent component analysis. IEEE Trans Neural Networks 10(3):626–634

Olorunfemi MO (1985) Statistical relationships among some formation parameters for sherwood sandstone, England. Math Geol 17:845–852

Pawlowsky-Glahn V (2004) Geostatistical analysis of compositional data—studies in mathematical geology. Oxford University Press, Oxford

Pearson K (1901) On lines and planes of closest fit to a system of points in space. Philosophical Magazine and Journal of Science 6(2):559–572

Reyment RA, Savazzi E (1999) Aspects of multivariate statistical analysis in geology. Elsevier Science, Amsterdam

Streckeisen A (1974) Classification and Nomenclature of Plutonic Rocks. Geologische Rundschau 63:773–786

Streckeisen A (1976) To each plutonic rock its proper name. Earth Sci Rev 12:1–33

Swan ARH, Sandilands M (1995) Introduction to geological data analysis. Blackwell Sciences, Oxford

Trauth MH, Bookhagen B, Mueller A, Strecker MR (2003) Erosion and climate change in the Santa Maria Basin, NW Argentina during the last 40,000 yrs. J Sediment Res 73(1):82–90

Weltje GJ, Tjallingii R (2008) Calibration of XRF core scanners for quantitative geochemical logging of sediment cores: Theory and application. Earth Planet Sci Lett 274:423–438

Westgate JA, Shane PAR, Pearce NJG, Perkins WT, Korisettar R, Chesner CA, Williams MAJ, Acharyya SK (1998) All Toba Tephra occurrences across peninsular India belong to the 75,000 yr BP Eruption. Quatern Res 50:107–112

Zadeh L (1965) Fuzzy sets. information. Control 8:338–353

Directional Data

10

10.1 Introduction

Methods for analyzing *circular and spherical data* are widely used in the earth sciences. Structural geologists, for example, measure and analyze the orientation of striated slickensides on fault planes. Circular statistics is also common in paleomagnetic applications. Microstructural investigations include analyzing grain shapes and quartz *c*-axis orientations in thin sections, and paleoenvironmentalists reconstruct paleocurrent directions from the alignments of elongated fossil shells (Fig. 10.1). There are two types of directional data in the earth sciences: directional data sensu stricto and oriented data. Directional data (e.g., the paleocurrent direction of a river as documented by flute marks, or the flow direction of a glacier as indicated by glacial striae) have a true polarity, whereas oriented data describe axial data and lines (e.g., the orientation of joints) without any sense of direction.

A number of useful publications are available on the statistical analysis of directional data, such as the books by Fisher (1993) and by Mardia and Jupp (1999) as well as chapters on the subject in the books by Swan and Sandilands (1995), by Davis (2002), and by Borradaile (2003). Furthermore, a chapter in the book by Middleton (2000), two journal articles by Jones (2006a, b), and Chap. 10 in the book by Marques de Sá (2007) discuss the use of Python for statistically analyzing directional data. Python is not the first choice for analyzing directional data since it does not provide the relevant functions, such as algorithms that can be used to compute the probability density function of a *von Mises distribution* or to run Rayleigh's test for the significance of a mean direction. Earth scientists

Supplementary Information The online version contains supplementary material available at https://doi.org/10.1007/978-3-031-07719-7_10.

M. H. Trauth, *Python Recipes for Earth Sciences*, Springer
Textbooks in Earth Sciences, Geography and Environment,
https://doi.org/10.1007/978-3-031-07719-7_10

Fig. 10.1 *Orthoceras* fossils from an outcrop at Neptuni Acrar near Byxelkrok on Öland, Sweden. *Orthoceras* is a cephalopod with a straight shell that lived in the Ordovician era about 450 Mio. years ago. Such elongated, asymmetric objects tend to orient themselves in the most hydrodynamically stable position and therefore indicate paleocurrent directions. The statistical analysis of cephalopod orientations at Neptuni Acrar reveals a significant southerly paleocurrent direction, which is in agreement with paleogeographic reconstructions.

have therefore developed numerous stand-alone programs with which to analyze such data, including the excellent desktop and smartphone software developed by Richard W. Allmendinger, structural geologist and professor emeritus at Cornell University (Allmendinger et al. 2017; Allmendinger 2019):

```
http://www.geo.cornell.edu/geology/faculty/RWA/programs.html
```

The following tutorial on the analysis of directional data is independent of these tools. It provides simple Python codes for displaying directional data, for computing the von Mises distribution, and for running simple statistical tests. Sect. 10.2 introduces rose diagrams as the most widely used method for displaying directional data. Subsequently, Sects. 10.3 and 10.4 have a similar concept to Chap. 3 (on univariate statistics) and cover the use of empirical and theoretical distributions for describing directional data. Then, Sects. 10.5 to 10.7 describe the three most important tests for directional data: namely tests for the randomness of circular data, for the significance of a mean direction, and for the difference between two sets of circular data. Finally, Sects. 10.8 and 10.9 describe the graphical representation and statistics of spherical data. For the analysis of directional data, we use the *NumPy* (https://numpy.org), *Matplotlib* (https://matplotlib.org), and *SciPy* packages (https://scipy.org), which contain all the necessary routines.

10.2   Graphical Representation of Circular Data

The classic way of displaying directional data is in a rose diagram, which is a histogram for measurements of angles. In contrast to a bar histogram, in which with the height of the bars is proportional to the frequency, a rose diagram comprises segments of a circle in which the radius of each sector is proportional to the frequency. We use synthetic data to illustrate two types of rose diagrams that are used to display directional data. We first load a set of directional data from the file *directional_1.txt*:

```
%reset -f

import numpy as np
import matplotlib.pyplot as plt

data_degrees_1 = np.loadtxt('directional_1.txt')
```

The data set contains two hundred measurements of angles (in degrees, or °, for short). We need to convert all measurements to radians before plotting the data. We then bin the data using `histogram()` and display the counts in a rose diagram. This is achieved by first selecting the `polar` projection in `subplot()` and then plotting a bar chart in the polar coordinate system:

```
data_radians_1 = np.pi*data_degrees_1/180

e = np.linspace(0,2*np.pi,13)
v = np.linspace(2*np.pi/24,2*np.pi-2*np.pi/24,12)
n,e = np.histogram(data_radians_1,bins=e)

plt.figure()
ax = plt.subplot(projection='polar')
ax.bar(v,n,width=2*np.pi/12)
ax.set_xticks(e[0:12])
plt.show()
```

This rose diagram counts in a counterclockwise direction, with zero degrees lying along the *x*-axis of the coordinate graph. In the geosciences, the angles increase in a clockwise direction, with 0° indicating due north, 90° indicating due east, and so on. We use `set_theta_offset(np.pi/2)` to rotate the rose diagram by 90° (or $\pi/2$) and `set_theta_direction(-1)` to reverse the direction of the rose diagram. We also select the ticks at the values of the class boundaries e (i.e., 0°, 30°, 60°, ...) using `set_xticks()` (Fig. 10.2):

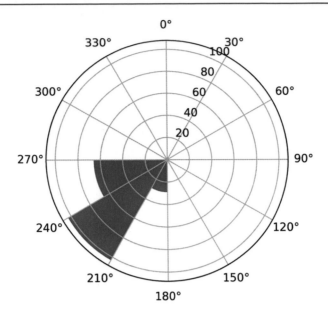

Fig. 10.2 Rose diagram for displaying directional data. The radii of the area segments are proportional to the frequencies for each class. We use `bar` with `set_theta_offset(np.pi/2)` and `set_theta_direction(-1)` such that 0° points due north and 90° points due east, i.e., the angles increase in a clockwise direction.

```
plt.figure()
ax = plt.subplot(projection='polar')
ax.bar(v,n,width=2*np.pi/12)
ax.set_xticks(e[0:12])
ax.set_theta_offset(np.pi/2)
ax.set_theta_direction(-1)
plt.show()
```

The area of the arc segments increases with the increasing square of the frequency, and directions that are twice as frequent as others thus appear to be four times as prominent as the others. In a modification, the rose diagram is therefore scaled to the square root of the class frequency (Fig. 10.3):

```
plt.figure()
ax = plt.subplot(projection='polar')
ax.bar(v,np.sqrt(n),width=2*np.pi/12)
ax.set_xticks(e[0:12])
ax.set_theta_offset(np.pi/2)
ax.set_theta_direction(-1)
plt.show()
```

This plot satisfies all conventions in the geosciences.

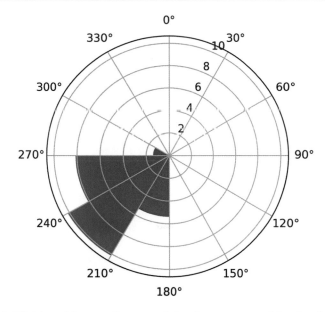

Fig. 10.3 Modification of the rose diagram scaled to the square root of the class frequency. This modification avoids increasing the area of the arc segments with the increasing square of the frequency and thus also avoids causing directions that are twice as frequent as others to appear to be four times as prominent as the others.

10.3   Empirical Distributions of Circular Data

This section introduces the statistical measures used to describe empirical distributions of directional data. The characteristics of directional data are described by measures of central tendency and dispersion, which is similar to the statistical characterization of univariate data sets (Chap. 3). Assume that we have collected a number of angular measurements, such as fossil alignments. The collection of data can be written as

$$\theta_1, \theta_2, \theta_3, \ldots, \theta_n$$

which contains n observations θ_i. Sine and cosine values are computed for each direction θ_i and are then used to compute the *resultant* (or *mean*) *direction* for the set of angular data (Fisher 1993; Mardia and Jupp 1999):

$$x_r = \sum \sin \theta_i$$

$$y_r = \sum \cos \theta_i$$

The resultant direction of the data set is

$$\overline{\theta} = \tan{(x_r/y_r)}$$

and the resultant length is

$$R = \sqrt{\left(x_r^2 + y_r^2\right)}$$

The resultant length clearly depends on the dispersion of the data. Normalizing the resultant length to the number of observations yields the mean resultant length:

$$\overline{R} = R/n$$

The value of the mean resultant length decreases with increasing dispersion (Fig. 10.4). The *circular variance* is then the difference between one and the mean resultant length and is therefore often used as a measure of dispersion for directional data:

$$\sigma_0 = 1 - \overline{R}$$

The following example illustrates the use of these parameters by means of synthetic directional data. We first load the data from the file *directional_1.txt* and convert all measurements to radians:

```
%reset -f

import numpy as np
import matplotlib.pyplot as plt

data_degrees_1 = np.loadtxt('directional_1.txt')
data_radians_1 = np.pi*data_degrees_1/180
```

We now calculate the resultant vector R. First, we compute the x and y components of the resultant vector,

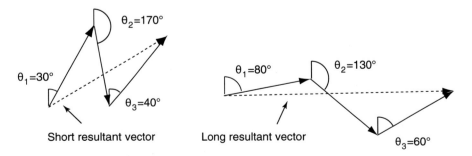

Fig. 10.4 The resultant length R of a sample decreases with increasing dispersion of the data θ_i. Modified from Swan and Sandilands (1995).

```
x_1 = np.sum(np.sin(data_radians_1))
y_1 = np.sum(np.cos(data_radians_1))
print(x_1)
print(y_1)
```

which yields

```
-146.54800998477873
-120.34536799675672
```

The mean direction is the inverse tangent `arctan()` of the ratio of x and y,

```
mean_radians_1 = np.arctan(x_1/y_1)
mean_degrees_1 = 180*mean_radians_1/np.pi
print(mean_radians_1)
print(mean_degrees_1)
```

which yields

```
0.883260987889874
50.6071268146455
```

This result suggests that the resultant vector R is around 0.88 radians, or ~51°. However, since both x and y are negative, the true value of mean_degrees is located in the third quadrant, and we therefore add 180°,

```
mean_degrees_1 = mean_degrees_1 + 180
print(mean_degrees_1)
```

which yields

```
230.60712681464548
```

which results in a mean direction of around 231°. The length of this vector is the absolute value of the vector, which is

```
R_1=np.sqrt(x_1**2+y_1**2)
print(R_1)
```

which yields

```
189.6294461015314
```

The resultant length depends on the dispersion of the directional data. Normalizing the resultant length to the sample size yields the mean resultant length Rm of

```
Rm_1 = R_1 / len(data_radians_1)
print(Rm_1)
```

which yields

```
0.948147230507657
```

A high Rm value suggests less variance. We then compute the circular variance sigma, which is

```
sigma_1 = 1 - Rm_1
print(sigma_1)
```

which yields

```
0.051852769492342965
```

10.4   Theoretical Distributions of Circular Data

As in Chap. 3, the next step in a statistical analysis is to find a suitable theoretical distribution that we then fit to the empirical distribution that was visualized and described in the previous section. The classic theoretical distribution for describing directional data is the *von Mises distribution*, which is named after Austrian mathematician Richard Edler von Mises (1883–1953). The probability density function of a von Mises distribution is

$$f(\theta) = \frac{1}{2\pi I_0(\kappa)} e^{\kappa \cos(\theta - \mu)}$$

where μ is the mean direction and κ is the concentration parameter (Mardia and Jupp 1999) (Fig. 10.5). The term $I_0(\kappa)$ is the zero-order modified Bessel function of the first kind. Bessel functions are solutions of a second-order differential equation (Bessel's differential equation) and are important in many problems involving wave propagation in a cylindrical waveguide or in problems involving heat conduction in a cylindrical object. The von Mises distribution is also known as the circular normal distribution since it has similar characteristics to a normal distribution (Sect. 3.4). This distribution is used when the mean direction is the most frequent direction. The probability of deviations is equal on either side of the mean direction and decreases with increasing distance from the mean direction.

As an example, let us assume a mean direction of mu=0 and five different values for the concentration parameter kappa:

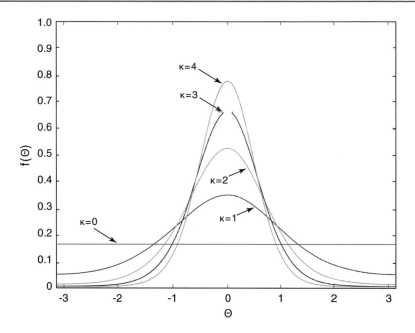

Fig. 10.5 Probability density function *f(x)* of a von Mises distribution with $\mu = 0$ and five different values for κ.

```
%reset -f

import numpy as np
from scipy import special
import matplotlib.pyplot as plt
from numpy.random import default_rng

mu = 0
kappa = np.array([0,1,2,3,4])
```

We first define an angle scale for a plot that runs from −180° to 180° with intervals of 1°:

```
theta = np.linspace(-180,180,361)
```

All angles are converted from degrees to radians:

```
mu_radians = np.pi*mu/180
theta_radians = np.pi*theta/180
```

We then compute the von Mises distribution for these values. The formula uses the zero-order modified Bessel function of the first kind, which can be calculated using `special.iv()`. We next compute the probability density function for the five values of `kappa`:

```
mises = np.zeros((len(theta),5))
theta_radians_array = np.zeros((len(theta),5))
for i in range(0,5):
    mises[:,i] = (1/(2*np.pi*special.iv(0,kappa[i]))) * \
    np.exp(kappa[i]*np.cos(theta_radians-mu_radians))
    theta_radians_array[:,i] = theta_radians
```

The results are plotted by

```
plt.figure()
for i in range(0,5):
    plt.plot(theta_radians_array[:,i],mises[:,i])
plt.legend(['$\kappa $=0',
    '$\kappa $=1',
    '$\kappa $=2',
    '$\kappa $=3',
    '$\kappa $=4'])
plt.show()
```

The mean direction and concentration parameters of such theoretical distributions can be easily modified for comparison with empirical distributions.

We can create a Python code that yields a pseudorandom number generator for von Mises-distributed data from the algorithm described on Page 155 of Best and Fisher (1979). We first define the sample size n, the mean direction mu (converted to radians), and the concentration parameter kappa:

```
n = 200
mu = 230
mu = np.pi*mu/180
kappa = 5
```

We then reset the random number generator using `random.seed(0)` and calculate n pseudorandom numbers (in °) using the algorithm created by Best and Fisher (1979):

```
np.random.seed(0)
tau = 1 + (1 + 4*kappa**2)**0.5
rho = (tau - (2*tau)**0.5)/(2*kappa)
r = (1 + rho**2)/(2*rho)
i = 1
```

```
theta = np.zeros(n)
while i < n:
    u1 = np.random.rand(1)
    u2 = np.random.rand(1)
    u3 = np.random.rand(1)
    z = np.cos(np.pi*u1)
    f = (1+r*z)/(r+z)
    c = kappa*(r-f)
    if c*(2-c)-u2 > 0:
        theta[i]=(np.sign(u3-0.5)*np.arccos(f))
        i = i+1
    elif np.log(c/u2)+1-c >= 0:
        i = i
    # Nothing happens, return to
    # beginning of while loop

data_radians_2 = theta + mu
data_degrees_2 = 180*data_radians_2/np.pi
```

Please note that the `elif` condition does not contain any conditional expressions. In the algorithm by Best and Fisher (1979), the condition simply says *return to step 1* (which is the beginning of the `while` loop) without evaluating any line of code. Instead of using this code, we can also use `vmises()`, which is included in the electronic supplement of this book:

```
from vmises import vmises
help(vmises)

data_radians_3 = vmises(n,mu,kappa)
data_degrees_3 = 180*data_radians_3/np.pi
```

We can display the data in a rose diagram using the code from the previous section:

```
ee = np.linspace(0,2*np.pi,13)
vv = np.linspace(2*np.pi/24,2*np.pi-2*np.pi/24,12)
nn,ee = np.histogram(data_radians_3,bins=ee)

plt.figure()
ax = plt.subplot(projection='polar')
ax.bar(vv,nn,width=2*np.pi/12)
ax.set_xticks(ee[0:12])
ax.set_theta_offset(np.pi/2)
ax.set_theta_direction(-1)
plt.show()
```

10.5   Test for Randomness of Circular Data

The first test for directional data compares the data set with a uniform distribution. Directional data that follow a uniform distribution are purely random (i.e., there is no preference for any one direction). We use the χ^2-test (Sect. 3.8) to compare the empirical frequency distribution with the theoretical uniform distribution. We first load our sample data:

```
%reset -f

import numpy as np
from scipy import stats
import matplotlib.pyplot as plt

data_degrees_1 = np.loadtxt('directional_1.txt')
```

We then use `histogram()` to count the number of observations within 12 classes, each of which has a width of 30°:

```
e = np.linspace(0,360,13)
counts, e = np.histogram(data_degrees_1,bins=e)
```

The expected number of observations is 200/12, where 200 is the total number of observations and 12 is the number of classes:

```
expect = 200/12*np.ones(12)
```

The χ^2-test explores the squared differences between the observed and expected frequencies. The quantity χ^2 is defined as the sum of these squared differences divided by the expected frequencies,

```
chi2cal = np.sum((counts-expect)**2/expect)
print(chi2cal)
```

which yields

```
748.5999999999998
```

The critical χ^2-value can be calculated using `chi2.ppf()`. The χ^2-test requires the degrees of freedom Φ. In our example, we test the hypothesis that the data are uniformly distributed (i.e., we estimate one parameter). Since the number of classes is 12, the number of degrees of freedom is $\Phi = 12-1-1 = 10$. We test our hypothesis at a $p = 5$ % significance level. The function `chi2.ppf()` computes the inverse of the cumulative distribution function (CDF) of the χ^2-distribution, with numbers of parameters specified by Φ for the corresponding probabilities in p,

```
chi2crit = stats.chi2.ppf(1-0.05,12-1-1)
print(chi2crit)
```

which yields

```
18.307038053275146
```

Since the critical χ^2 of 18.3070 is well below the measured χ^2 of 748.6000, we can reject the null hypothesis and conclude that our data do not follow a uniform distribution (i.e., they are not randomly distributed).

10.6   Test for the Significance of a Mean Direction

Once we have measured a set of directional data in the field, we may wish to know whether the data document a prevailing direction. We can find out using Rayleigh's test for the significance of a mean direction (Mardia 1972; Mardia and Jupp 1999). This test uses the mean resultant length introduced in Sect. 10.3, which increases as the preferred direction increases:

$$\overline{R} = \frac{1}{n} \sqrt{\left(\sum \sin \theta_i \right)^2 + \left(\sum \cos \theta_i \right)^2}$$

The data show a preferred direction if the calculated mean resultant length is below a critical value (Mardia 1972) (Tab. 10.1). As an example, we again load the data contained in the file *directional_1.txt*:

```
%reset -f

import numpy as np
import matplotlib.pyplot as plt

data_degrees_1 = np.loadtxt('directional_1.txt')
data_radians_1 = np.pi*data_degrees_1/180;
```

We then calculate the mean resultant vector Rm_1_calc by typing

```
x_1 = np.sum(np.sin(data_radians_1))
y_1 = np.sum(np.cos(data_radians_1))
```

Tab. 10.1 Critical values of the mean resultant length for Rayleigh's test for the significance of a mean direction of n measurements (Mardia 1972).

Level of Significance α					
n	0.100	0.050	0.025	0.010	0.001
5	0.677	0.754	0.816	0.879	0.991
6	0.618	0.690	0.753	0.825	0.940
7	0.572	0.642	0.702	0.771	0.891
8	0.535	0.602	0.660	0.725	0.847
9	0.504	0.569	0.624	0.687	0.808
10	0.478	0.540	0.594	0.655	0.775
11	0.456	0.516	0.567	0.627	0.743
12	0.437	0.494	0.544	0.602	0.716
13	0.420	0.475	0.524	0.580	0.692
14	0.405	0.458	0.505	0.560	0.669
15	0.391	0.443	0.489	0.542	0.649
16	0.379	0.429	0.474	0.525	0.630
17	0.367	0.417	0.460	0.510	0.613
18	0.357	0.405	0.447	0.496	0.597
19	0.348	0.394	0.436	0.484	0.583
20	0.339	0.385	0.425	0.472	0.569
21	0.331	0.375	0.415	0.461	0.556
22	0.323	0.367	0.405	0.451	0.544
23	0.316	0.359	0.397	0.441	0.533
24	0.309	0.351	0.389	0.432	0.522
25	0.303	0.344	0.381	0.423	0.512
30	0.277	0.315	0.348	0.387	0.470
35	0.256	0.292	0.323	0.359	0.436
40	0.240	0.273	0.302	0.336	0.409
45	0.226	0.257	0.285	0.318	0.386
50	0.214	0.244	0.270	0.301	0.367
100	0.150	0.170	0.190	0.210	0.260

```
mean_radians_1 = np.arctan(x_1/y_1)
mean_degrees_1 = 180*mean_radians_1/np.pi
mean_degrees_1 = mean_degrees_1 + 180

Rm_1_calc = 1/len(data_degrees_1) * (x_1**2+y_1**2)**0.5
print(Rm_1_calc)
```

which yields

```
0.948147230507657
```

The mean resultant length `Rm_1_calc` in our example is 0.9481. There are a number of approximations involved in calculating the critical `Rm_1_crit`, such as Wilkie's approximation (Wilkie 1983), according to which `Rm_1_crit` can be calculated by typing

```
n = len(data_degrees_1)

alpha = 0.050

Rm_1_crit = np.sqrt((-np.log(alpha)-(2*np.log(alpha)+ \
    (np.log(alpha))**2)/(4*n))/n)
print(Rm_1_crit)
```

which yields

```
0.12231115217794654
```

The critical `Rm` for $\alpha = 0.05$ and $n = 200$ is 0.1223. Since this value is lower than the `Rm_1_calc` from the data, we can reject the null hypothesis and conclude that there is a preferred single direction, which is

```
theta_1 = 180 * np.arctan(x_1/y_1) / np.pi
print(theta_1)
```

which yields

```
50.6071
```

However, the negative signs of the sine and cosine imply that the true result is in the third sector (180° to 270°), and the correct result is therefore $180 + 50.6071 = 230.6071$,

```
theta_1 = 180 + theta_1
print(theta_1)
```

which yields

```
230.60712681464548
```

10.7   Test for the Difference Between Two Sets of Directions

Let us consider two sets of measurements in two files, *directional_1.txt* and *directional_2.txt*. We wish to compare the two sets of directions and test the null hypothesis that they are not significantly different. We use the Watson–Williams test to test the similarity between two mean directions,

$$
F = \left(1 + \frac{3}{8\kappa}\right) \frac{(n-2)(R_A + R_B - R_T)}{n - R_A - R_B}
$$

where κ is the concentration parameter, R_A and R_B are the resultant lengths of samples A and B, respectively, and R_T is the resultant length of the combined samples (Watson and Williams 1956; Mardia and Jupp 1999). The concentration parameter can be obtained from tables using R_T (Batschelet 1981, Gumbel et al. 1953, Tab. 10.2). The calculated F is compared with critical values from the standard F tables (Sect. 3.8). The two mean directions are not significantly different if the calculated F-value is less than the critical F-value, which depends on the degrees of freedom $\Phi_a = 1$ and $\Phi_b = n-2$ as well as on the significance level α. Both samples must follow a von Mises distribution (Sect. 10.4).

We use two synthetic data sets of directional data to illustrate how this test is applied. We first load the data and convert the degrees to radians:

```
%reset -f

import numpy as np
from scipy import stats
import matplotlib.pyplot as plt

data_degrees_1 = np.loadtxt('directional_1.txt')
data_degrees_2 = np.loadtxt('directional_2.txt')

data_radians_1 = np.pi*data_degrees_1/180;
data_radians_2 = np.pi*data_degrees_2/180;
```

We then compute the lengths of the resultant vectors,

```
x_1 = np.sum(np.sin(data_radians_1))
y_1 = np.sum(np.cos(data_radians_1))
x_2 = np.sum(np.sin(data_radians_2))
y_2 = np.sum(np.cos(data_radians_2))

mean_radians_1 = np.arctan(x_1/y_1)
mean_degrees_1 = 180*mean_radians_1/np.pi
```

Tab. 10.2 Maximum likelihood estimates of the concentration parameter κ for calculated mean resultant length. Adapted from Batschelet (1981) and from Gumbel et al. (1953).

R	κ	R	κ	R	κ	R	κ
0.000	0.000	0.260	0.539	0.520	1.224	0.780	2.646
0.010	0.020	0.270	0.561	0.530	1.257	0.790	2.754
0.020	0.040	0.280	0.584	0.540	1.291	0.800	2.871
0.030	0.060	0.290	0.606	0.550	1.326	0.810	3.000
0.040	0.080	0.300	0.629	0.560	1.362	0.820	3.143
0.050	0.100	0.310	0.652	0.570	1.398	0.830	3.301
0.060	0.120	0.320	0.676	0.580	1.436	0.840	3.479
0.070	0.140	0.330	0.700	0.590	1.475	0.850	3.680
0.080	0.161	0.340	0.724	0.600	1.516	0.860	3.911
0.090	0.181	0.350	0.748	0.610	1.557	0.870	4.177
0.100	0.201	0.360	0.772	0.620	1.600	0.880	4.489
0.110	0.221	0.370	0.797	0.630	1.645	0.890	4.859
0.120	0.242	0.380	0.823	0.640	1.691	0.900	5.305
0.130	0.262	0.390	0.848	0.650	1.740	0.910	5.852
0.140	0.283	0.400	0.874	0.660	1.790	0.920	6.539
0.150	0.303	0.410	0.900	0.670	1.842	0.930	7.426
0.160	0.324	0.420	0.927	0.680	1.896	0.940	8.610
0.170	0.345	0.430	0.954	0.690	1.954	0.950	10.272
0.180	0.366	0.440	0.982	0.700	2.014	0.960	12.766
0.190	0.387	0.450	1.010	0.710	2.077	0.970	16.927
0.200	0.408	0.460	1.039	0.720	2.144	0.980	25.252
0.210	0.430	0.470	1.068	0.730	2.214	0.990	50.242
0.220	0.451	0.480	1.098	0.740	2.289	0.995	100.000
0.230	0.473	0.490	1.128	0.750	2.369	0.999	500.000
0.240	0.495	0.500	1.159	0.760	2.455	1.000	5000.000
0.250	0.516	0.510	1.191	0.770	2.547		

```
mean_radians_2 = np.arctan(x_2/y_2)
mean_degrees_2 = 180*mean_radians_2/np.pi

mean_degrees_1 = mean_degrees_1 + 180
mean_degrees_2 = mean_degrees_2 + 180
print(mean_degrees_1)
print(mean_degrees_2)
```

```
R_1 = np.sqrt(x_1**2+y_1**2)
R_2 = np.sqrt(x_2**2+y_2**2)
```

which yields

```
230.60712681464548
212.18856343005913
```

The orientations of the resultant vectors are approximately $223°$ and $201°$. We also need the resultant length for both samples combined; thus, we combine both data sets and again compute the resultant length,

```
data_radians_T = np.concatenate([data_radians_1, \
        data_radians_2])

x_T = np.sum(np.sin(data_radians_T))
y_T = np.sum(np.cos(data_radians_T))

mean_radians_T = np.arctan(x_T/y_T)
mean_degrees_T = 180*mean_radians_T/np.pi

mean_degrees_T = mean_degrees_T + 180

R_T = np.sqrt(x_T**2 + y_T**2)
print(R_T)
Rm_T = R_T / (len(data_radians_T))
print(Rm_T)
```

which yields

```
372.4449200527211
0.9311123001318028
```

We then apply the F-statistic to the data for `kappa=7.426` and for `Rm_T=0.9311` (Tab. 10.2). The computed value for `F` is

```
n = len(data_radians_T)

Fcalc = (1+3/(8*7.426)) * (((n-2)*(R_1+R_2-R_T))/(n-R_1-R_2))
print(Fcalc)
```

which yields

```
89.60541051467561
```

Using the F-statistic, we find that for 1 and $400-2=398$ degrees of freedom and $\alpha = 0.05$, the critical value is

```
Fcrit = stats.f.ppf(1-0.05,1,n-2)
print(Fcrit)
```

which yields

```
3.8649292212017365
```

which is well below the observed value of F = 89.6054 (Sect. 3.8). We can therefore reject the null hypothesis and conclude that the two samples could not have been drawn from populations with the same mean direction.

10.8   Graphical Representation of Spherical Data

There are many ways to display spherical data. However, since we always present data in two dimensions (i.e., either on a sheet of paper or on a computer screen), we need to convert them to one of many available projections. Python does not offer many functions for displaying or processing spherical data, especially not in a form commonly used in the geosciences. The only exception lies with functions such as those used to display geographic data on a globe, some of which we use here.

We begin by using synthetic data for longitude lon and latitude lat, which are calculated using vmises(). We first reset the random number generator using random.seed(0) and define the sample size n. We then define the mean of the lon and lat values in degrees, convert both values to radians, define the concentration parameter kappa, and use vmises() (which is included in the electronic supplement of this book) to calculate n = 100 values for lon and lat. Finally, we convert the lon and lat values to degrees:

```
%reset -f

import numpy as np
from vmises import vmises
import scipy.stats as stats
import matplotlib.pyplot as plt
from numpy.random import default_rng

n = 100;
np.random.seed(0)
mu_lon_degrees = 10
mu_lat_degrees = 35
```

```
mu_lon_radians = np.pi*mu_lon_degrees/180
mu_lat_radians = np.pi*mu_lat_degrees/180
kappa_lon = 10
kappa_lat = 10
lon_radians = vmises(n,mu_lon_radians,kappa_lon)
lat_radians = vmises(n,mu_lat_radians,kappa_lat)
lon = 180*lon_radians/np.pi
lat = 180*lat_radians/np.pi
```

Next, we calculate the data density c with gaussian_kde():

```
c = stats.gaussian_kde([lon,lat])
```

We then calculate the grid coordinates grid_coords and a color value zc for each data point (lon,lat) according to the kernel density c:

```
grid_coords = np.append(lon.reshape(-1,1), \
    lat.reshape(-1,1),axis=1)
zc = 256*c(grid_coords.T) / np.max(c(grid_coords.T))
```

We subsequently display the data points in a geographic density plot with axes that use a *Lambert* projection (Fig. 10.6):

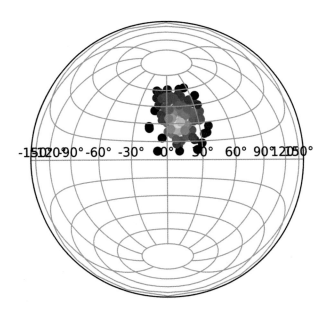

Fig. 10.6 Display of spherical data using the lambert map projection with subplot().

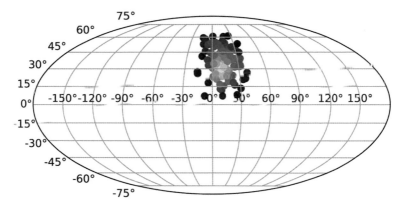

Fig. 10.7 Display of spherical data using the `mollweide` map projection with `subplot()`.

```
plt.figure()
ax = plt.subplot(projection='lambert')
ax.scatter(lon_radians,lat_radians,c=zc)
ax.grid()
plt.show()
```

Alternatively, we can also use a *Mollweide* projection (Fig. 10.7):

```
plt.figure()
ax = plt.subplot(projection='mollweide')
ax.scatter(lon_radians,lat_radians,c=zc)
ax.grid()
plt.show()
```

Finally, we display the data in a rectilinear coordinate system (Fig. 10.8):

```
plt.figure()
plt.scatter(lon,lat,c=zc)
plt.show()
```

This simple method of presentation might be the preferred option in some applications.

10.9   Statistics of Spherical Data

The statistical analysis of spherical data is a vast topic. Since Python offers very little support for analyzing this type of data, we remain very brief here. More detailed articles on this topic can be found in the books by Mardia and Jupp (1999) and by Borradaile (2003). The following section demonstrates how to

Fig. 10.8 Three-dimensional histogram using `bar3d()` to show the frequency distribution of geographic data using colors that correspond to the counts. The median is marked as a filled red circle above the bar plot.

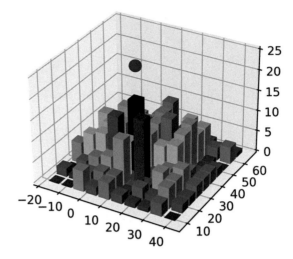

calculate the mean, the standard deviation, and the standard distance of geographic points. It also introduces a Python function for calculating a frequency distribution for geographic points using equirectangular bins.

First, we again create synthetic data using `vmises()`, which is included in the electronic supplement of this book:

```
%reset -f

import numpy as np
from vmises import vmises
import scipy.stats as stats
import matplotlib.pyplot as plt
from numpy.random import default_rng

n = 500;
np.random.seed(0)
mu_lon_degrees = 10
mu_lat_degrees = 35
mu_lon_radians = np.pi*mu_lon_degrees/180
mu_lat_radians = np.pi*mu_lat_degrees/180
kappa_lon = 10
kappa_lat = 10
lon_radians = vmises(n,mu_lon_radians,kappa_lon)
lat_radians = vmises(n,mu_lat_radians,kappa_lat)
lon = 180*lon_radians/np.pi
lat = 180*lat_radians/np.pi
```

Next, we calculate the data density c with `gaussian_kde()`:

```
c = stats.gaussian_kde([lon,lat])
```

We then calculate the grid coordinates grid_coords and a color value zc for each data point (lon,lat) according to the kernel density c:

```
grid_coords = np.append(lon.reshape(-1,1), \
    lat.reshape(-1,1),axis=1)
zc = 256*c(grid_coords.T) / np.max(c(grid_coords.T))
```

We subsequently calculate the median of the (lon,lat) coordinates using the Earth's arithmetic mean radius r of 6371 km by typing

```
r = 6371
x = r*np.cos(lat_radians)*np.cos(lon_radians)
y = r*np.cos(lat_radians)*np.sin(lon_radians)
z = r*np.sin(lat_radians)

x_median = np.median(x)
y_median = np.median(y)
z_median = np.median(z)

lon_radians_median = np.arctan2(y_median,x_median)
lat_radians_median = np.arcsin(z_median/r)

lon_degrees_median = 180*lon_radians_median/np.pi
lat_degrees_median = 180*lat_radians_median/np.pi

print(lon_degrees_median)
print(lat_degrees_median)
```

which yields

```
9.448245726147533
33.297480017287604
```

We find that the mean values for the longitudes (10°) and latitudes (30°) are very similar to the values we used to create the synthetic data with `vmises()`:

```
plt.figure()
plt.scatter(lon,lat,c=zc)
plt.plot(lon_degrees_median,lat_degrees_median,
    marker='o',
    markersize = 10,
```

```
    markeredgecolor=(0.9,0.3,0.1),
    markerfacecolor=(0.9,0.3,0.1))
plt.show()
```

We then calculate the histogram and display the counts in a 3D bar plot using colors that correspond to the counts, and we mark the median as a filled red circle above the bar plot (Fig. 10.8):

```
h,xedges,yedges = np.histogram2d(lon,lat)
vx = (xedges[:-1] + xedges[1:]) / 2
vy = (yedges[:-1] + yedges[1:]) / 2
vxx,vyy = np.meshgrid(vx,vy)
VX = vxx.flatten()
VY = vyy.flatten()
H = h.flatten()
bottom = np.zeros_like(H)
width = depth = 5


zc = plt.cm.coolwarm(H/float(np.max(H)))


fig = plt.figure()
ax = fig.add_subplot(projection='3d')
ax.bar3d(VX,VY,bottom,width,depth,H,color=zc)
plt.plot(lon_degrees_median,lat_degrees_median,25,
    marker='o',
    markersize = 10,
    markeredgecolor=(0.9,0.3,0.1),
    markerfacecolor=(0.9,0.3,0.1))
plt.show()
```

Recommended Reading

Allmendinger RW, Cardozo N, Fisher DM (2012) Structural geology algorithms: vectors and tensors. Cambridge University Press, New York

Allmendinger RW, Siron CR, Scott CP (2017) Structural data collection with mobile devices: accuracy, redundancy, and best practices. J Struct Geol 102:98–112

Allmendinger RW (2019) Modern structural practice, a structural geology laboratory manual for the 21st century. http://www.geo.cornell.edu/geology/faculty/RWA/structure-lab-manual/downloads.html

Batschelet E (1981) Circular statistics in biology. Academic, Cambridge, Massachusetts

Best DJ, Fisher NI (1979) Efficient simulation of the von mises distribution. Appl Stat 28:152–157

Borradaile G (2003) Statistics of earth science data – their distribution in time, space and orientation. Springer, Berlin

Davis JC (2002) Statistics and data analysis in geology, 3rd edn. Wiley, New York

Fisher NI (1993) Statistical analysis of circular data. Cambridge University Press, New York

Gumbel EJ, Greenwood JA, Durand D (1953) The circular normal distribution: tables and theory. J Am Stat Assoc 48:131–152

Jones TA (2006a) Python functions to analyze directional (azimuthal) data–I: single sample inference. Comput Geosci 32:166–175

Jones TA (2006b) Python functions to analyze directional (azimuthal) data–II: correlation. Comput Geosci 32:176–183

Mardia KV (1972) Statistics of directional data. Academic, London

Mardia KV (1975) Statistics of directional data. J Roy Stat Soc B (Methodological) 37:349–393

Mardia KV, Jupp PE (1999) Directional statistics. Wiley, Chichester

Marques de Sá JP (2007) Applied statistics using SPSS, STATISTICA, Python and R. Springer, Berlin

Middleton GV (2000) Data analysis in the earth sciences using python. Prentice Hall, Upper Saddle River, New Jersey

Swan ARH, Sandilands M (1995) Introduction to geological data analysis. Blackwell Sciences, Oxford

Watson GS, Williams EJ (1956) On the construction of significance tests on a circle and the sphere. Biometrika 43:344–352

Wilkie D (1983) Rayleigh test for randomness of circular data. Appl Stat 32:311–312

Printed in the United States
by Baker & Taylor Publisher Services